UNITED NATIONS CONFERENCE ON TRADE AND

REVIEW OF MARITIME TRANSPORT

2020

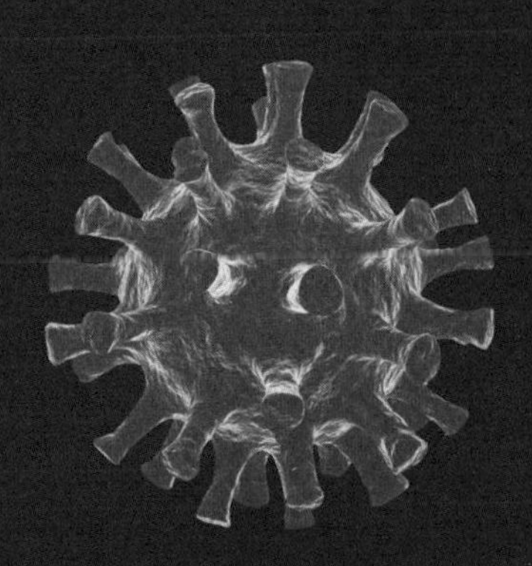

UNITED NATIONS

Geneva, 2020

United Nations Publications
300 East 42nd Street
New York, New York 10017
United States of America
Email: publications@un.org
Website: un.org/publications

United Nations publication issued by the United Nations Conference on Trade and Development

UNCTAD/RMT/2020

ISBN 978-92-1-112993-9
eISBN 978-92-1-005271-9
ISSN 0566-7682
eISSN 2225-3459
Sales No. E.20.II.D.31

ACKNOWLEDGEMENTS

The *Review of Maritime Transport 2020* was prepared by UNCTAD under the overall guidance of Shamika N. Sirimanne, Director of the Division on Technology and Logistics of UNCTAD, and under the coordination of Jan Hoffmann, Chief of the Trade Logistics Branch. Administrative and editorial support was provided by Wendy Juan. Regina Asariotis, Mark Assaf, Gonzalo Ayala, Ahmed Ayoub, Hassiba Benamara, Dominique Chantrel, Jan Hoffmann, Alexandre Larouche-Maltais, Anila Premti, Luisa Rodríguez, Sijia Sun and Frida Youssef were contributing authors.

The *Review* was edited by the Intergovernmental Support Service of UNCTAD. Magali Studer designed the publication, and Carlos Bragunde López did the formatting.

Comments and inputs from the following reviewers are gratefully acknowledged: Hashim Abbas, Gail Bradford, Pierre Cariou, Trevor Crowe, Neil Davidson, Juan Manuel Díez Orejas, Goran Dominioni, Azhar Jaimurzina Ducrest, Mahin Faghfouri, Fredrik Haag, Morten Ingebrigtsen, Eleni Kontou, Andy Lane, Mikael Lind, Amparo Mestre Alcover, James Milne, Turloch Mooney, Theo Notteboom, Athanasios A. Pallis, Ricardo Sánchez, Alastair Stevenson, Stellios Stratidakis, Antonella Teodoro and Patrick Verhoeven. Experts from the International Chamber of Shipping contributed to chapter 2.

Thanks are due to the following experts for their contributions to the case studies and testimonials included in chapter 4: Alison Newell, Member of the Chartered Institute of Ecology and Environmental Management, Chartered Environmentalist and Member of the management team of Sailing for Sustainability (Fiji), and Peter Nuttall, Scientific and Technical Adviser, Micronesian Centre for Sustainable Transport. Contributions from the Permanent Secretariat of the Northern Corridor Transit and Transport Coordination Authority; Port Authority of Valencia; Mediterranean Shipping Company; and Panama Canal Authority, in particular Ilya Espino de Marotta, Deputy Administrator, and Silvia de Marucci, Manager Market Analysis and Customer Relations, are also gratefully acknowledged.

Comments received from other UNCTAD divisions as part of the internal peer review process, as well as comments from the Office of the Secretary-General, are acknowledged with appreciation.

Thanks are also due to Vladislav Shuvalov for reviewing the publication in full.

TABLE OF CONTENTS

Tables

Figures

Boxes

ABBREVIATIONS

BIMCO	Baltic and International Maritime Council
COSCO	China Ocean Shipping Company
dwt	dead-weight ton(s)
e-commerce	electronic commerce
FEU	40-foot equivalent unit
GDP	gross domestic product
ICT	information and communications technology
IMO	International Maritime Organization
MARPOL	International Convention for the Prevention of Pollution from Ships, 1973, as modified by the Protocol of 1978 relating thereto
OECD	Organization for Economic Cooperation and Development
TEU	20-foot equivalent unit
United Nations Comtrade	United Nations International Trade Statistics Database
WTO	World Trade Organization

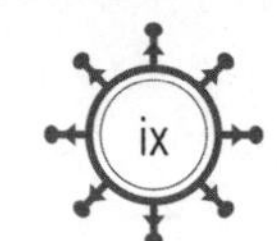

NOTE

The *Review of Maritime Transport* is a recurrent publication prepared by the UNCTAD secretariat since 1968 with the aim of fostering the transparency of maritime markets and analysing relevant developments. Any factual or editorial corrections that may prove necessary, based on comments made by Governments, will be reflected in a corrigendum to be issued subsequently.

This edition of the *Review* covers data and events from January 2019 until June 2020. Where possible, every effort has been made to reflect more recent developments.

All references to dollars ($) are to United States dollars, unless otherwise stated.

"Ton" means metric ton (1,000 kg) and "mile" means nautical mile, unless otherwise stated.

Because of rounding, details and percentages presented in tables do not necessarily add up to the totals.

Two dots (..) in a statistical table indicate that data are not available or are not reported separately.

All websites were accessed in September 2020.

The terms "countries" and "economies" refer to countries, territories or areas.

Since 2014, the *Review of Maritime Transport* does not include printed statistical annexes. Instead, UNCTAD has expanded the coverage of statistical data online via the following links:

Overview: http://stats.unctad.org/maritime

Seaborne trade: http://stats.unctad.org/seabornetrade

Merchant fleet by flag of registration: http://stats.unctad.org/fleet

Merchant fleet by country of ownership: http://stats.unctad.org/fleetownership

National maritime country profiles: http://unctadstat.unctad.org/CountryProfile/en-GB/index.html

Number of port calls, annual: http://stats.unctad.org/portcalls_number_a

Seafarer supply: http://stats.unctad.org/seafarersupply

Share of the world merchant fleet value by country of beneficial ownership: http://stats.unctad.org/vesselvalue_ownership

Share of the world merchant fleet value by flag of registration: http://stats.unctad.org/vesselvalue_registration

Shipbuilding by country in which built: http://stats.unctad.org/shipbuilding

Ship scrapping by country of demolition: http://stats.unctad.org/shipscrapping

Liner shipping connectivity index: http://stats.unctad.org/lsci

Liner shipping bilateral connectivity index: http://stats.unctad.org/lsbci

Container port throughput: http://stats.unctad.org/teu

Vessel groupings used in the *Review of Maritime Transport*

Group	Constituent ship types
Oil tankers	Oil tankers
Bulk carriers	Bulk carriers, combination carriers
General cargo ships	Multi-purpose and project vessels, roll-on roll-off cargo ships, general cargo ships
Container ships	Fully cellular container ships
Other ships	Liquefied petroleum gas carriers, liquefied natural gas carriers, parcel (chemical) tankers, specialized tankers, refrigerated container ships, offshore supply vessels, tugboats, dredgers, cruise, ferries, other non-cargo ships
Total all ships	Includes all the above-mentioned vessel types

Approximate vessel-size groups according to commonly used shipping terminology

Crude oil tankers	
Ultralarge crude carrier	320,000 dead-weight tons (dwt) and above
Very large crude carrier	200,000–319,999 dwt
Suezmax crude tanker	125,000–199,999 dwt
Aframax/longe-range 2 crude tanker	85,000–124,999 dwt
Panamax/long-range 1 crude tanker	55,000–84,999 dwt
Medium-range tankers	40,000–54,999 dwt
Short-range/Handy tankers	25,000–39,000 dwt
Dry bulk and ore carriers	
Capesize bulk carrier	100,000 dwt and above
Panamax bulk carrier	65,000–99,999 dwt
Handymax bulk carrier	40,000–64,999 dwt
Handysize bulk carrier	10,000–39,999 dwt
Container ships	
Neo-Panamax	Container ships that can transit the expanded locks of the Panama Canal with up to a maximum 49 m beam and 366 m length overall; fleets with a capacity of 12,000–14,999 20-foot equivalent units (TEUs) include some ships that are too large to transit the expanded locks of the Panama Canal based on current dimension restrictions.
Panamax	Container ships above 3,000 TEUs with a beam below 33.2 m, i.e. the largest size vessels that can transit the old locks of the Panama Canal.
Post Panamax	Fleets with a capacity greater than 15,000 TEUs include some ships that are able to transit the expanded locks.

Source: Clarksons Research.

Note: Unless otherwise indicated, the ships mentioned in the *Review of Maritime Transport* include all propelled seagoing merchant vessels of 100 gross tons and above, excluding inland waterway vessels, fishing vessels, military vessels, yachts, and fixed and mobile offshore platforms and barges (with the exception of floating production storage, offloading units and drillships).

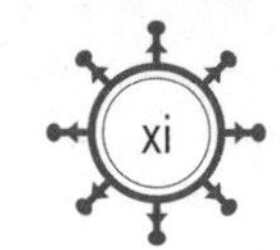

EXECUTIVE SUMMARY

The coronavirus disease (COVID-19) pandemic has underscored the global interdependency of nations and set in motion new trends that will reshape the maritime transport landscape. The sector is at a pivotal moment facing not only immediate concerns resulting from the pandemic but also longer-term considerations, ranging from shifts in supply-chain design and globalization patterns to changes in consumption and spending habits, a growing focus on risk assessment and resilience-building, as well as a heightened global sustainability and low-carbon agenda. The sector is also dealing with the knock-on effects of growing trade protectionism and inward-looking policies.

The pandemic has brought to the fore the importance of maritime transport as an essential sector for the continued delivery of critical supplies and global trade in time of crisis, during the recovery stage and when resuming normality. Many, including UNCTAD and other international bodies, issued recommendations and guidance emphasizing the need to ensure business continuity in the sector, while protecting port workers and seafarers from the pandemic. They underscored the need for ships to meet international requirements, including sanitary restrictions, and for ports to remain open for shipping and intermodal transport operations.

International maritime trade under severe pressure

The global health and economic crisis triggered by the pandemic has upended the landscape for maritime transport and trade and significantly affected growth prospects. UNCTAD projects the volume of international maritime trade to fall by 4.1 per cent in 2020. Amid supply-chain disruptions, demand contractions and global economic uncertainty caused by the pandemic, the global economy was severely affected by a twin supply and demand shock.

These trends unfolded against the backdrop of an already weaker 2019 that saw international maritime trade lose further momentum. Lingering trade tensions and high policy uncertainty undermined growth in global economic output and merchandise trade. Volumes expanded by 0.5 per cent in 2019, down from 2.8 per cent in 2018, and reached 11.08 billion tons in 2019. In tandem, global container port traffic decelerated to 2 per cent growth, down from 5.1 per cent in 2018.

Trade tensions caused trade patterns to shift, as the search for alternative markets and suppliers resulted in a redirection of flows away from China towards other markets, especially in South-East Asian countries. The United States of America increased its merchandise exports to the rest of the world, which helped to somewhat offset its reduced exports to China. New additional tariffs are estimated to have cut maritime trade by 0.5 per cent in 2019, with the overall impact being mitigated by increased trading opportunities in alternative markets.

Increased supply capacity remains a concern for the container shipping industry

At the beginning of 2020, the total world fleet amounted to 98,140 commercial ships of 100 gross tons and above, equivalent to a capacity of 2.06 billion dwt. In 2019, the global commercial shipping fleet grew by 4.1 per cent, representing the highest growth rate since 2014, but still below levels observed during the 2004–2012 period.

Gas carriers experienced the fasted growth, followed by oil tankers, bulk carriers and container ships. The size of the largest container vessel in terms of capacity went up by 10.9 per cent. The largest container ships are now as big as the largest oil tankers and bigger than the largest dry bulk and cruise ships. Experience from other ship types and limitations affecting access channels, port infrastructure and shipyards, suggest that container ship sizes have probably reached a peak.

Economies of scale primarily of benefit to shipping carriers

Larger ports, with more ship calls and bigger vessels, also report better performance and connectivity indicators. Increasing the number of calls by 1 per cent in container ports for example, is associated with a decrease of the time a ship spends in port per container by 0.18 per cent. Similarly, increasing the average vessel size of port calls by 1 per cent decreases the time a ship spends in port per container by 0.52 per cent.

Gains from the economies of scale resulting from the deployment of larger vessels do not necessarily benefit ports and inland transport service providers, as they often increase total transport costs across the logistics chain. A rise in the average call or ship size often leads to peak demand for trucks, yard space and intermodal connections, as well as additional investment requirements for dredging and bigger cranes.

The concentration of cargo in bigger ships and fewer ports often implies business for a smaller number of companies. The cost savings made on the seaside are not always passed on to clients in the form of lower freight rates. This is more evident in markets such as small island developing States, where only few service providers are in operation. These additional costs will have to be borne by shippers, ports and inland transport providers. Thus, economies of scale arising from the deployment of larger vessels accrue mainly to carriers.

Positive performance of freight rates despite the pandemic

As structural container shipping market imbalances remained a concern, liner shipping carriers closely monitored and adjusted ship supply capacity to match the lower demand levels in 2020. Suppressed demand forced container shipping companies to adopt more stringent strategies to manage capacity and reduce costs. Carriers started to significantly reduce capacity in the second quarter of 2020. Capacity management strategies such as suspending services, blanking scheduled sailings and re-routing vessels have all been used. From the perspective of shippers, service cuts and reduced supply capacity meant space limitations to transport goods and delays in delivery dates, affecting supply chains.

In the first half of 2020, freight rates were higher compared with 2019 for most routes, with reported profits of many carriers exceeding 2019 levels. While keeping freight rates at levels that ensure economic viability for the sector may have been justified as a crisis-mitigation strategy, sustained cuts in ship supply capacity for longer periods and during the recovery phase will be problematic for maritime transport and trade, including shippers and ports.

High freight rate volatility in dry and wet bulk segments

Tanker rates surged in March and April 2020, reflecting growing demand for floating storage. The oil market was in a state of super contango where front-month prices were much lower than prices in future months, making storing oil for future sales profitable. Traders chartered tankers to store low-cost crude oil, thereby reducing the availability of vessels for transport and supporting tanker rates. Freight rates declined sharply in May 2020, with about a third of total vessels locked in floating storage returning to active trade and inflating oil supply.

Dry bulk freight rates continued to be shaped by supply and demand imbalances, which increased with the disruptions caused by the pandemic. As a result, rates have shown high volatility especially among the larger vessel categories.

Seafarers and international cooperation: Essential and critical

Due to restrictions relating to the outbreak of COVID-19, large numbers of seafarers had their service extended on board ships after many months at sea, unable to be replaced or repatriated after long tours of duty – unsustainable, both for the safety and well-being of seafarers and the safe operation of ships. Others who had been on break could not return to work, with dire implications for their personal income. UNCTAD and others have issued calls to designate seafarers and other marine personnel, regardless of nationality, as key workers, and exempt them from travel restrictions, to ensure that crew changes can be carried out. In addition, temporary guidance was developed for flag States, enabling the extension of the validity of seafarers and ship licences and certificates under mandatory instruments of the International Maritime Organization (IMO) and the International Labour Organization.

Sustainable shipping, decarbonization and ship pollution control remain priorities

More stringent environmental requirements continue to shape the maritime transport sector. Carriers need to maintain service levels and reduce costs, and at the same time ensure sustainability in operations. Greenhouse gas emissions from international shipping continue to rank high on the international policy agenda. Progress was made at IMO towards the ambition set out in its initial strategy on reduction of greenhouse gas emissions from ships. These include ship energy efficiency, alternative fuels and the development of national action plans to address greenhouse gas emissions from international shipping.

The increase in vessel size, combined with multiple efficiency gains and the recycling of less efficient vessels, have constrained growth in carbon dioxide emissions, despite growth in total fleet tonnage. Some further gains can reasonably be expected over the next decade, as modern eco-designs continue to replace older and less efficient ships. However, these marginal improvements will not be sufficient to meaningfully decrease overall carbon-dioxide emissions as specified in the IMO target of reducing total annual greenhouse gas emissions by at least 50 per cent by 2050 compared with levels in 2008. Achieving these targets will require radical engine and fuel technology changes.

With regard to the protection of the marine environment and the conservation and sustainable use of marine biodiversity, there are several areas where regulatory action has recently been taken or is under way. These include the implementation of the IMO 2020 sulphur limit, ballast-water management, measures to address biofouling, the reduction of pollution from plastics and microplastics, safety considerations of new fuel blends and alternative marine fuels, and the conservation and sustainable use of marine biodiversity of areas beyond national jurisdiction.

The implementation of the IMO sulphur cap regulation as of 1 January 2020 had been considered relatively smooth at the outset. However, difficulties arose in relation to disruptions caused by the COVID-19 pandemic. In March 2020, the ban on the carriage of non-compliant fuel oil entered into force to support the implementation of the sulphur cap. Its enforcement by port State control authorities was limited, due to measures put in place to reduce the number of inspections and contain the risk of spreading the coronavirus. It will be important to ensure

that any delay will not have a negative impact on the long-term implementation of the sulphur cap regulation.

Sustainability and resilience take on their full meaning in small island developing States

Wide-ranging economic impacts of the COVID-19 crisis on small island developing States are likely to exacerbate existing vulnerabilities, making sustainable and resilient transport systems in those States ever more crucial. These States already face unique transport and logistical challenges that derive from their inherent size and geographical, topographical and climate features. These include a significantly lower transport connectivity, a narrow export base and low cargo volumes, limited economies of scale, higher transport costs and exposure to external shocks – as also evidenced by the pandemic.

Some small island developing States are among those with the longest port ship turnaround times and lowest service frequencies. Such States are thus confronted with diseconomies of scale as well as low levels of competition and limited choice for their importers and exporters. On the other hand, some small island developing States can attract trans-shipment services and use the additional fleet deployed to service national trade, as illustrated by the Bahamas, Jamaica and Mauritius. By serving as hub ports handling other countries' trade, these island countries have increased their own liner-shipping connectivity levels, which in turn benefits their respective importers and exporters.

The inherent vulnerabilities of small island developing States put them at the forefront of shocks and disruptions, including from pandemics and climate-change factors. Enabling a sustainable and resilient maritime transportation system in these States requires immediate actions and investment plans that promote low-carbon interregional and domestic shipping solutions and transport connectivity. They also require measures that anticipate and mitigate disruption risks and enable the adaptation of coastal transport infrastructure to climate change impacts and other stressors.

The pandemic's legacy

Maritime transport, as reiterated in the reflections by selected stakeholders showcased in this publication, is essential to keep trade flowing and supply chains connected during and outside crises. While experiences may vary depending on pre-existing conditions and levels of preparedness, all in all, maritime transport and logistics kept essential goods and trade flows moving during the pandemic. However, a number of key trends with wide-ranging policy implications for maritime transport and trade have been observed due to the disruption. These include the following:

A paradigm shift – risk management and resilience-building are becoming new policy and business mantras. Business continuity plans and emergency-response mechanisms have never been as vital as in the case of the COVID-19 crisis. This experience has underscored the need for the maritime transport of the future to be calibrated to risk exposure and for enhanced risk management and resilience-building capabilities to be ensured. Understanding exposure, vulnerabilities and potential losses is key to informing resilience-building in the sector. Industry players and policymakers are expected to increasingly focus on developing emergency-response guidelines and contingency plans to deal with future disruptions. Criteria and metrics on risk assessment and management, digitalization, and harmonized disaster and emergency-response mechanisms are likely to be mainstreamed into relevant national and regional transport policies. Early warning systems, scenario planning, improved forecasts, information sharing, end-to-end transparency, data analytics, business continuity plans and risk management skills will need to feature more prominently on policy agendas and the industry's business plans.

Accelerated shift in globalization patterns and supply chain designs. The slowdown of globalization reflected in lower trade-to-gross domestic product (GDP) ratios observed since the 2008 financial crisis and the regionalization of trade are likely to accelerate, with the post-pandemic world featuring an element of shortened supply chains (near shoring, reshoring) and redundancy (excess stocks and inventory). Investing in warehousing and storage will become more important to ensure sufficient safety stocks and inventories. The established just-in-time supply chain model will be reassessed to include considerations such as resilience and robustness. Diversification in sourcing, routing and distribution channels will grow in importance. Moving away from single country-centric location sourcing to multiple location sourcing that is not only focused on cutting costs and delays but also on risk management and resilience will evolve further.

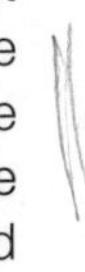

New consumer spending and behaviour. As tastes, consumption and shopping patterns continue to evolve, changes in production and transport requirements are likely to follow. Examples include a further rise in online shopping in the post-pandemic world and a requirement for more customized goods. These trends are likely to emphasize the last-mile transport leg and promote shorter supply chains though the use of three-dimensional printing and robotics. These trends will trigger more demand for warehousing and space for stocks, a move away from established patterns that promoted lean inventory and storage.

A strengthened case for digitalization and dematerialization. Technology, digitalization and innovation will further permeate supply chains and their distribution networks, including transport and logistics.

Adopting technological solutions and keeping abreast of the most recent advances in the field will become a requisite and no longer an option. The pandemic has demonstrated that first movers in terms of technological uptake have been able to better weather the storm (for example, commerce and online platforms, those using blockchain solutions and information technology-enabled third-party logistics companies). The digitalization of interactions and information-sharing has been critical to the continuity of maritime transport operations during the pandemic. It has helped to maintain continuity in transport operations and trade processes while reducing the risk of contagion. Quick deployment of technological solutions has ensured the continuity of business activities and government processes. This has been more evident in the case of cross-border trade and when responding to new consumer expectations in an environment characterized by supply-chain disruption, remote working and increased engagement through business-to-consumer electronic commerce (e-commerce) for business operations.

A significant increase in the use of electronic trade documentation. Governments have made notable efforts to keep their ports operational and speed up the use of new technologies and digitalization. In addition, industry associations have been working to promote the use of electronic equivalents to negotiable bills of lading and their increased acceptance by government authorities, banks and insurers. International cooperation and coordination will be required to ensure that commercial parties across the world readily accept and use electronic records and that legal systems are adequately prepared. Capacity-building may be required, particularly for small and medium-sized enterprises from developing countries that may lack access to the necessary technology or means of implementation.

Standards and interoperability becoming more important. For ports and shipping companies to benefit from benchmarking, data should be comparable, and ship types, key performance indicators, definitions and parameters need to be standardized. For instance, in the long run, the UNCTAD port performance scorecard has the potential to become an industry standard and thus a globally accepted benchmark, helping the port sector to continuously improve its efficiency. UNCTAD seeks to include more port entities and countries from the TrainForTrade network that are not yet reporting in the port performance scorecard component.

Cybersecurity becoming a major concern. Increased cyberattacks in shipping during the COVID-19 crisis were exacerbated by the limited ability of companies to sufficiently protect themselves, including because of travel restrictions, social distancing measures and economic recession. With ships and ports becoming better connected and further integrated into information technology networks, the implementation and strengthening of cybersecurity measures are becoming essential priorities. New IMO resolutions encourage administrations to ensure that cybersecurity risks are appropriately addressed in safety-management systems. Owners who fail to do so are not only exposed to such risks but may have their ships detained by port State control authorities that need to enforce this requirement. Cybersecurity risks are likely to continue to grow significantly as a result of greater reliance on electronic trading and an increasing shift to virtual interactions at all levels. This deepens vulnerabilities across the globe, with a potential to produce crippling effects on critical supply chains and services.

Adjustments in maritime transport to adapt to the new operating landscape. In addition to the oversupply of ship capacity, which remains a concern for carriers, the pandemic and its fallout will heighten competitive pressures and drive stakeholders in the maritime transport sector to increasingly tap new business opportunities to ensure relevance, profitability and business continuity. Some shipping lines and port operators have been taking greater interest in potential business opportunities that may exist in the supply chain through inland logistics. The aim is to be closer to shippers and emerge as reliable end-to-end logistics service providers. Concerns over market concentration and oligopolistic market structures require close monitoring of trends that promote rationalization, consolidation and integration of services to ensure adequate competition levels.

A greater need for systemic and coordinated policy responses at the global level. The pandemic has highlighted the importance of coordinated action when dealing with cross-border disruptions with broad-ranging ripple effects. This has been recognized widely, as illustrated by a call to action by the COVID-19 Task Force on Geopolitical Risks and Responses of the Sustainable Ocean Business Action Platform of the United Nations Global Compact. The document sets out recommendations for urgent political action to keep global ocean-related supply chains moving, stating that "the scale, complexity and urgency of the problem call for a comprehensive, systemic and coordinated approach at the global level."[1] These issues cannot be effectively dealt with on a case-by-case basis, bilaterally or between a limited number of countries.

Six policy actions to prepare for a post-pandemic world

There are six priority areas for policy action to be taken in response to the COVID-19 pandemic and the persistent challenges facing the maritime transport and trade of developing countries.

[1] See www.unglobalcompact.org/news/4534-05-05-2020 and https://ungc-communications-assets.s3.amazonaws.com/docs/publications/Call-To-Action_Imminent-Threats-to-the-Integrity-of-Global-Supply-Chains.pdf.

1. **Support trade so it can effectively sustain growth and development.** Trade tensions, protectionism, export restrictions, particularly for essential goods in times of crisis, bring economic and social costs. These should, to the extent possible, be avoided. Further, non-tariff measures and other obstacles to trade should be addressed, including by stepping up trade facilitation action and customs automation.

2. **Help reshape globalization for sustainability and resilience.** Disruptions caused by the COVID-19 outbreak have re-ignited the debate on the risks associated with international manufacturing production and extended supply chains. It will be important to carefully assess the varied options when it comes to changes in supply-chain design and outcomes that are aligned with the Sustainable Development Goals and the 2030 Agenda for Sustainable Development. For example, a shortening of supply chains through re-shoring or near shoring may reduce transport costs and fuel consumption, but it does not necessarily future-proof supply chains against disruptions that could take place, regardless of the location. Multi-sourcing approaches may guarantee greater resilience than approaches that concentrate production in a single location, whether at home or abroad. The debate on globalization should focus on identifying ways in which unsustainable globalization patterns could be mitigated to generate more value to a wider range of economies.

3. **Promote greater technology uptake and digitalization.** Polices should support a digital transformation that improves the resilience of supply chains and their supporting transportation networks. For maritime transport to play its role in linking global economies and supply chains, it should leverage the crisis by investing in technology and adopting solutions that meet the needs of the supply chains of the future and support resilience efforts. Digitalization efforts should enable enhanced efficiencies, including energy efficiency, and productivity in transport (for example, smart ports and shipping). It should also help countries tap e-commerce capabilities and transport facilitation benefits that boost trade. For more impact, cybersecurity should be strengthened at all levels.

4. **Harness data for monitoring and policy responses.** The use of fast-evolving data capabilities can support efforts to forecast growth and monitor recovery trends. New sources of data and enhanced possibilities emanating from digitalization provide ample opportunities to analyse and improve policies. The pandemic has highlighted the potential for real-time data on ship movement and port traffic, as well as information on shipping schedules to generate early warning systems for economic growth and seaborne trade.

5. **Enable agile and resilient maritime transport systems.** There is a need to invest in risk management and emergency response preparedness beyond pandemics. Future-proofing the maritime supply chain and risk management require greater visibility of door-to-door transport operations. To do so, it is necessary to formulate plans setting out key actions and protocols to be implemented in response to crises while ensuring business continuity. Special consideration is needed to address seafarers' concerns, most of whom come from developing countries. Collaboration across port States and among different actors within countries remains key to improving crew changeover processes and ensuring standardized procedure and risk-management protocols.

6. **Maintain the momentum on sustainability, climate-change adaptation and resilience-building.** Current efforts to deal with carbon emissions from shipping and the ongoing energy transition away from fossil fuels should remain a priority. Governments could direct stimulus packages to support recovery while promoting other priorities such as climate-change mitigation and adaptation action. Thus, policies adopted in the context of a post-pandemic world should support further progress in the shipping industry's transition to greening and sustainability. Meanwhile, sustainability and resilience concerns, such as connectivity among small island developing States and climate-change adaptation, remain key priorities. In these States, critical coastal transport infrastructure is a lifeline for external trade, tourism, and food and energy security. The generation and dissemination of tailored data and information plays an important role in risk assessment, the improvement of connectivity levels, the development of effective adaptation measures, the preparation of targeted studies and effective multidisciplinary and multi-stakeholder collaboration. In addition, progress towards the realization of target 8.1 of the Sustainable Development Goals – sustainable economic growth in the least developed countries – is ever more important to strengthen the resilience of the least developed countries and their ability to cope with future disruptions.

The COVID-19 pandemic is a litmus test, not only for globalization but for global solidarity and collaboration as well. The success of the above-mentioned policy measures will depend on effective international collaboration to ensure coordinated policy responses. Coordinated efforts are also necessary for the standardization of data, tracking of port performance and development of protection mechanisms against cybercrime. In facing the challenges ahead, policymakers should ensure that financial support, technical cooperation and capacity-building are provided to developing countries, in particular the most vulnerable groups of countries, including the least developed countries, landlocked developing countries and small island developing States.

1 INTERNATIONAL MARITIME TRADE AND PORT TRAFFIC

Growth in international maritime trade stalled in 2019, reaching its lowest level since the global financial crisis of 2008–2009. Lingering trade tensions and high policy uncertainty undermined growth in global economic output and merchandise trade and by extension, maritime trade. Maritime trade volumes expanded by 0.5 per cent, down from 2.8 per cent in 2018 and reached a total of 11.08 billion tons in 2019. Growth in world gross domestic product slowed to 2.5 per cent, down from 3.1 per cent in 2018 and 1.1 percentage point below the historical average over the 2001–2008 period. In tandem, global merchandise trade contracted by 0.5 per cent, as manufacturing activity came under pressure and the negative impact of trade tensions between the two largest world economies took a toll on investment and trade.

Against the backdrop of a weaker 2019, the short-term prospects of maritime transport and trade darkened in early 2020. While initial expectations were that 2020 would bring moderate improvements in the economy and trade, the unprecedented global health and economic crisis triggered by the COVID-19 pandemic severely affected the outlook. The fallout on maritime transport and trade was dramatic, with all economic indicators pointing downward. Taking into account the prevailing and persistent uncertainty, UNCTAD estimates that the volume of international maritime trade will fall by 4.1 per cent in 2020. Predicting the timing and scale of the recovery is also challenging, as many factors can significantly influence the outlook. Bearing this in mind, UNCTAD projections indicate that maritime trade will recover in 2021 and expand by 4.8 per cent.

As the debate on the recovery continues to evolve, it is becoming clear that disruptions caused by the COVID-19 pandemic will have a lasting impact on shipping and trade. These disruptions may trigger deep shifts in the overall operating landscape, together with a heightened sustainability and resilience-building imperative. Potential shifts range from changes in globalization patterns to alterations in supply-chain design, just-in-time production models, technology uptake and consumer spending habits. Depending on how these patterns unfold and interact, the implications for maritime transport can be transformational. Further, risk assessment and management, as well as resilience-building to future-proof supply chains and maritime transport, are likely to feature more prominently on policy and business agendas. While maritime transport could emerge as a catalyst supporting some of these trends, it will also need to brace itself for change and adapt and ensure that it is also well prepared to enter the post-COVID-19 pandemic world.

The *Review of Maritime Transport 2020* is structured around five substantive chapters. Chapter 1 considers the demand for maritime transport services. Chapter 2 considers the factors that shape maritime transport infrastructure and services supply, including ship-carrying capacity, ports and related maritime businesses. Chapter 3 assesses the sector's performance using a set of indicators on port calls, port-waiting times, connectivity and the environmental sustainability of ships. Chapter 4 provides an overview of selected contributions received from various stakeholders, including government and industry, sharing experiences and lessons learned in connection with the pandemic. Chapter 5, the final chapter, presents key legal and regulatory developments, as well as trends in technology and innovation affecting maritime transport and trade.

The present chapter on international maritime trade and port traffic reviews major developments in the world economy, merchandise trade, industrial activity and manufacturing supply chains that underpin demand for maritime transport infrastructure and services. Section A discusses volumes of international maritime trade and port traffic and outlines key trends affecting maritime trade in 2019. Section B focuses on the unprecedented health and economic global crisis triggered by the pandemic and considers its immediate impacts and its fallout on the varied shipping segments and ports, as well as its implications for the outlook of maritime transport and trade. Section C concludes with some priority action areas with a view to ensuring the longer-term sustainability and resilience of maritime transport networks and supply chains.

MARITIME TRADE AND PORT CARGO TRAFFIC

IN 2019

SEABORNE TRADE

Growth in maritime trade stalled **+0.5%**

- below 2.8% in 2018
- lowest level since 2008–2009 downturn

Volumes reached **11.08 billion tons**

IMPACTS OF TRADE TENSIONS

Trade tensions and great policy uncertainty undermined growth in maritime trade

Trade diversion and re-routing

United States

China

East As

Brazil

CONTAINER PORT TRAFFIC

Growth in global traffic **+2%**

- down from 5.1% in 2018

811.2 million TEUs handled in container ports worldwide

IN 2020

COVID-19 DISRUPTION

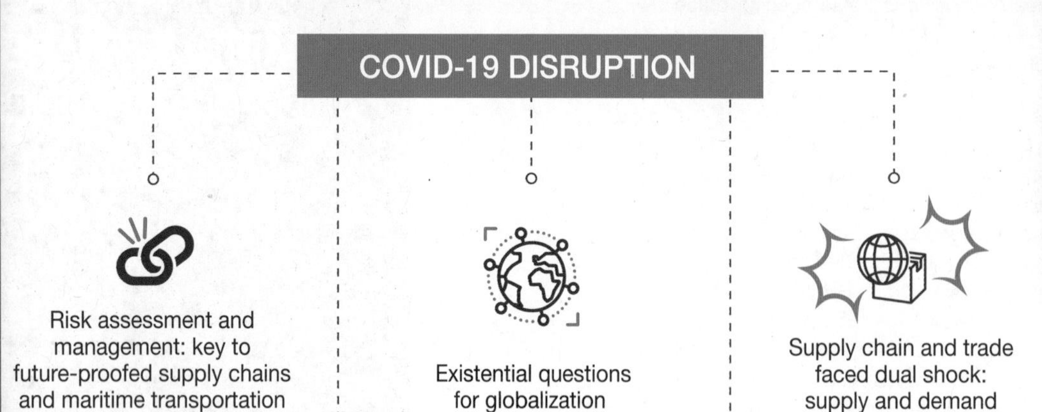

Risk assessment and management: key to future-proofed supply chains and maritime transportation

Existential questions for globalization

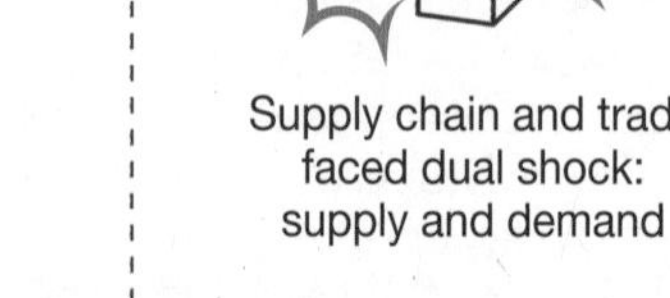

Supply chain and trade faced dual shock: supply and demand

Shockwaves through supply chains, shipping and ports

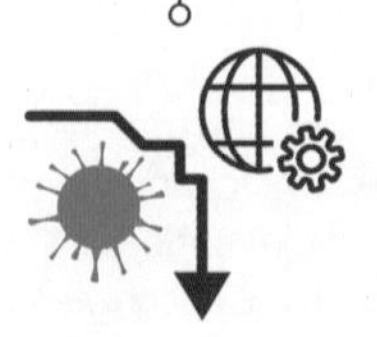

International maritime trade projected to fall by **4.1% in 2020**

LOOKING AHEAD

International maritime trade projected to recover and expand by **4.8% in 2021**

A. VOLUME OF INTERNATIONAL MARITIME TRADE AND PORT TRAFFIC

1. Maritime trade lost momentum in 2019 and came under pressure in 2020

Owing to the slowdown in the world economy and trade, growth in international maritime trade stalled in 2019 and reached its lowest level since the financial crisis of 2008–2009. After rising moderately (2.8 per cent) in 2018, volumes expanded at a marginal 0.5 per cent in 2019. A number of factors weighed on the performance of maritime trade. These included trade policy tensions; adverse economic conditions and social unrest in some countries; sanctions; supply-side disruptions, such as the Vale dam collapse in Brazil and *Cyclone Veronica* in Australia; and low oil demand growth. UNCTAD estimates the total volume of maritime trade in 2019 at 11.08 billion tons (tables 1.1 and 1.2).

As shown in figure 1.1, growth in maritime trade decelerated in line with the slowdown in world GDP growth. Data also point to a negative outlook for 2020, with world GDP and maritime trade projected to contract by 4.1 per cent. The onset of the pandemic in early 2020 and its fallout on world economies, travel, transport and consumption patterns, as well as manufacturing activity and supply chains, are causing a global recession in 2020. See section C for a more detailed discussion on the pandemic and its implications for maritime transport and trade.

2. Negative trends in the world economy and trade put a dent in international maritime trade

Shipping is a derived demand largely determined by developments in the world economy and trade. Therefore, negative economic and trade trends affected maritime trade growth in 2019. Global economic growth decelerated in 2019 against a backdrop of lingering trade tensions and high policy uncertainty. Growth in world GDP slowed down to 2.5 per cent, below 3.1 per cent in 2018 and 1.1 percentage point below the historical average in 2001–2008 (table 1.3). Developed and developing economies alike were affected, reflecting the continued trade tensions between China and the United States and the overall weakening of

Figure 1.1 **Development of international maritime trade and global output, 2006–2020**
(Annual percentage change)

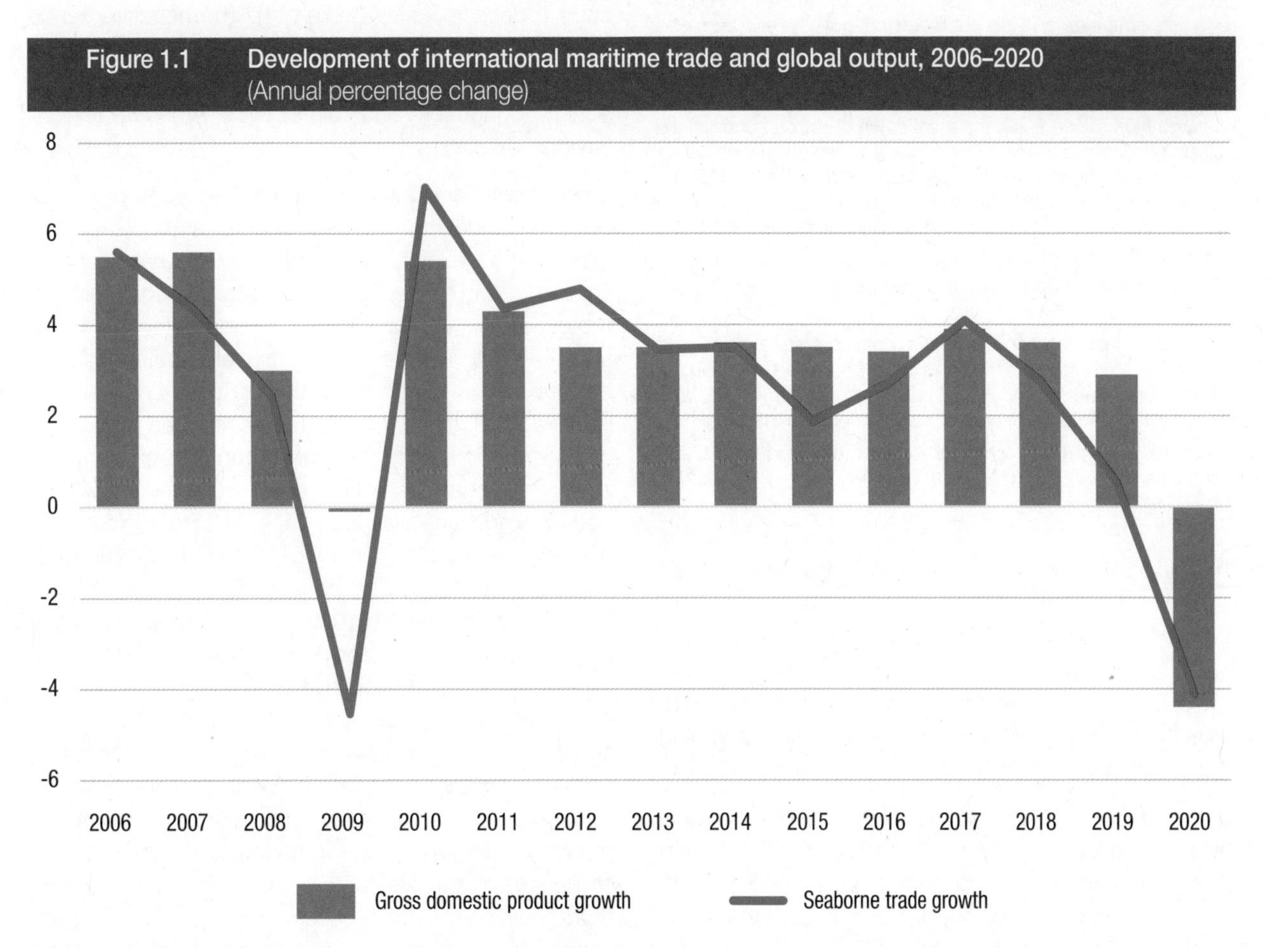

Source: UNCTAD calculations, based on the *Review of Maritime Transport*, various issues, data from UNCTADstat and table 1.12 of this report.

Table 1.1 Development of international maritime trade, selected years (Million tons loaded)

Year	Tanker trader[a]	Main bulk[b]	Other dry cargo[c]	Total (all cargo)
1970	1 440	448	717	2 605
1980	1 871	608	1 225	3 704
1990	1 755	988	1 265	4 008
2000	2 163	1 186	2 635	5 984
2005	2 422	1 579	3 108	7 109
2006	2 698	1 676	3 328	7 702
2007	2 747	1 811	3 478	8 036
2008	2 742	1 911	3 578	8 231
2009	2 641	1 998	3 218	7 857
2010	2 752	2 232	3 423	8 408
2011	2 785	2 364	3 626	8 775
2012	2 840	2 564	3 791	9 195
2013	2 828	2 734	3 951	9 513
2014	2 825	2 964	4 054	9 842
2015	2 932	2 930	4 161	10 023
2016	3 058	3 009	4 228	10 295
2017	3 146	3 151	4 419	10 716
2018	3 201	3 215	4 603	11 019
2019	3 169	3 225	4 682	11 076

Sources: UNCTAD calculations, based on data supplied by reporting countries and as published on government and port industry websites, and by specialist sources. Dry cargo data for 2006 onwards were revised and updated to reflect improved reporting, including more recent figures and a better breakdown by cargo type. Since 2006, the breakdown of dry cargo into main bulk and dry cargo other than main bulk is based on various issues of the *Shipping Review* and *Outlook and Seaborne Trade Monitor,* produced by Clarksons Research. Estimates of total maritime trade figures for 2019 are based on preliminary data or on the last year for which data were available.

[a] Tanker trade includes crude oil, refined petroleum products, gas and chemicals.

[b] Main bulk includes iron ore, grain, coal, bauxite/alumina and phosphate. With regard to data as of 2006, main bulk includes iron ore, grain and coal only. Data relating to bauxite/alumina and phosphate are included under dry cargo other than main bulk.

[c] Includes minor bulk commodities, containerized trade and general cargo.

the world economy. In developed countries, GDP growth decelerated to 1.8 per cent, down from 2.3 per cent in 2018, while developing regions expanded by 3.5 per cent, a relatively higher rate in comparison, but below the 4.3 per cent growth recorded in 2018. Growth in transition economies also stalled, expanding at 2.2. per cent in 2019 against 2.8 per cent in 2018.

In the United States, the supportive effect of fiscal stimulus measures (*New York Times*, 2018) and strong domestic demand that underpinned growth in 2018 diminished slightly in 2019. Growth in the European Union fell to 1.5 per cent, the lowest rate since 2013. Concerns in Europe and the uncertainty surrounding a potential “no-deal” departure from the European Union by the United Kingdom of Great Britain and Northern Ireland (Brexit) had a negative impact on the economy. While the economy of China continued to gradually mature and diversify, trade tensions seem to have contributed to weaker GDP expansion in 2019. Growth slowed to 6.1 per cent, the country’s weakest performance since the early 1990s. Economic growth decelerated across East Asia, South Asia and South-East Asia in varying degrees. In particular, the economy of India slowed down to 4.2 per cent GDP growth in 2019, down from 6.8 per cent in 2018.

In the developing Americas, economic growth was hindered by adverse domestic and global conditions. In 2019, GDP growth in the region contracted by 0.3 per cent. Subdued growth (0.9 per cent) in Western Asia reflected weaker oil prices and geopolitical tensions in the region, including those arising from the sanctions placed on the Islamic Republic of Iran. Growth in Africa remained relatively steady, increasing by 3.1 per cent.

Global merchandise trade contracted in 2019 as manufacturing activity slowed over the course of the year. Rising tariffs have heightened policy uncertainty, undermined investment and weighed on global trade. In 2019, world merchandise trade volumes declined and fell by 0.5 per cent, its lowest level since the financial crisis a decade earlier (table 1.4). The negative trends were mainly driven by a contraction in imports from developing countries, including China, other emerging Asian economies and developing America (United Nations, 2020a).

Global trade tensions increased in 2019 and extended beyond China, the United States and Brexit. For example, complaints were made by several countries against Indian tariffs, reciprocal allegations of protectionism were put forward by the European Union and the United States, and a trade dispute occurred between Japan and the Republic of Korea. For example, in June 2020, the United States announced that it was considering imposing more tariffs on European goods in view of the contention over subsidies to Airbus and Boeing. The new list of goods that may face duties of up to 100 per cent, potentially doubling the price of certain goods, caused European stocks to fall, particularly those of beverage companies, luxury goods manufacturers and truck makers (Whitten and Ben-Moussa, 2020). Such developments, together with rising nationalist sentiment (MDS Transmodal, 2020a) and inward-looking policies, added to the uncertainty, caused business confidence to deteriorate, affected investment growth in many countries and undermined global trade. This environment also amplified the challenges in the electronics and automotive sectors, both of which have large international production value chains. These two sectors were hit particularly hard. However, some countries gained export market shares

Table 1.2 International maritime trade in 2018–2019 (Type of cargo, country group and region)

Designation	Year	Goods loaded				Goods unloaded			
		Total	Crude oil	Other tanker trade[a]	Dry cargo	Total	Crude oil	Other tanker trade[a]	Dry cargo
Millions of tons									
World	2018	11 019.0	1 881.0	1 319.7	7 818.3	11 016.8	2 048.8	1 338.6	7 629.4
	2019	11 075.9	1 860.2	1 308.4	7 907.3	11 083.0	2 033.4	1329.3	7 720.3
Developed economies	2018	3 862.8	206.2	507.5	3 149.1	3 844	931.9	494.8	2 417.8
	2019	3 935.2	242.9	506.9	3 185.4	3 780	913.6	472.6	2 394.0
Transition economies	2018	713.0	203.8	37.6	471.6	99.4	0.3	4.8	94.3
	2019	715.8	193.9	41.1	480.8	102.0	0.8	5.4	95.8
Developing economies	2018	6 443.4	1 471.1	774.6	4 197.6	7 072.9	1 116.6	839.0	5 117.3
	2019	6 424.8	1 423.3	760.3	4 241.2	7 200.7	1 118.9	851.3	5 230.5
Africa	2018	763.0	297.4	70.4	395.2	501.8	39.0	99.9	362.8
	2019	762.1	293.5	69.9	398.7	504.5	39.2	99.3	365.9
America	2018	1 385.4	200.6	88.7	1 096.1	638.1	47.1	149.3	441.8
	2019	1 386.3	204.2	82.3	1 099.8	621.7	47.8	138.8	435.1
Asia	2018	4 280.4	971.3	607.8	2 701.3	5 918.9	1 029.7	584.7	4 304.5
	2019	4 261.8	923.9	600.5	2 737.5	6 059.1	1 031.1	607.7	4 420.3
Oceania	2018	14.5	1.7	7.8	5.1	14.1	0.8	5.0	8.2
	2019	14.6	1.8	7.7	5.1	15.4	0.7	5.5	9.1
Designation	Year	Goods loaded				Goods unloaded			
		Total	Crude oil	Other tanker trade[a]	Dry cargo	Total	Crude oil	Other tanker trade[a]	Dry cargo
Percentage share									
World	2018	100.0	17.1	12.0	71.0	100.0	18.6	12.2	69.3
	2019	100.0	16.8	11.8	71.4	100.0	18.3	12.0	69.7
Developed economies	2018	35.1	11.0	38.5	40.3	34.9	45.5	37.0	31.7
	2019	35.5	13.1	38.7	40.3	34.1	44.9	35.5	31.0
Transition economies	2018	6.5	10.8	2.8	6.0	0.9	0.0	0.4	1.2
	2019	6.5	10.4	3.1	6.1	0.9	0.0	0.4	1.2
Developing economies	2018	58.5	78.2	58.7	53.7	64.2	54.5	62.7	67.1
	2019	58.0	76.5	58.1	53.6	65.0	55.0	64.0	67.8
Africa	2018	6.9	15.8	5.3	5.1	4.6	1.9	7.5	4.8
	2019	6.9	15.8	5.3	5.0	4.6	1.9	7.5	4.7
America	2018	12.6	10.7	6.7	14.0	5.8	2.3	11.1	5.8
	2019	12.5	11.0	6.3	13.9	5.6	2.4	10.4	5.6
Asia	2018	38.8	51.6	46.1	34.6	53.7	50.3	43.7	56.4
	2019	38.5	49.7	45.9	34.6	54.7	50.7	45.7	57.3
Oceania	2018	0.1	0.1	0.6	0.1	0.1	0.0	0.4	0.1
	2019	0.1	0.1	0.6	0.1	0.1	0.0	0.4	0.1

Source: UNCTAD calculations, based on data supplied by reporting countries and as published on government and port industry websites, and by specialist sources. Dry cargo data for 2006 onwards were revised and updated to reflect improved reporting, including more recent figures and a better breakdown by cargo type. Estimates of total maritime trade figures for 2019 are based on preliminary data or on the last year for which data were available.

Note: For longer time series and data prior to 2019, see UNCTADstat Data Centre (http://unctadstat.unctad.org/wds/TableViewer/tableView.aspx?ReportId=32363).

[a] Includes refined petroleum products, gas and chemicals.

Table 1.3 World economic growth, 2018–2021
(Annual percentage change)

Region or country	Average 2001–2008	2018	2019	2020[a]	2021[a]
World	**3.6**	**3.1**	**2.5**	**-4.3**	**4.1**
Developed countries	**2.3**	**2.3**	**1.8**	**-5.8**	**3.1**
of which:					
European Union (27)	2.1	2.1	1.5	-7.3	3.5
Japan	1.2	0.3	0.6	-4.5	1.9
United States	2.6	2.9	2.3	-5.4	2.8
Developing countries	**6.6**	**4.3**	**3.5**	**-2.1**	**5.7**
of which:					
Africa	5.8	3.1	3.1	-3.0	3.5
East Asia	9.2	5.9	5.4	1.0	7.4
of which:					
China	10.9	6.6	6.1	1.3	8.1
South Asia	6.7	5.1	2.8	-4.8	3.9
of which:					
India	7.6	6.8	4.2	-5.9	3.9
South-East Asia	5.7	5.1	4.4	-2.2	4.3
Western Asia	5.5	2.0	0.9	-4.5	3.6
Latin American and the Caribbean	**3.9**	**0.6**	**-0.3**	**-7.6**	**3.0**
of which:					
Brazil	3.7	1.3	1.1	-5.7	3.1
Caribbean	5.0	3.5	1.9	-6.4	2.3
Transition economies	**7.2**	**2.8**	**2.2**	**-4.3**	**3.5**
of which:					
Russian Federation	6.8	2.3	1.3	-4.2	3.4

Source: UNCTAD calculations, based on UNCTAD, 2020a, *Trade and Development Report 2020: From Global Pandemic to Prosperity for All – Avoiding Another Lost Decade*, chapter 1.

[a] Forecast.

Table 1.4 Volumes of exported and imported goods, selected group of countries, 2018–2020
(Annual percentage change)

Group/country	Volume of exports (percentage change)			Volume of imports (percentage change)		
	2018	2019	2020[a]	2018	2019	2020[a]
World	**3.1**	**-0.5**	**-9.0**	**3.8**	**-0.4**	**-8.8**
Developed countries *of which:*	**2.6**	**0.0**	**-12.4**	**2.5**	**0.2**	**-10.9**
Euro area	1.9	-0.2	-13.3	2.2	0.0	-12.1
Japan	2.6	-1.6	-11.3	3.1	0.9	-4.9
United States	4.2	-0.5	-13.3	5.2	-0.3	-9.8
Other developed countries	**2.9**	**1.1**	**-10.8**	**0.5**	**0.6**	**-11.6**
Developed countries *of which:*	**3.7**	**-1.7**	**-4.7**	**5.7**	**-1.2**	**-5.7**
China	5.4	0.5	-4.5	6.9	-0.4	-0.9
Africa and the Middle East	1.0	-3.9	-5.2	0.8	-0.2	-2.8
Asia (not including China)	3.7	-1.7	-3.9	6.9	-2.3	-7.1
Latin America	3.0	0.5	-7.0	4.8	-1.6	-12.8
Transition economies	**3.9**	**-1.3**	**-4.1**	**2.2**	**3.1**	**-5.9**

Source: UNCTAD calculations, based on CPB World Trade Monitor, August 2020. Data source and methodology are aligned with UNCTAD, 2020a, *Trade and Development Report 2020: From Global Pandemic to Prosperity for All – Avoiding Another Lost Decade*.

Note: Country coverage in the aggregated country groupings is not comprehensive.

[a] Percentage change between the average for the period January to June 2020 and January to June 2019.

as companies looked for new suppliers from countries that were not directly affected by the rising tariffs.

In December 2019, China and the United States agreed on the first phase of a trade agreement to help de-escalate the tensions between the two economies. On 15 January 2020, both countries signed the agreement on the understanding that China would increase its merchandise imports from the United States by $200 billion (United Nations, 2020a). In return, the United States would cut by half its 15 per cent tariffs on $120 billion of imports from China. In Europe, reduced uncertainty over Brexit was a welcome development, although the European Union and the United Kingdom still needed to define a new trading relationship before January 2021 (United Nations, 2020a). In June 2020, the United Kingdom outlined new customs and border arrangements for 2021 and indicated its commitment to introducing a three-phase plan of import changes, building new border facilities for carrying out required checks and providing targeted support to ports to build new infrastructure (Lloyd's Loading List, 2020a). Further, the European Union is expected to impose full customs controls and checks on goods from the United Kingdom starting 1 January 2021 (United Nations, 2020a).

3. Regional and country grouping contribution to maritime trade

In 2019, developing economies continued to account for the lion's share of goods being loaded (58 per cent) and unloaded (65 per cent) in seaports worldwide (figure 1.2). Together, developed economies and economies in transition generated 42 per cent of global merchandise exports by sea (goods loaded) and imported 35 per cent (goods unloaded) of such global trade. While the role of developing regions as a source and destination for maritime trade is significant, developing economies are not a homogenous group. The grouping includes countries and economies in

Figure 1.2 Participation of developing economies in international maritime trade, selected years
(Percentage share in total tonnage)

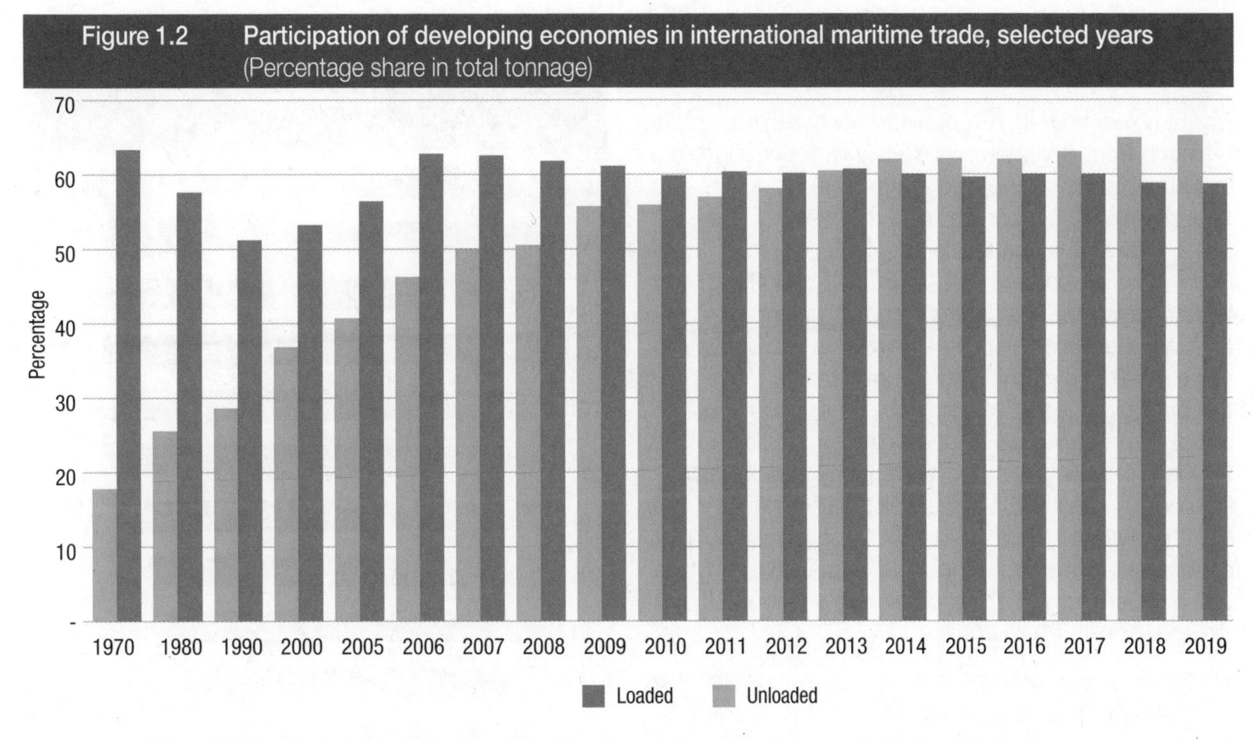

Source: UNCTAD calculations, based on the *Review of Maritime Transport*, various issues and table 1.2 of this report.

varying stages of development and degrees of integration in the world's manufacturing and trading networks. Much of the growth recorded over the past decade is largely driven by fast growing emerging developing countries, most notably China. These countries have also been driving the structural shift in trade patterns observed since 2013, whereby volumes unloaded in developing countries exceeded volumes loaded. The shift is a reversal of a historical pattern where developing countries acted as suppliers of large-volume low-value raw materials imported by developed countries.

There is a predominance of Asian and intra-Asian trade in globalized production processes and value chain growth. A closer look at this trend indicates that the globalization of manufacturing processes has never been truly global. There is scope for other developing regions within and outside Asia to diversify their economies, expand their maritime transport capacity and participate more effectively in regional and international production processes. The marginal contribution of these economies to global value chains is reflected in their relatively limited contribution to container trade flows and global container port throughput. Maritime transport, combined with supportive trade and industrial policies, can be a catalyst for growth and greater integration in the world economy for a broader range of such developing countries.

In 2019, 41 per cent of the total goods loaded (exported) were sourced from Asia and 62 per cent of total goods unloaded (imported) were received in this same region (figure 1.3). The contribution of developing America and Africa to maritime trade flows remained marginal. In comparison, and as previously noted, Asia has benefited from a greater integration into global manufacturing and trading networks, promoting intraregional trade. Capitalizing on the fragmentation of globalized production processes, Asia has become a maritime hub that brings together over 50 per cent of global maritime trade volumes.

Figure 1.3 International maritime trade, by region, 2019
(Percentage share in total tonnage)

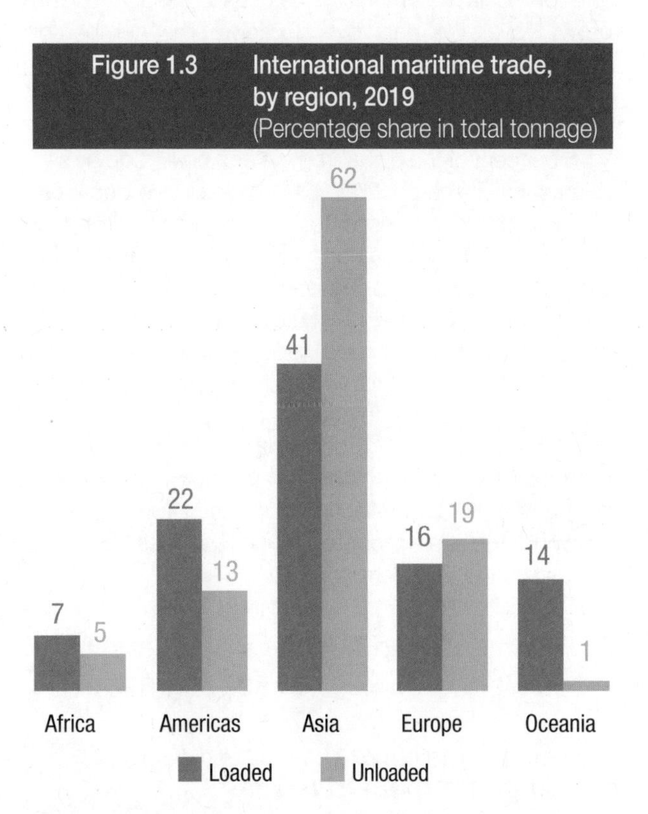

Source: UNCTAD calculations, based on data supplied by reporting countries and as published on government and port industry websites, and by specialist sources.

Note: Estimated figures are based on preliminary data or on the last year for which data were available.

4. Maritime trade underperformed across market segments

Dry cargo continued to account for over two thirds of total maritime trade volumes, while liquid bulk commodities, including crude oil, refined petroleum products, gas and chemicals, accounted for the remaining share. In 2019, growth in all market segments decelerated. Trade in dry cargo expanded at 1.1 per cent over 2018, and tanker trade volumes contracted by 1 per cent. A look at how the various market segments have evolved since 1990 shows that growth in maritime trade over the past three decades has been sustained by bullish trends in containerized trade volumes starting in the 2000s, coinciding with the wave of hyperglobalization (figures 1.4 and 1.5). It was also supported by the swift growth of trade in dry bulk commodities that accompanied the rapid industrial expansion of China that accelerated with its accession to the World Trade Organization (WTO) in 2001.

When adjusted for distances travelled, international maritime trade grew at a slightly faster rate of 1 per cent in 2019, supported by growing long-haul oil exports from Brazil and the United States to Asia. Clarksons Research estimates seaborne trade in ton-miles to have reached 59,503 billion ton-miles in 2019 (figure 1.6).

Figure 1.7 shows that trade in ton-miles by cargo expanded in varying degrees. Trade in container and dry bulk commodities has fuelled much of the growth over the past two decades. The number of cargo ton-miles generated by dry cargo has been rising steadily over the years. In 2002, China imported 121.7 million tons of iron ore and coal, accounting for 11.8 per cent of the global iron ore and coal trade by sea (Clarksons Research, 2006). In less than two decades, these volumes increased to 1.3 billion tons, bringing the country' market share to nearly 50 per cent of the world total (Clarksons Research, 2020b). Gas trade in ton-miles expanded swiftly to 9.9 per cent in 2019. Other segments recorded relatively smaller growth; ton-miles generated by trade in chemicals expanded by 3.2 per cent, followed by container trade (1.9 per cent) and other dry cargo (1.6 per cent). Growth in ton-miles produced by trade in oil and major bulk commodities contracted in 2019, reflecting declines in iron ore trade following the disruption to mining activities in Brazil caused by the Vale dam collapse.

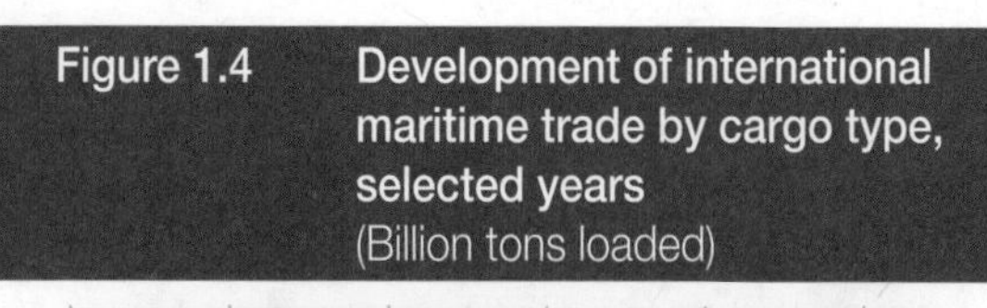

Figure 1.4 Development of international maritime trade by cargo type, selected years
(Billion tons loaded)

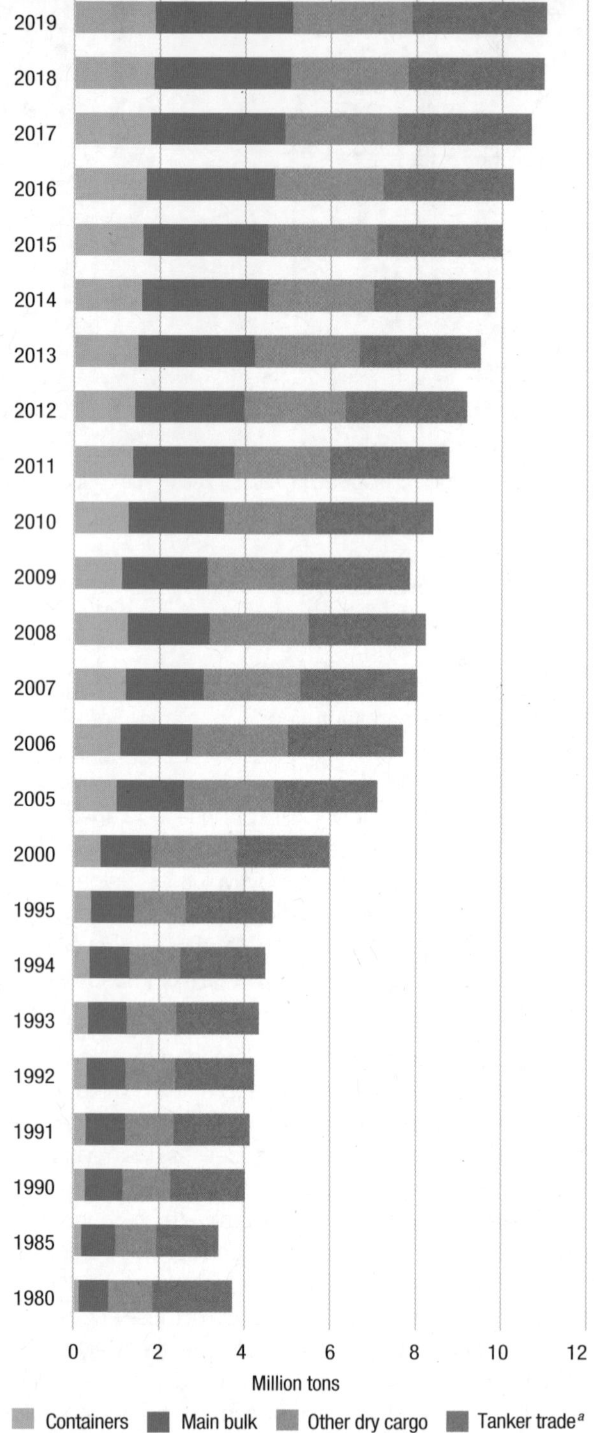

Source: UNCTAD, Review of Maritime Transport, various issues. For 2006–2019, the breakdown by cargo type is based on Clarksons Research, 2020a, *Shipping Review and Outlook*, spring 2020 and *Seaborne Trade Monitor, various issues*.

Note: 1980–2005 figures for main bulk include iron ore, grain, coal, bauxite/alumina and phosphate. With regard to data starting in 2006, main bulk figures include iron ore, grain and coal only. Data relating to bauxite/alumina and phosphate are included under "other dry cargo".

[a] Tanker trade includes crude oil, refined petroleum products, gas and chemicals.

5. Demand and supply-side pressures weighed on key market segments

Trade in oil weakened, while trade in gas remained robust

Since the onset of the shale revolution in the United States, developments in the country's energy sector have played a significant role in shaping global tanker trade. This was apparent throughout 2019, with a

Figure 1.5 Development of international maritime trade by cargo type, selected years
(Index: 1990 = 100)

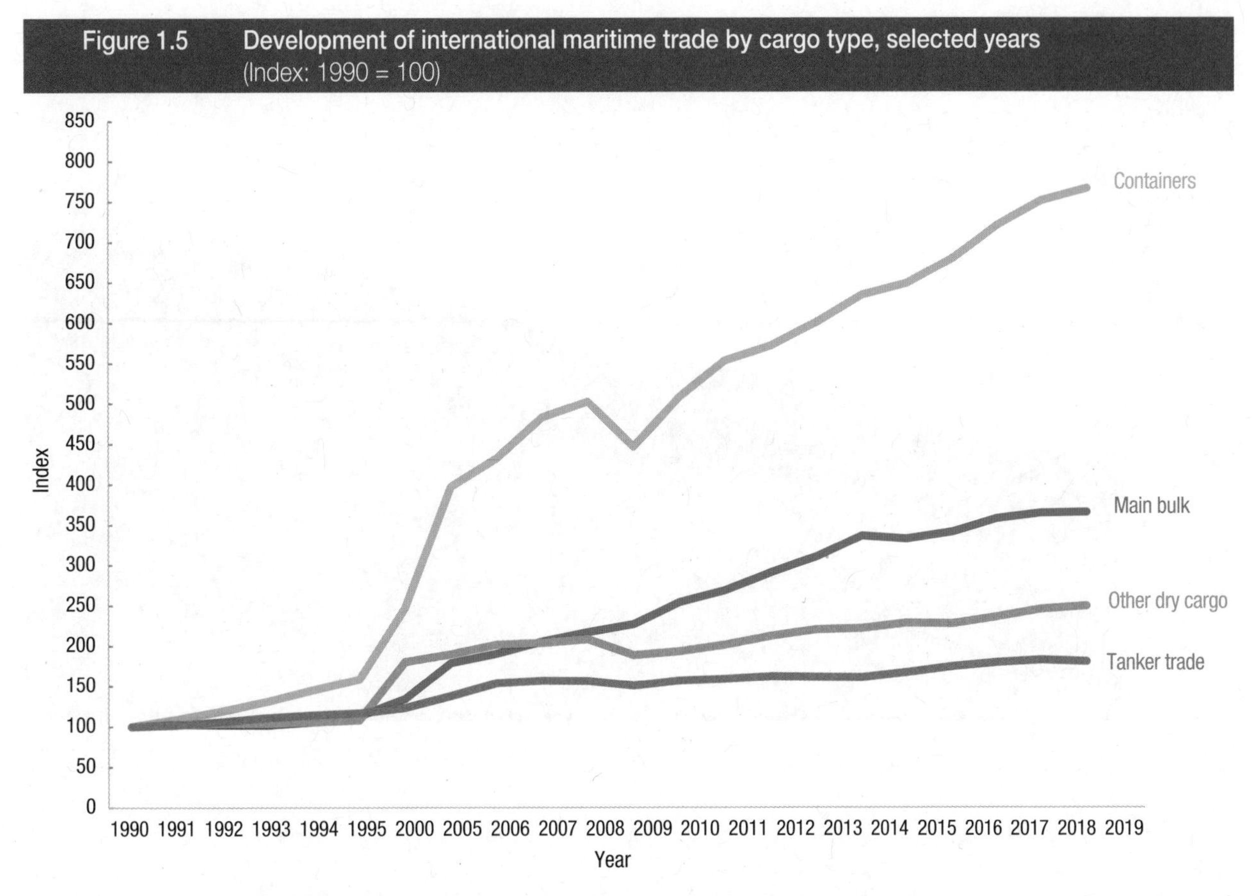

Source: UNCTAD, *Review of Maritime Transport*, various issues. For 2006–2019, the breakdown by cargo type is based on Clarksons Research, 2020a, *Shipping Review and Outlook,* spring 2020 *and Seaborne Trade Monitor*, various issues.

Note: 1980–2005 figures for main bulk include iron ore, grain, coal, bauxite/alumina and phosphate. Since 2006, main bulk figures include iron ore, grain and coal only. Data relating to bauxite/alumina and phosphate are included under "other dry cargo". Tanker trade includes crude oil, refined petroleum products, gas and chemicals.

decline in United States crude oil imports and a rise in its long-haul exports. Overall tanker trade contracted by 1 per cent in 2019, owing to lower volumes of crude oil and refined petroleum products (table 1.5). An overview of global players in the oil and gas sector is presented in table 1.6.

Crude oil trade decreased by 1.1. per cent in 2019. Downside factors include the cuts in supply by members of the Organization of the Petroleum Exporting Countries aimed at supporting oil prices, as well as disruptions affecting exports from the Islamic Republic of Iran and the Bolivarian Republic of Venezuela. The impact on exports from Western Asia resulting from the attacks on Saudi oil infrastructure was limited. Pressure on the demand side include lower global oil demand, a sharp reduction in United States imports and a decline in global refinery activity. However, expansion in exports from Brazil and the United States have supported long-haul journeys from the Atlantic to Asia. Crude oil imports to China increased by 10.6 per cent in 2019, compared with the previous year, while imports to the United States declined (Clarksons Research, 2020c). In Asia, extended refinery maintenance and smaller refining margins contributed to limiting import growth (Clarksons Research, 2020d).

Other tanker trade experienced difficulty in 2019, contracting by nearly 1 per cent. Major setbacks included slower global economic growth and extended refinery maintenance periods, with many refiners adjusting production in preparation for the coming into force on 1 January 2020 of the IMO 2020 regulation on a sulphur cap for marine fuels. In addition, naphtha faced competition from liquefied petroleum gas as a petrochemical feedstock, arbitrage opportunities were limited (Clarksons Research, 2020e) and fuel oil trade declined. The latter accounts for over 20 per cent of trade in seaborne refined petroleum products (Clarksons Research, 2020d).

Mexican imports, a key driver of global trade growth in recent years, dropped in 2019 as domestic supply increased. Growth in imports to Latin America and rising exports from China provided support to product tanker demand.

Trade in gas remained strong, with volumes expanding by nearly 11 per cent in 2019. Trade in liquefied natural gas increased by 11.9 per cent, supported by project start-ups in Australia and the United States. In comparison, trade in liquefied petroleum gas grew by 6 per cent, driven largely by growing supply in Australia, Canada and the United States (Clarksons Research, 2020c). Despite the

Figure 1.6 International maritime trade in cargo ton-miles, 2000–2020
(Billion ton-miles)

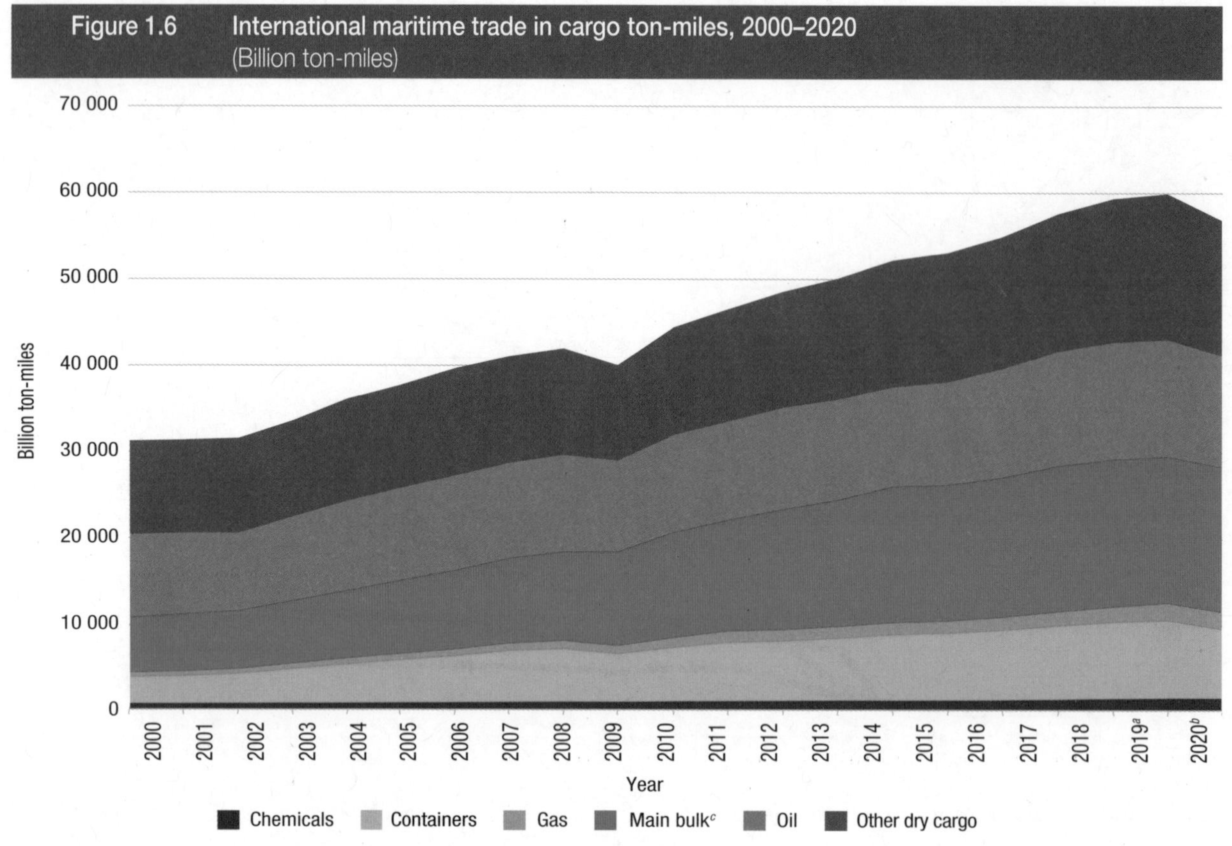

Source: Clarksons Research, 2020a, *Shipping Review and Outlook*, spring.

Note: Seaborne trade data in ton-miles are estimated by Clarksons Research. Given methodological differences, containerized trade data in tons sourced from Clarksons Research are not comparable with data in TEUs sourced from MDS Transmodal.

[a] Estimated.

[b] Forecast.

[c] Includes iron ore, grain and coal.

Figure 1.7 International maritime trade in cargo ton-miles, 1999–2020
(Billion ton-miles; index: 1999 = 100)

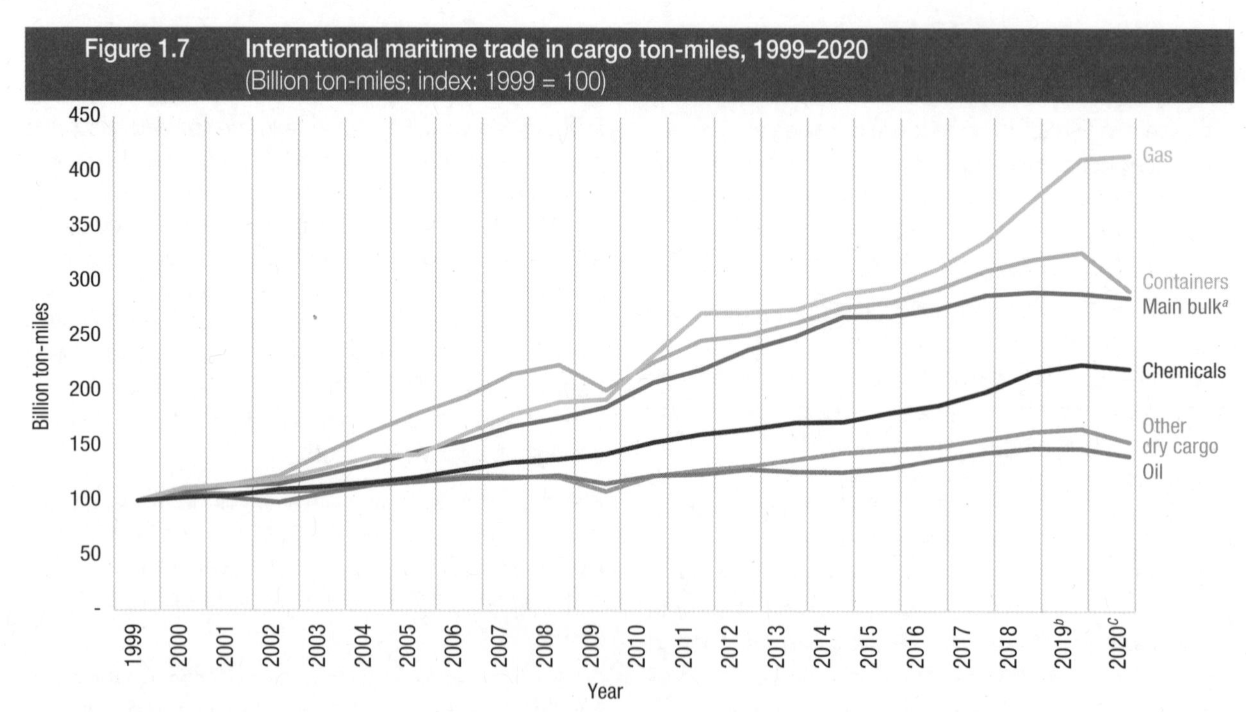

Source: Clarksons Research, 2020a, *Shipping Review and Outlook*, spring.

Note: Seaborne trade data in ton-miles are estimated by Clarksons Research. Given methodological differences, containerized trade data in tons sourced from Clarksons Research are not comparable with data in TEUs sourced from MDS Transmodal.

[a] Includes iron ore, grain and coal.

[b] Estimated.

[c] Forecast.

Table 1.5	Tanker trade, 2018–2019 (Million tons and annual percentage change)		
Tanker trade[a]	2018	2019	Percentage change 2018–2019
Crude oil	1 881	1 860	-1.1
Other tanker trade[a] *of which:*	1 320	1 308	-0.9
Gas	416	461	10.8
Total tanker trade	**3 201**	**3 169**	**-1.0**

Sources: UNCTAD calculations, derived from table 1.2 of this report.

Note: Gas figures are derived from Clarksons Research, 2020c, *Seaborne Trade Monitor*, Volume 7, No. 6, June.

[a] Includes refined petroleum products, gas and chemicals.

Table 1.6	Major producers and consumers of oil and natural gas, 2019 (World market share in percentage)		
World oil production	Percentage	World oil consumption	Percentage
Western Asia	32	Asia and the Pacific	36
North America	23	North America	23
Transition economies	16	Europe	15
Developing America	9	Western Asia	9
Africa	9	Developing America	9
Asia and the Pacific	8	Transition economies	4
Europe	3	Africa	4
Oil refinery capacities		**Oil refinery throughput**	
Asia and the Pacific	35	Asia and the Pacific	37
North America	21	North America	22
Europe	15	Europe	15
Western Asia	11	Western Asia	11
Transition economies	8	Transition economies	8
Developing America	7	Developing America	5
Africa	3	Africa	2
World natural gas production		**World natural gas consumption**	
North America	27	North America	25
Transition economies	21	Asia and the Pacific	22
Western Asia	17	Transition economies	15
Asia and the Pacific	17	Western Asia	15
Europe	6	Europe	13
Developing America	6	Developing America	6
Africa	6	Africa	4

Source: UNCTAD calculations, based on data published in British Petroleum 2020, *BP [British Petroleum] Statistical Review of World Energy 2020*, June 2020.

Note: Oil includes crude oil, shale oil, oil sands and natural gas liquids. The latter term excludes liquid fuels from other sources such as biomass and coal derivatives.

trade tensions, long-haul United States exports to Asia continued to expand steadily due to substitution trends and limited growth in Western Asian exports stemming from sanctions and supply cuts. With regard to imports, China and India remained key markets. Imports into China picked up speed in 2019 compared with 2018, supported by its petrochemical sector demand and the coming online of new propane dehydrogenation capacity. Reduced shipments from the United States were offset by increased imports from Africa, Australia and Western Asia. In India, import demand for liquefied petroleum gas was supported by the continued rollout of liquefied petroleum gas infrastructure in rural areas under a government subsidy programme.

While trade in chemicals rose rapidly in 2018, there was little growth in the segment in 2019, reflecting pressure on demand. In China, demand for palm oil soared in 2019, given higher domestic soybean oil prices as a consequence of the trade tensions and the African swine fever affecting pig farming in China, causing a reduction in soymeal feed. Strong demand in India for palm oil, following a decline in import taxes in January 2020, supported growth in this segment. Trade in palm oil remains highly sensitive to policy shifts, such as the rise in Indian import duties on Malaysian palm oil (The Indian Express, 2020), the decision by the European Union to phase out palm oil-based biofuel by 2030 and higher taxes on Indonesian biofuel and liquefied petroleum gas.

The mainstay of maritime trade, growth in dry bulk commodity trade, faltered in 2019

Major bulk

Dry bulk commodities, in particular minerals and ores, are closely linked to industrial and steel production, as well as manufacturing and construction.[1] With many relevant indicators pointing downward in 2019, global trade in dry bulk lost momentum during the year and grew marginally, (0.5 per cent), bringing the total to 5.3 billion tons (table 1.7) (Clarksons Research, 2020f). An overview of global players in the dry bulk commodities and steel trade sector is presented in table 1.8.

For the first time in 20 years, iron ore trade fell by 1.5 per cent due to severe supply-side disruptions caused by the Vale dam collapse in Brazil and Cyclone Veronica in Australia. Other factors at play include a shift in the make-up of steel production in China, which favours scrap steel over imported iron ore. As China represented 72 per cent of global seaborne iron ore imports in 2019 (Clarksons Research, 2020f), changes affecting its import demand could have a strong impact on trade in global dry bulk commodities. Australia and Brazil are major suppliers of iron ore to China. However, growing Chinese investments in Guinea are likely to make this

[1] Detailed figures on dry bulk commodities are derived from Clarksons Research, 2020f.

Table 1.7 Dry bulk trade, 2018–2019
(Million tons and annual percentage change)

	2018	2019	Percentage change 2018–2019
Major bulks[a] *of which:*	3 215.0	3 225.0	0.3
Coal	1 263.0	1 293.0	2.4
Grain	475.0	477.0	0.4
Iron ore	1 477.0	1 455.0	-1.5
Minor bulk *of which:*	2 010.0	2 028.0	0.9
Forest products	380.0	382.0	0.5
Steel products	388.0	371.0	-4.4
Total dry bulk	**5 225.0**	**5 253.0**	**0.5**

Source: UNCTAD calculations, based on Clarksons Research, 2019d, *Dry Bulk Trade Outlook,* Volume 26, No. 6, June.

[a] Includes iron ore, coal (steam and coking) and grains (wheat, coarse grain and soybean).

Table 1.8 Major dry bulk commodities and steel: Producers, users, exporters and importers, 2019
(World market shares in percentage)

Steel producers		Steel users	
China	53	China	51
India	6	India	6
Japan	5	United States	6
United States	5	Japan	4
Russian Federation	4	Republic of Korea	3
Republic of Korea	4	Russian Federation	2
Germany	2	Germany	2
Turkey	2	Turkey	1
Brazil	2	Italy	1
Other	17	Other	24
Iron ore exporters		**Iron ore importers**	
Australia	57	China	72
Brazil	23	Japan	8
South Africa	5	Europe	7
Canada	4	Republic of Korea	5
India	2	Other	8
Sweden	2		
Other	7		
Coal exporters		**Coal importers**	
Indonesia	35	China	19
Australia	30	India	18
Russian Federation	12	Japan	15
United States	6	European Union	11
South Africa	6	Republic of Korea	11
Colombia	6	Taiwan Province of China	5
Canada	3	Malaysia	3
Other	2	Other	18
Grain exporters		**Grain importers**	
Brazil	25	East and South Asia	46
United States	22	Western Asia	14
Argentina	13	Africa	13
Ukraine	12	South and Central America	12
European Union	8	European Union	10
Russian Federation	7	North America	1
Canada	6	Other	4
Australia	3		
Other	4		

Sources: UNCTAD calculations, based on data from Clarksons Research, 2020f, *Dry Bulk Trade Outlook*, Volume 26, No. 6, June; World Steel Association, 2019, World Steel short range outlook October 2019, 14 October; World Steel Association, 2020, *2020 World Steel in Figures*.

country an important alternative source of supply that may capture part of the Chinese market (Drewry, 2020a). Although growth in the economy of China continued to decelerate, its steel demand expanded by 7.8 per cent in 2019, largely driven by real estate investment (World Steel Association, 2019). By contrast, steel demand was low in the rest of the world. The Chinese manufacturing sector, similarly to that of many other countries, came under pressure due to the slowing economy and the effect of trade tensions, particularly on the manufacturing and automotive industries.

In 2019, growth in coal (coking and thermal) trade slowed to 2.4 per cent, reflecting fewer thermal coal imports into Europe and lower coking coal demand in China. With regard to exports of thermal and coking coal, Indonesia remained in the top position, with a share of 35.3 per cent, followed by Australia with 29.7 per cent (Clarksons Research, 2020g). In China, seaborne thermal coal imports increased by 9.2 per cent, supported by lower coal prices and government efforts to stimulate industrial activity and growth. The country's topping up of its domestic coal supply with imports is a key risk factor for global seaborne coal trade. Its import demand varies according to domestic output, prices and government policies, including decarbonization and air pollution control efforts. In India and countries of South-East Asia, imports continued to rise, given new coal-fired power generation capacities. India, the world's largest seaborne coking coal importer, and Viet Nam, which is becoming a major steel producer, increased their coking coal imports in 2019 to support growth in their steel sectors.

Agricultural bulk commodities, notably grains, are an important issue in trade tensions between China and the United States. In 2019, grain volumes expanded by 0.4 per cent. Soybean imports into China, which accounted for about 60 per cent of global soybean imports, continued to be affected by the new tariffs and the spread of swine fever in the country's pig population. In this context and through a substitution effect, Brazil overtook the United States as the world's largest seaborne grain exporter. The United States has long been the world's largest grain exporter and, if fully implemented, the first phase of a trade agreement between China and the United States could potentially support increased

soybean and other grain exports from the United States. Shipping can benefit from this development, with the two exporters complementing each other, since the grain export season in the United States runs from September to February, and that of Brazil, from March to September.

Minor bulk

A contraction of 4.4 per cent in the trade of steel products detracted from the overall growth in seaborne shipments of minor bulk commodities. In 2019, minor bulk volumes expanded by 0.9 per cent, down from 3.8 per cent in 2018 (Clarksons Research, 2020g). Exports from China, Japan, the Republic of Korea and the Russian Federation came under pressure as demand from Europe and the United States lessened. Imports into China of some minor bulk commodities, namely nickel ore, bauxite and cement, continued to support this type of trade. An important development with a potential impact on this segment is a ban placed by Indonesia on nickel ore exports that came into force in January 2020. However, exports from the Philippines and New Caledonia may help to partially bolster trade in these commodities.

Other dry cargo: Containerized trade

In 2019, global containerized trade expanded at a slower rate of 1.1 per cent, down from 3.8 per cent in 2018 bringing the total to 152 TEUs (figure 1.8). Much of the growth was driven by activity on non-mainlane East–West, South–South and intraregional trade routes. Excluding intraregional flows, global containerized trade increased by 0.4 per cent in 2019. The challenges facing the global car industry and motor manufacturing in 2019 have had some impact, as trade in automotive-related goods is an important sector for some individual trade lanes. Global car sales decreased for the first time by about 1.5 per cent in 2018, after steady growth for over a decade. Sales continued to decline in 2019. China, the largest market, recorded a double-digit drop. In addition to the slowdown in the economy, other factors came into play: new emissions standards, a shift towards electrification, greater durability of cars with an extended life cycle and the growing popularity of used cars and ridesharing (Drewry, 2019).

Mainlane East–West containerized trade routes, namely Asia–Europe, the trans-Pacific and the transatlantic, handled 39.1 per cent of worldwide containerized trade flows in 2019. Trade on other routes, which involves greater participation from developing countries, has gained in importance over time, as these countries accounted for 60.9 per cent of containerized trade in 2019 (figure 1.9 and table 1.9). Together, intraregional trade, principally intra-Asian flows, and South–South trade represented over 39.9 per cent of the total in 2019.

Figure 1.8 Global containerized trade, 1996–2020
(Million 20-foot equivalent units and annual percentage change)

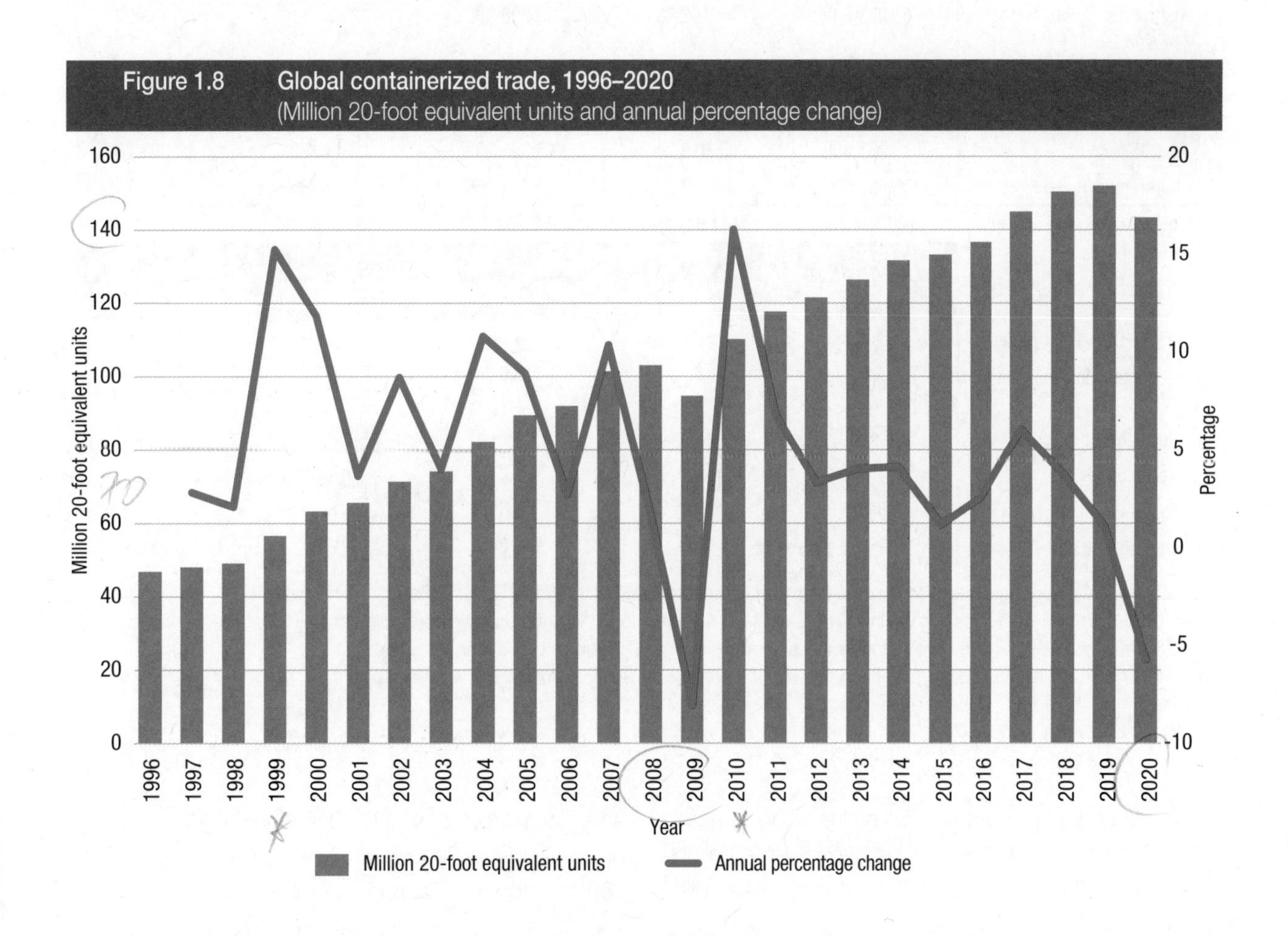

Source: UNCTAD calculations, based on data from MDS Transmodal, 2020b, 19 August.

Figure 1.9 Market share of global containerized trade by route, 2019
(Percentage)

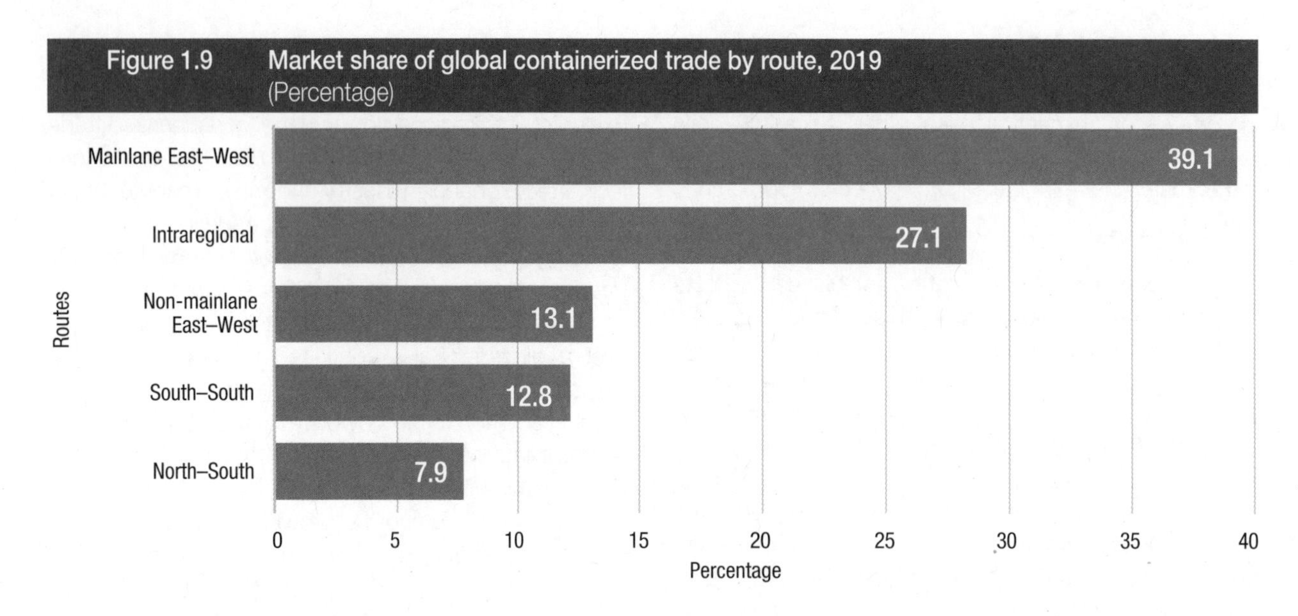

Source: UNCTAD calculations, based on data from MDS Transmodal, 2020b, World Cargo Database, 19 August.

Table 1.9 Containerized trade on mainlane East–West routes and other routes, 2016–2020
(20-foot equivalent units and annual percentage change)

	2016	2017	2018	2019	2020[a]
	20-foot equivalent units				
Mainlane East–West routes	54 610 793	57 695 035	60 512 411	59 451 778	55 529 706
Other routes *of which:*	81 973 339	87 152 831	89 796 992	92 439 115	87 733 977
Non-mainlane East–West	17 928 632	18 977 780	18 961 472	19 869 413	18 099 717
North–South	11 108 989	11 753 235	11 963 148	12 018 424	11 576 259
South–South	16 251 689	17 619 241	18 898 303	19 433 908	18 007 289
Intraregional	36 684 030	38 802 575	39 974 069	41 117 369	40 050 711
World total	**136 584 133**	**144 847 866**	**150 309 403**	**151 890 894**	**143 263 682**
	Percentage change				
	2016	**2017**	**2018**	**2019**	**2020[a]**
Mainlane East–West routes	4.06	5.6	4.9	-1.8	-6.6
Other routes (non-mainlane) *of which:*	1.59	6.3	3.0	2.9	-5.1
Non-mainlane East–West	2.7	5.9	-0.1	4.8	-8.9
North–South	-0.31	5.8	1.8	0.5	-3.7
South–South	-0.98	8.4	7.3	2.8	-7.3
Intraregional	2.83	5.8	3.0	2.9	-2.6

Source: UNCTAD calculations, based on data from MDS Transmodal, 2020b, World Cargo Database, 19 August.
Notes: Non-mainlane East–West: Trade involving East Asia, Europe, North America and Western Asia and the Indian subcontinent.
North–South: Trade involving Europe, Latin America, North America, Oceania and sub-Saharan Africa.
South–South: Trade involving East Asia, Latin America, Oceania, sub-Saharan Africa and Western Asia.
[a] Forecast.

The continued prominence of Asia as the world's factory continued to boost expansion in intra-Asian container trade, with a growing contribution from South-East Asia.

Non-mainlane, or secondary East–West trade routes and North–South routes accounted for 13.1 per cent and 7.9 per cent of the market, respectively. Trade on the non-mainlane East–West routes involves flows between the Far East and Western Asia, the Far East and South Asia, South Asia and Europe, and Western Asia and Europe, for example. Sanctions on the Islamic

Republic of Iran and geopolitical tensions in the region create volatility on these types of trade. Cargo bound for Saudi Arabia and the United Arab Emirates make up over 50 per cent of the containers carried from the Far East to Western Asia. In 2019, trade on the westbound leg of this route increased, reflecting the gradual economic recovery in these two countries. Imports into Iraq also improved, which may reflect an element of diverted trade away from the Islamic Republic of Iran. The number of imports on the Eastern Asia–South Asia lane diminished in 2019 in line with poor economic performance in India. Lower consumption demand, as well as bans on waste imports, and declining vehicle sales and car manufacturing contributed to lower growth. It appears at the time of writing (September 2020) that India, unlike Viet Nam, has not yet capitalized on the trade tensions between China and the United States to attract the production moving away from China (Drewry, 2019).

In 2019, main East–West trade lanes contracted by 1.8 per cent, compared with positive growth on other routes (+2.9 per cent growth). Trade tensions and escalating tariffs between China and the United States took a toll on trans-Pacific containerized trade. Volumes on this key East–West lane contracted by 4.7 per cent in 2019. This reflected a decrease of 7.4 per cent on the peak leg, East Asia–North America, on the one hand, and a 3.8 per cent drop on the return leg from North America to East Asia, on the other (table 1.10). Although significant, the slump in trade flows was moderated by the substitution of Chinese volumes by exports to the United States from other Asian economies. The substitution impact became clear: the number of shipments from China and Hong Kong, China declined, while those from from several other countries rose (Malaysia, Thailand and Viet Nam – and to a lesser extent – Indonesia, Japan, the Republic of Korea and Taiwan Province of China).

Table 1.10 Containerized trade on major East–West trade routes, 2014–2020
(Million 20-foot equivalent units and annual percentage change)

	Trans-Pacific			Asia–Europe			Transatlantic		
	Eastbound	Westbound		Eastbound	Westbound		Eastbound	Westbound	
Year	East Asia–North America	North America–East Asia	Trans-Pacific	Northern Europe and Mediterranean to East Asia	East Asia–Northern Europe and Mediterranean	Total Asia–Europe	North America–Northern Europe and Mediterranean	Northern Europe and Mediterranean–North America	Transatlantic
2014	16.2	7.0	**23.2**	6.3	15.5	**21.8**	2.8	3.9	**6.7**
2015	17.4	6.9	**24.3**	6.4	15.0	**21.3**	2.7	4.1	**6.8**
2016	18.2	7.3	**25.5**	6.8	15.3	**22.1**	2.7	4.3	**7.0**
2017	19.4	7.3	**26.7**	7.1	16.4	**23.4**	3.0	4.6	**7.5**
2018	20.8	7.4	**28.2**	7.0	17.3	**24.3**	3.1	4.9	**8.0**
2019	20.0	6.8	**26.8**	7.2	17.5	**24.7**	2.9	4.9	**7.9**
2020	18.1	7.0	**25.1**	6.9	16.1	**23.0**	2.8	4.7	**7.4**
Annual percentage change									
2014–2015	7.9	-2.0	**4.9**	1.4	-2.6	**-1.4**	-2.4	5.6	**2.2**
2015–2016	4.4	6.6	**5.1**	6.3	2.5	**3.6**	0.4	2.9	**1.9**
2016–2017	6.7	-0.5	**4.7**	4.1	6.9	**6.0**	7.9	8.3	**8.1**
2017–2018	7.0	0.9	**5.4**	-1.3	5.7	**3.6**	5.8	6.8	**6.4**
2018–2019	-3.8	-7.4	**-4.7**	2.9	1.4	**1.8**	-5.0	-0.2	**-2.1**
2019–2020	-9.7	2.6	**-6.6**	-3.6	-8.3	**-6.9**	-5.3	-5.8	**-5.6**

Source: UNCTAD calculations, based on MDS Transmodal, 2020b, World Cargo Database, 19 August.

Volumes on the Asia–Europe trade lane grew by 1.8 per cent. Volumes on the westbound leg expanded by 1.4 per cent, supported by the replenishment by European importers of their own stocks, inventory building in the United Kingdom before Brexit and an increased export focus by China on Europe (Clarksons Research, 2020h). Eastbound volumes from Europe to Asia rose by 2.9 per cent, strengthened by an uplift in refrigerated pork shipments in response to the outbreak of African swine fever in China (Drewry, 2019). Shipments of wastepaper and plastic also increased in 2019, as loads destined for recycling in China reflected greater compliance with the country's new regulations on waste contamination levels or, alternatively, were redirected to markets outside China, such as Indonesia and Malaysia.

Transatlantic trade volumes declined by 2.1 per cent in 2019. Volumes on the eastbound journey from North America to Europe contracted at 5.0 per cent. On the westbound leg, the number of imports into the United States fell slightly (0.2 per cent), reflecting a reduced need to ship parts and components for motor vehicle manufacturing in the United States. The potentially negative impact of escalating trade tensions between the European Union and the United States remained a major reason for concern. In October 2019, WTO authorized the United States to apply new tariffs of 25 per cent on $7.5 billion worth of imports from the European Union, following a 15-year dispute over subsidies granted to Airbus. The European Union has since threatened to also apply tariffs to the United States, and WTO is expected to make a decision regarding the United States subsidies to Boeing (Drewry, 2019). The possibility that tariffs may be applied to European exports of cars and motor vehicle parts to the United States remains a concern.

6. Trade tensions curbed maritime shipments and caused trade patterns to shift

In 2019, the United States increased its merchandise exports to the rest of the world, which helped offset to a certain extent reduced exports to China. Less than 2 per cent of world maritime trade in metric tons and 7 per cent of containerized cargo are estimated to be subject to the new tariffs introduced by China and the United States between 2018 and 2019 (Clarksons Research, 2020a). It is estimated that additional tariffs curbed maritime trade by 0.5 per cent in 2019, the overall impact of which was mitigated by substitution trends, that is to say, by exporting and/or importing from alternative markets, and the extent to which demand for tariffed goods is sensitive to increased tariff levels. The quest for alternative markets and suppliers resulted in changing trade patterns and a redirection of flows away from China towards other markets, especially in South-East Asia, thereby promoting the deployment of smaller vessels in intra-Asian trade (Clarksons Research, 2020a).

Between 2017 and 2019, all major shipping segments experienced declines in exports of tariffed goods. Although United States exports of such goods were redirected to new markets, they failed to fully compensate for the volumes lost to China. This is the case for dry bulk commodities exports, for example. A greater number of exports to the rest of the world may have added volumes but did not support maritime trade in ton-miles, as countries importing more dry bulk commodities from the United States were at a shorter distance, compared with China.

Viet Nam benefited the most from the changing trade patterns triggered by trade tensions. Although there has been some migration in sourcing to other countries in South-East Asia since 2018, the market shares of Cambodia, Indonesia, Malaysia, the Philippines, Singapore and Thailand did not increase at the same pace as those of Viet Nam. The share of China in United States imports from Asia dropped to 63.8 per cent in 2019, down from 69.1 per cent in 2018 (JOC.com, 2020a). The production of some goods, such as electronics and footwear, had already been delocalized to Viet Nam as the country continued to boost its capacity to receive new business by developing port and inland transportation infrastructure and upgrading manufacturing skills. In a parallel development, carriers added trans-Pacific services at ports in Viet Nam. Other South-East Asian nations were also expanding their manufacturing base, but at a slower pace. Different patterns are associated with each of the containerized and bulk trades. In general, the bulk commodities and raw material cargoes sectors seek different markets, while the containerized and manufactured goods sectors seek alternative suppliers.

7. Slower growth in port traffic in 2019 and shifts in port-call patterns

UNCTAD estimates that growth in global container port throughput decelerated to 2 per cent in 2019, down from 5.1 per cent in 2018. In 2019, some 811.2 million TEUs were handled in container ports worldwide, reflecting an additional 16.0 million TEUs over 2018 (table 1.11).

Table 1.11 World container port throughput by region, 2018–2019
(Million 20-foot equivalent units and annual percentage change)

	20-foot equivalent units		Annual percentage change 2018–2019
	2018	2019	
Asia	514.9	526.7	2.3
Europe	121.7	123.6	1.5
North America	61.6	62.5	1.6
Latin America and the Caribbean	52.3	52.6	0.7
Africa	31.3	32.5	3.9
Oceania	13.5	13.2	-2.2
Small island developing States			
Oceania	13.5	13.2	-2.2
World total	**795.3**	**811.2**	**2.0**

Sources: UNCTAD calculations, based on data collected by various sources, including Lloyd's List Intelligence, Dynamar B. V., Drewry, as well as information published on the websites of port authorities and container port terminals.

Note: Data are reported in the format available. In some cases, estimates of country volumes are based on secondary source information, reported growth rates and estimates based on correlations with other variables, such as the liner shipping connectivity index of UNCTAD. Country totals may conceal the fact that minor ports may not be included. Therefore, in some cases, data in the table may differ from actual figures.

In 2019, nearly 65 per cent of global port-container cargo handling was concentrated in Asia – the share of China alone exceeded 50 per cent (figure 1.10). Europe ranked second in terms of container port-handling volumes,

Figure 1.10 Estimated world container port throughput by region, 2019
(Percentage share in total 20-foot equivalent units)

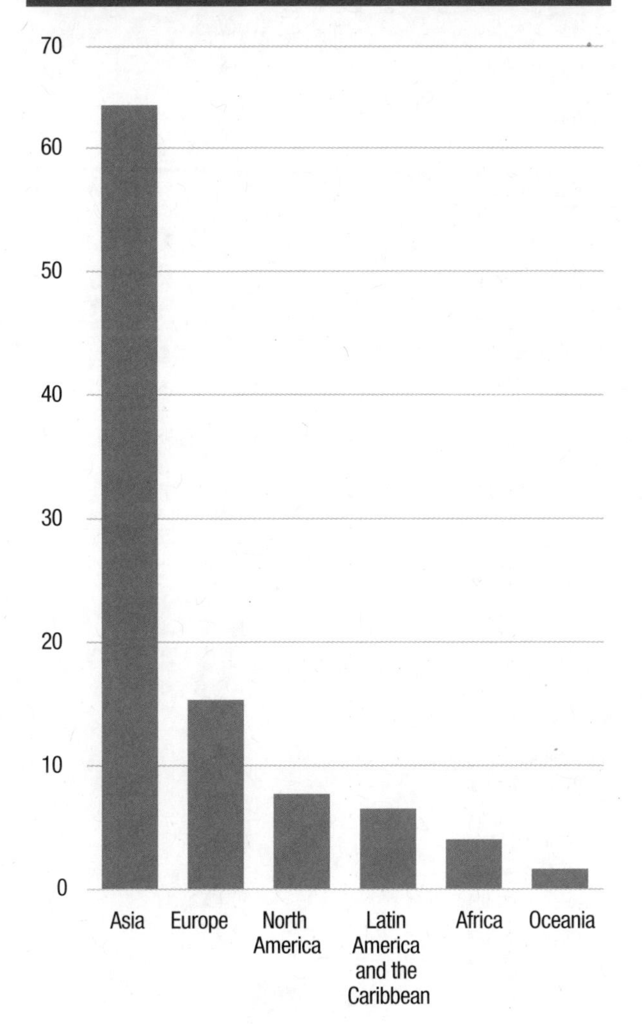

Sources: UNCTAD calculations, derived from table 1.11 of this report.

behind Asia, whose share was more than four times greater. Other regions in descending order are North America (7.7 per cent), Latin America and the Caribbean (6.5 per cent), Africa (4 per cent) and Oceania (1.6 per cent).

Although the rankings of the world's top 20 container ports in 2019 changed little compared with 2018, slower growth in the world economy and trade translated into moderated growth in global container port throughput. As shown in figure 1.11(a) and (b), there were reductions in volumes handled in some ports such as Dalian, China; Dubai, United Arab Emirates; Hong Kong, China; and Long Beach, United States. In comparison, container port activity continued to grow in other ports such as Antwerp, Belgium; Hamburg, Germany; Klang, Malaysia; Qingdao, China; and Tianjin, China (Lloyd's List, 2020).

In China, growth in Shanghai lagged behind that of Ningbo in 2019, as the latter benefited from feeder and rail traffic growth. During the year, six new rail connections came into operation and helped attract more traffic from neighbouring provinces, reflecting government policy to concentrate container trade in selected ports to prevent unhealthy port competition. Volumes in Hong Kong, China dipped by 6.3 per cent, as the political crisis had a negative impact on the economy. The port has also been losing market share to ports in mainland China. Qingdao and Tianjin, China have seen more domestic traffic move by sea as a result of government anti-pollution measures to restrict trucking operations.

In South-East Asia, the port of Klang, Malaysia continued to capture more trans-shipment market share. However, this was not sufficient to recover the entire volumes that had been moving to Singapore for some time. Cargo handled by the port of Tanjung Pelepas, Malaysia increased by 1.55 per cent, while growth in Singapore remained at 1.63 per cent.

European ports recorded less volume growth, reflecting the persistent weakness that had plagued the manufacturing sector and importers drawing from stocks and inventories. Rotterdam, the Netherlands expanded volumes by 2.1 per cent compared with 2018, while Antwerp, Belgium achieved 6.8 per cent growth. The move of THE Alliance's Atlantic services in Germany from Bremerhaven to Hamburg, is reflected in the 2019 throughput of these ports. Hamburg recorded an increase of 6.1 per cent in volumes handled, supported by the addition of new connections to Baltic services, while Bremerhaven recorded a decline in volumes (Drewry, 2020b).

Container port throughput at North American ports moderated in 2019. West coast ports performed poorly, compared with the east coast and the coast of the Gulf of Mexico. Ports on the United States west coast lost market share in the combined import-export market. While the trend accelerated with the trade tensions, there was already a tendency for cargo to move away from the west coast of North America. In 2019, the share of the ports of Los Angeles and Long Beach, United States dropped to 22.9 per cent, down from 26.5 per cent in 2015. Cargo migration has also had an impact on the west coast ports of Canada and Mexico, in particular, the ports of Vancouver, Lázaro Cárdenas and Manzanillo, which also lost some market share.

In the United States, exporters looked for other export markets to avoid the increased reciprocal tariffs imposed by China (JOC.com, 2020b). As previously noted, trade tensions required shippers to find alternative markets and source imports from locations outside China, such as South-East Asia. Thailand and Viet Nam benefited from the change in trade patterns and routing, while the market share of China shrank. Ports on the Atlantic and Gulf coasts are better positioned to handle shipments arriving from other parts of Asia. The performance of the ports of Houston and Savannah, United States, for example, whose market share increased, is a case in point.

Figure 1.11 Leading 20 global container ports, 2018–2019
in (a) million 20-foot equivalent units and (b) annual percentage change

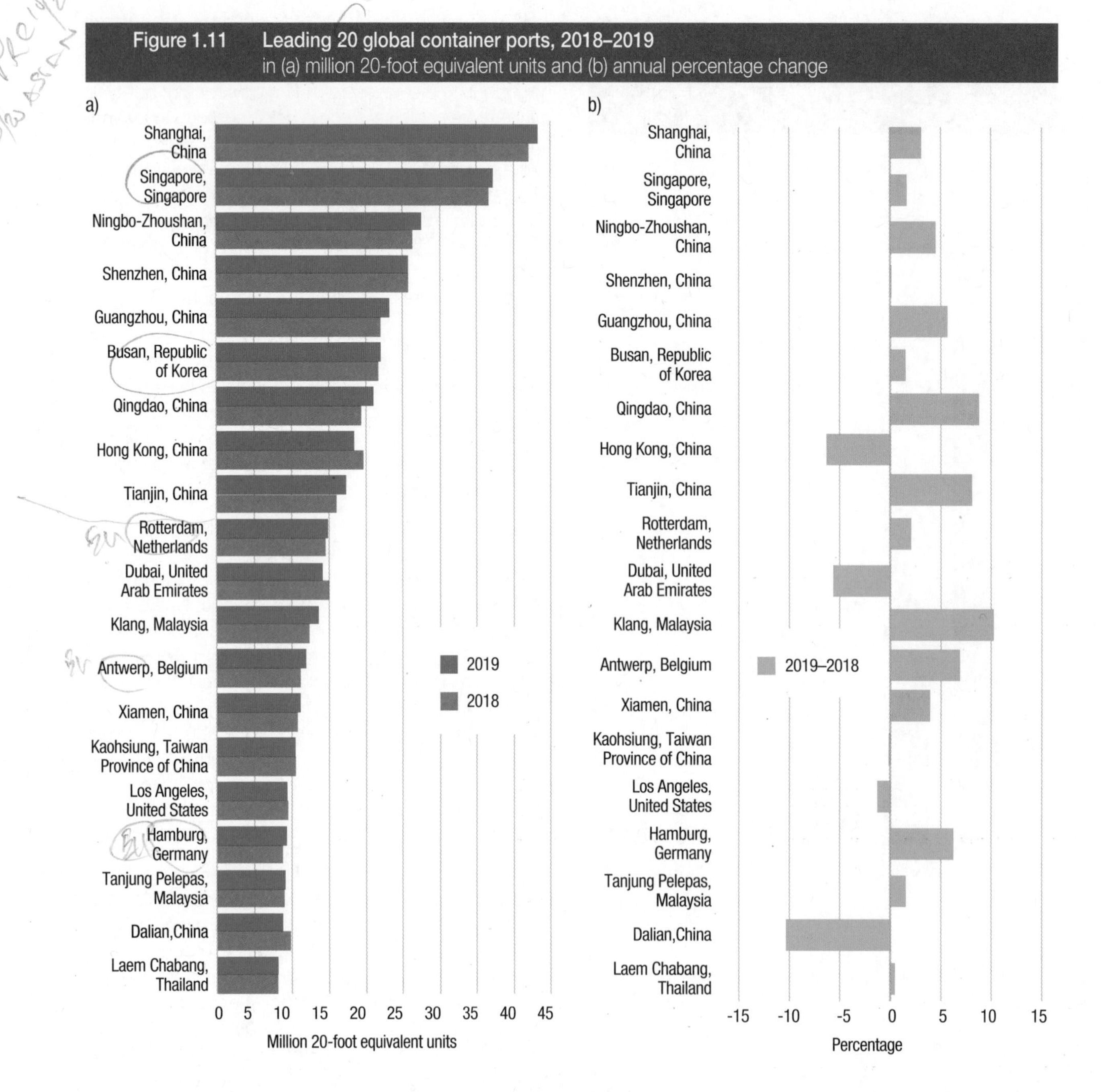

Sources: UNCTAD calculations, based on Lloyd's List, 2020a, *One Hundred Ports*.

Challenging economic trends in Argentina, recession in Brazil and social unrest in Chile constrained cargo volumes in ports of Latin America and the Caribbean. However, some ports such as Freeport in the Bahamas; Itajaí, Sao Francisco do Sul and Paranaguá in Brazil; and two Panama Pacific terminals recorded positive growth. In Western Asia, container port volumes continued to be affected by sanctions and political tensions. In 2019, the gradual recovery of the economies of Saudi Arabia and the United Arab Emirates provided some support to port-handling activity, while in the Islamic Republic of Iran, volumes in Bandar Abbas decreased. In the United Arab Emirates, Khalifa port activity rose, as both the China Ocean Shipping Company (COSCO) and the Mediterranean Shipping Company moved more business over to their respective terminals, away from Jebel Ali (Drewry, 2020b).

Growth in container activity in South Asia stalled in 2019, reflecting slower economic growth in India and austerity measures in Pakistan. While the ports of Jawaharlal Nehru and Mundra reported some growth, Chennai port continued to lose traffic to newer east coast ports such as Kattupalli. Other Indian ports such as Visakhapatnam and Krishnapatnam are benefiting from increased trans-shipment and coastal traffic generated by a relaxation of the country's cabotage rules. In Sri Lanka, subdued growth in Colombo reflected a declining trend in gateway traffic and some erosion in trans-shipment

cargo because of the amended cabotage rules in India. In Africa, a weakening in the economies of Nigeria and South Africa constrained container port volume growth. In Oceania, container port activity declined by 2.2 per cent as the economy of Australia slowed down and consumer confidence fell (Drewry, 2019).

8. Adapting port strategies and seeking new opportunities

Today, ports are showing more interest in strengthening connections with the hinterland to get closer to the shippers and tap the cargo volumes that could be committed. Providing intermodal access, warehousing and other logistics services illustrates the type of actions that may help ports capture local market volumes. For example, the port of Savannah, United States has, for three decades, been a pioneer in driving port centric logistics and is growing as a hub for retail import distribution. In the Republic of Korea, the port of Busan is investing in port-distribution centres ("distriparks") to strengthen its position as a regional logistics centre. In Egypt, the port of Damietta is focusing more and more on its gateway market as opposed to the trans-shipment business. This is illustrated by the development of recent dry port and rail connection projects (Drewry, 2019). This change in strategy, as well as a gradual shift towards further mergers and acquisitions, as opposed to the development of new projects, reflects the uncertainty surrounding the outlook for port growth and the need to diversify business strategies and respond to the evolving landscape (Drewry, 2020b). For example, China Merchants Port Holdings concluded an agreement with CMA CGM to transfer 10 terminal assets to Terminal Link.

The South Asian company Adani acquired 75 per cent of shares in Krishnapatnam Port Company in India. With regard to future developments, ports will need to expand environmental facilities in line with the accelerated environmental sustainability agenda. Similarly to ports, shipping companies such as Maersk, for example, are also showing increasing interest in integrating their services with ports and inland logistics (The Loadstar, 2019).

9. Challenges ahead for the sector with the onset of the pandemic

All in all, 2019 was a weak year for shipping and maritime trade. On the upside, a hard Brexit was avoided or delayed, as it remains to be seen how the new trade relations between the European Union and the United Kingdom will evolve. There was also an apparent easing in the trade tensions between China and the United States that may be associated with the first phase of a trade agreement between the two countries signed in January 2020.

Initial expectations were that a moderate improvement in global economic conditions would occur in 2020. However, the unprecedented global health and economic crisis triggered by the COVID-19 pandemic in early 2020 undermined the growth prospects for maritime transport and trade. A black swan event that is extremely rare and unpredictable, with potentially severe consequences (Drewry, 2020c), the pandemic and its global fallout transformed the world. While making a precise assessment of the immediate impacts and longer-term implications is a challenging task, there is no doubt, however, that the outlook has significantly deteriorated and has become more uncertain.

B. MARITIME TRADE IN THE ERA OF PANDEMIC

Initially localized in China, the pandemic evolved rapidly and became a global game changer by the first quarter of 2020. The spread of the disease worldwide and the consequent disruptions to societies and economies have far-reaching implications, including for transport and trade. Amid supply-chain disruptions, falling global demand and global economic uncertainty caused by the pandemic, the global economy has suffered dislocation, first at the supply end, then at the demand end.

While disruptions such as natural disasters, conflicts, strikes and security incidents are common in maritime transport, the pandemic is exceptional, given its scale, speed and direct impact on global supply chains, transport and trade. Historically, no disruption has ever resulted in a global lockdown of people and business. Restrictions on mobility, travel and economic activities worldwide, although in varying degrees, are unprecedented. By mid-April 2020, nearly 90 per cent of the world economy had been affected by some form of lockdown (United Nations, 2020b), and by month's end, about 4.2 billion people or 54 per cent of the global population (International Energy Agency, 2020). As many as 100 countries closed their national borders, disrupting supply and supressing global demand for goods and services. No country was prepared to face the combined health and economic crisis.

Risk assessment and management are common practice in business and policymaking processes, especially with the emergence of various risks – security threats, environmental risks, changing weather patterns and rising social unrest. However, it would appear that the likelihood of a disruption of the type and scale of the COVID-19 outbreak was not foreseen or it was underestimated. Many factors may be at play, including competing policy priorities, immediate versus longer-term concerns, budget pressures and institutional capacity constraints. However, research on behavioural economics suggests that limitations inherent to human minds may also be interfering with relevant risk assessment and decision-making processes (see box).

By June 2020, it appeared as if the brunt of the economic shock was going to be concentrated in the first half of 2020 and that impacts were going to vary by region in line with the gradual geographical spread of the pandemic. Breaking out in stages and gradually moving from one region to another, the pandemic has had a particular impact on supply chains. These have been affected multiple times as goods cross borders and in different ways, depending on where the pathway of the pandemic is in each region. As a result, instead of managing the pandemic response based on a single location, responses had to take into account multiple locations.

Box 1.1 Blind spots in risk assessment and management

The frequency and severity of supply-chain disruptions is on the rise. Supply chains are vulnerable to a broad range of threats, including pandemics, extreme weather events, cyberattacks and political crises. Risk management has become more widely known in recent years, given events such as the terror attacks of 11 September 2001 in the United States, tsunamis and the 2008–2009 global financial crisis. Yet the COVID-19-induced disruptions revealed the extent to which the world was ill-prepared in the face of a rapidly evolving global pandemic. This calls into question the effectiveness of relevant risk assessment and management plans, especially in the current context of highly interdependent and interconnected world economies. Paradoxically, there is no lack of pandemic plans. However, they generally failed to account for the full importance and ramifications of global supply chains. Research on behavioural economics, pioneered by Nobel Prize winner Daniel Kahneman, suggests that when it comes to evaluating risks, biases inherent to the human mind often interfere. Thinking critically is important when assessing risks. However, humans are prone to making errors in reasoning, as many fallacies and cognitive illusions clutter the thinking. Examples of such cognitive blind spots include relying on intuition to evaluate evidence, assess probabilities and take risks; being on autopilot – that is to say, being primed by certain social and cultural conditions; making snap judgments; using shortcuts to make quick decisions based on trial and error, rule of thumb or educated guess; ignoring facts, hard data and statistics; being influenced by vivid mental images; and being motivated by emotional factors and gut feeling and not necessarily rational and objective thinking. Understanding these biases and how they shape judgments and decisions is therefore important when assessing risks and devising response measures and plans. To help overcome these limitations, policymakers and business executives could start by becoming aware of the various cognitive biases that may undermine sound policies and decisions, and adopt potential mitigation measures, as deemed appropriate.

Sources: Economic and Social Commission for Asia and the Pacific, 2013; Kahneman, 2011; Piattelli-Palmarini, 1994; Rodrigue, 2020.

Since more than 80 per cent of world merchandise trade by volume is carried by sea, the impact of the pandemic on maritime transport can have far-reaching implications. The impact is magnified by the role played by China in maritime trade, as prosperity within the shipping sector has long been strongly tied to that country. In 2003, amid the outbreak of severe acute respiratory syndrome, China made up 5 per cent of global GDP. Today this figure stands at 16 per cent. In 2019, China accounted for over 20 per cent of world imports by sea, up from less than 10 per cent in 2003. While its share of total exports has remained stable at 5 per cent of the world total since 2003, its share in global container exports has increased. In this context, its maritime trade has ripple effects on all shipping market segments, and supply-chain disruptions involving China naturally send shockwaves across shipping and ports worldwide.

As the pandemic weighed down on the maritime trade of China, especially during the first quarter of 2020, global maritime trade was bound to be affected. In addition to the sector's high exposure and sensitivity to developments in China, restrictions on vessels and crew in many ports, labour force shortages and restrictions on their movement, and operational challenges have sent shipping into unchartered waters. Impacts are being felt across the board, ranging from maritime trade flows to vessel movements, vessel crew changes, capacity deployed, port operations, warehousing capacity, hinterland connections and inland logistics.

By June 2020, leading economic and shipping indicators were showing resumed activity in China. However, this only partly helped the recovery, as consumers and business in export markets were still in lockdown. Even as major economies eased out of lockdown, the situation remained problematic and continued to evolve amid uncertainty about the pandemic and possible new spikes.

Against this background, the following section considers the implications of the pandemic for maritime transport and trade. While not exhaustive, the following four main issues highlight the type of challenges ahead and emphasize the need for maritime transport to act as a trade facilitator, supply-chain connector and key partner in promoting more resilient, robust and sustainable transport and trade patterns:

- The pandemic sent shockwaves through supply chains, shipping and ports.
- World output and merchandise trade are projected to fall in 2020.
- Global merchandise trade receives both supply and demand shocks.
- Disruptions caused by the pandemic raise existential questions for globalization.

With regard to the first issue, that of the pandemic's disruptions to supply chains, shipping and ports, these

disruptions inevitably invite comparisons with the global financial crisis of 2008–2009. The two crises are similar in certain respects but diverge in others. First, in both cases, governments intervened by injecting funds into the economy to stimulate recovery. Second, the two crises were accompanied by rising protectionist sentiment and scepticism about globalization. However, they differed in their type, scope, speed and scale. A crisis like no other, surpassing the 2008–2009 financial crisis, the COVID-19 crisis has been dubbed the "Great Lockdown" (International Monetary Fund, 2020a). The touch points of the financial crisis were more limited, whereas the pandemic swept the entire world in record time. The 2020 crisis was a double-hit disruption, which morphed from being a supply-side disruption in China to becoming a global cross-sectoral demand shock. Third, restrictions on economic activity and travel did not occur during the previous crisis. Fourth, the pre-existing trade and finance trends were different. Fifth, while the 2008–2009 crisis began in mid-2008, its worst effects became evident eight months later, while the impact of the 2020 crisis were almost immediate.

With regard to shipping and maritime trade, a fundamental difference was also the industry's response to suppressed demand. While carriers focused on safeguarding market shares during the months leading up to the outbreak of the pandemic, the focus shifted to managing supply to maintain rates. Also, in the case of the financial crisis, the size of the orderbook was much higher (see chapter 2). Although the precise impact on shipping and maritime trade is still difficult to gauge, the picture for 2020 is nonetheless not optimistic, given that key forecasting entities are predicting contractions in world GDP and merchandise trade.

With regard to the second issue, that world output and merchandise trade will most likely decline in 2020, existing estimates of the economic fallouts of the pandemic vary, given the high degree of uncertainty involved. Yet all converge and point to a global recession in the making. Bearing in mind the uncertain times, differences in forecasting techniques and assumptions, as well as the potential for revisions depending on how the pandemic continues to evolve and whether the various policy interventions have been effective in mitigating the pandemic and its effects, UNCTAD expects world GDP to fall by 4.3.per cent in 2020. The International Monetary Fund predicts a decline of 4.4 per cent (International Monetary Fund, 2020b) (figure 1.12). In comparison, UNCTAD analysis shows that world GDP contracted by 1.3 per cent in 2009. In both cases, GDP in all countries, developed and developing countries alike, is expected to decrease, except for East Asia, including China, which will see a marginal growth of 1.1 per cent. According to UNCTAD analysis, the pandemic-related recession is likely to translate into a $12 trillion loss in global income relative to the end of 2019. This is based on the UNCTAD baseline scenario for world GDP growth and takes into account that the average growth rate of the world

Figure 1.12 Varied forecasts of gross domestic product growth for 2020
(Percentage change)

Source: UNCTAD calculations, based on reports issued by the entities listed.

economy – the trend prior to the outbreak of the pandemic – was 3.0 per cent in 2017–2019 (UNCTAD, 2020a). Another estimate suggests that the cumulative output losses during 2020 and 2021 will approach $8.5 trillion (United Nations, 2020b).

Many developing countries will be affected by declining demand and export revenues, remittances, foreign direct investment and official development assistance. The least developed countries are hit hard, given their limited resources and exposure to supply-chain disruptions such as in exports of textiles and clothing products (for example, Bangladesh). For the economies of Africa, developing America and Western Asia, and transition economies, an added concern is the sharp fall in commodity prices. Commodity-dependent countries and small island developing States, which depend heavily on external flows, are particularly vulnerable to external shocks. For the latter, external flows account for nearly 35 per cent of GDP (United Nations, 2020b). Fiscal measures and stimulus packages introduced worldwide stand at $9 trillion, equivalent to over 10 per cent of global GDP in 2019. Further, several developing countries are also implementing limited fiscal stimulus, not exceeding

2 per cent of GDP. Many lack the fiscal resources to address the economic impact with large relief and stimulus measures (United Nations, 2020b).

With respect to the third issue, that global merchandise trade receives shocks to both supply and demand, trade is typically more volatile than output and tends to fall particularly sharply in times of crisis (World Bank, 2020). By mid-2020, the full impact of the outbreak of the pandemic on international trade remained uncertain, in line with projections for GDP growth. However, preliminary estimates and some leading indicators provide some useful pointers. While trade had already weakened in 2019, it became clear that disruptions brought by the pandemic had significantly suppressed trade and volumes had collapsed to record lows. Forecasts have varied with differences in assumptions, scenarios and models but all concur that international merchandise trade can be expected to decrease beyond the contraction levels of 2009.

UNCTAD estimates that the value of international merchandise trade declined by about 5 per cent in the first quarter of 2020 and that it will diminish further by 27 per cent in the second quarter (UNCTAD, 2020b). In the first quarter of 2020, the value of trade in textiles and apparel diminished by almost 12 per cent, and that of the office machinery and automotive sectors, by about 8 per cent. In April 2020, trade in energy and automotive products fell by about 40 per cent and 50 per cent in value, respectively. Significant declines were also observed in the value of trade in chemicals, machineries and precision instruments, with drops above 10 per cent. By contrast, trade in agrifood products and electronics fared comparatively better (WTO, 2020). For the full year, WTO projections point to reductions in world merchandise trade ranging from 13 to 32 per cent in 2020, depending on the scenario, before recovering at rates ranging from 21.3 to 24 per cent in 2021 (WTO, 2020). Overall, these numbers are do not bode well for maritime trade.

The fourth issue is that disruptions caused by the pandemic raise existential questions for globalization. This is because maritime transport is the backbone linking global supply chains, supporting trade and enabling participation in global value chains. When a pandemic of the magnitude of the COVID-19 crisis occurs, the sector works as a transmission channel that sends shockwaves across supply chains and regions. Restrictions introduced in response to the pandemic have raised obstacles that undermine the smooth movement of trade flows and supply-chain operations and can significantly erode the transport services trade liberalization and trade facilitation gains achieved over the years. In this context, the pandemic and its fallout have accelerated an existing debate on the benefits of globalization and extended supply chains. This debate was sparked by heightened trade tensions between China and the United States since 2018. The disruption caused by the pandemic has brought to the fore concerns regarding outsourcing production to distant locations and the need to diversify production and manufacturing sites and suppliers.

About 70 per cent of international trade is linked to global value chains (OECD, 2020b), with China predominating not only as a manufacturer and exporter of consumer products, but also as a supplier of intermediate inputs for manufacturing companies located in other countries. UNCTAD estimates intermediate products at half of the trade in world goods in 2018 – about $8.3 trillion (UNCTAD, 2020c). In 2020, an estimated 20 per cent of global trade in manufacturing intermediate products originated in China, up from 4 per cent in 2002 (UNCTAD, 2020d). The volume of intra-Asian containerized trade and its rapid growth over recent years reflect this trend. In this context, any disruption to supply chains in China is bound to affect production in the rest of the world, with wide-ranging impacts on machinery, automotive products, chemicals, communication equipment and precision instruments. Japan, the Republic of Korea, Taiwan Province of China, the United States and Viet Nam will be affected the most.

Preliminary analyses suggest that electronics and electrical equipment are the highest risk sector on a global scale. Although the automotive industry maintains low inventory levels, it does, however, depend less on China than the electronics industry (Aylor et al., 2020). Electronics manufacturing is global to a large degree, which adds to its complexity, as goods cross many borders. According to the OECD database of 2018 on trade in value added, the share of foreign value added in electronics exports was about 10 per cent for the United States, 25 per cent for China, 34 per cent for the Republic of Korea, 44 per cent for Singapore, more than 50 per cent for Malaysia and Mexico, and over 60 per cent for Viet Nam.

Constraints on transportation and logistics and lack of workers prevented timely delivery of components from China and other countries to factories in South-East Asia during the pandemic. As a result, response measures such as sourcing directly from Viet Nam, switching from land to air freight and rerouting shipping lanes that previously included stops at Chinese factories had to be taken (Aylor et al., 2020). For shipping, these measures translate into rerouting of vessels, changes in schedules and port calls, as well as variations in volumes. Further, they illustrate the challenges involved in the transport of time-sensitive trade when disruptions to supply chains occur and how the level of integration with the country's supply chain and level of inventories can change the outcomes.

Less sophisticated manufacturing in countries such as Bangladesh, Pakistan and Viet Nam, which have recently attracted factories to move their production away from China, is also highly exposed to COVID-19-induced disruptions. A case in point is Bangladesh, where about 85 per cent of its exports are composed of textile fibres,

textiles and made-up articles, clothing and accessories (categories of standard international trade classification) (UNCTAD, 2020e). The shock to this supply chain is demand driven and reflects cuts in spending on non-essential goods and store closures. One estimate expects global sales for fashion and luxury brands to drop by 25 to 35 per cent in 2020, compared with 2019 (Seara et al., 2020).

Factory closures, including in China and other East Asian countries, and lockdowns implemented worldwide, resulting in supply-chain disruptions, have revealed the shortcomings of extended and single-country-centric supply chains. They have rekindled the debate on the risks associated with an internationalization of production networks and overreliance on a few countries such as China for manufacturing production, as well as the predominance of low-inventory and just-in-time supply-chain models.

Some observers argue the need to revisit existing supply-chain patterns and reflect on strategies to shift away from the model that had been promoted by hyperglobalization (1999–2009). Others assert that the re-nationalizing of global value chains could, to some extent, insulate countries from the fallout of a pandemic (OECD, 2020b). In the United States, incentives to encourage companies to shift business away from China include tax breaks and a new reshoring fund (Lloyd's Loading List, 2020b). Japan announced that it will allocate $2.2 billion to attract Japanese manufacturers to shift production out of China, $2 billion of which will be earmarked for their relocation back to Japan. These developments could accelerate the move towards the China plus one[2] manufacturing hub model, which evolved amid rising labour costs in that country and has recently intensified trade tensions. The developments could also prompt further regionalization of supply chains and growth in intraregional containerized flows. It is likely that no single country can easily absorb the massive export manufacturing capacity of China.

Moving production home or closer to home is a complex process and should take into account factors other than labour costs. Analytical research suggests that the contraction of GDP would have been worse with re-nationalized global value chains, as government lockdowns also affect the supply of domestic inputs (OECD, 2020b). That said, it is becoming increasingly evident that a slowdown in globalization has taken place over the past decade. Prior to the pandemic, structural shifts, such as digitalization, the "servicification" of manufacturing (Haven and Van Der Marel, 2018), a growing sustainability imperative and the rise of protectionist sentiment, have been taking hold and increasingly re-shaping globalization trends. Companies have already been adding new operations to supplement current production.

Viet Nam is the largest country in the region to see new manufacturing growth from offshoring, as illustrated by agreements with Intel and Samsung. Others, such as Indonesia, Malaysia, the Philippines, Singapore and Thailand, are prime candidates. India is also contemplating a larger role and looking to establish itself as a regional manufacturing hub and to attract companies seeking to move their supply chains out of China (Bloomberg, 2020a). Tax incentives and easy access to land and other infrastructure are being considered. While these efforts pre-date the pandemic (Bloomberg, 2020b), trade tensions between China and the United States and the supply-chain vulnerabilities exposed by its outbreak will most probably accelerate the process.

Nonetheless, China is likely to remain a key player, given its strong supply-chain network and infrastructure and knowledge base, as well as its massive labour force, which has no match. For instance, even though Intel opened a new facility in Viet Nam, the company has maintained several assets in China. Viet Nam was simply added as an assembly and testing operation (Procurement Bulletin, 2020). This is further illustrated by the rise in United States imports from China in May 2020, reflecting the fact that retailers were rushing back to China for inventory replenishment and showing how difficult it would be to shift entire sourcing elsewhere (JOC.com, 2020c). The manufacturing activity that had already migrated to South-East Asia is tied to low-wage and low-skill workers who produce footwear and apparel. For higher-end products such as electronics, workers will require greater skills (JOC.com, 2020c). On the other hand, Chinese companies have also been shifting some of their production to neighbouring countries, reflecting in part the impact of tariff escalation since 2018.

The globalization process based on low labour-cost differentials and on an extensive outsourcing of production that stimulated trade may have reached its limits, with factors other than developments in the world economy and population likely to shape the maritime trade patterns of the future transport.

These include the global decarbonization agenda, which has implications for the two largest commodities transported at sea: crude oil and coal. Another driver would be the growing demand for smaller and low-value packages of physical goods that are increasingly bundled with services and require faster transit time. These shifts in demand patterns are expected to question the cost advantage of shipping compared with other means of transport (Port Economics, 2020).

In summary, the pandemic-induced disruption may trigger shifts in globalization patterns, supply-chain configuration and production models, with implications for transport and inventory decisions – all of which are of strategic importance for shipping. They have the potential to reshape the operational landscape, especially for

[2] A business strategy that aims to avoid investing and concentrating business only in China.

container shipping, including with regard to vessel size, capacity deployed and operations. For example, greater regionalization would lead to the increased fragmentation of trade flows which, in turn, would make the use of larger vessels more challenging (JOC.com, 2020d).

C. OUTLOOK

1. Poor short-term outlook for maritime trade

Uncertainty remains an overriding theme in 2020. Predicting the impact on maritime trade and the timing and scale of the recovery is fraught with uncertainty. Many factors are at play, significantly influencing the outlook. These include the pathway of the pandemic, the effectiveness of the efforts to control further outbreaks, continued shifts in spending patterns, trends in consumer and business confidence, developments in commodity prices and the ability of stimulus packages to give an impetus to growth and put the world economy back on track. Bearing this in mind and extrapolating from past trends, UNCTAD expects the volume of maritime trade to decline in 2020. Based on the maritime trade-to-GDP ratio for the period 1990–2019 and the forecast of GDP growth by the International Monetary Fund (October 2020), UNCTAD predicts that international maritime trade will fall by 4.1 per cent in 2020 (table 1.12). Seaborne trade forecasts for 2021 also depend on economic growth projections, and these vary.

For example, UNCTAD expects world GDP to rebound by 4.1 per cent in 2021 (see table 1.3 above), the Department of Economic and Social Affairs in its May 2020 forecast projects a global GDP expansion of 4.2 per cent and the International Monetary Fund in its June 2020 forecast predicts that growth will bounce back to 5.4 per cent in 2021. By contrast, the WTO forecast of April 2020 points to a recovery in world merchandise trade volume in 2021 ranging from 21.3 to 24 per cent, depending on the scenario (WTO, 2020). For 2021, UNCTAD estimates that maritime trade flows will recover by 4.8 per cent.

2. Falling containerized trade volumes and rising service cancellations in 2020

Container shipping is strongly affected by the disruptions induced by the pandemic, as containerized trade is closely linked to world economic developments, consumer activity and supply chains. Reflecting the negative impact of the combined demand and supply shocks, volumes are coming under pressure in 2020. The large share of ship capacity idled and the number of services cancelled are a good indication of the slowdown. To provide a general picture, 10 per cent of global vessel-carrying capacity was sitting idle in April 2020 (Drewry, 2020d).

As shown in figure 1.8 and tables 1.9 and 1.10, global containerized trade is projected to contract across all trade routes, with intra-regional trade faring relatively better than the others.

Data available for the first and second quarters of 2020 highlight the impact of the pandemic on containerized trade originating from China across the three main East–West containerized trade routes (figure 1.13 (a) and (b)). Journeys involving the Far East, especially the export leg (westbound Asia–Europe, eastbound trans-Pacific), contracted in the first quarter of 2020, compared with the same quarter in 2019. These numbers were more pronounced during the second quarter when the slump in demand in Europe and North America was felt. On the transatlantic route, where automotive goods are a staple of container flows, the outlook has also deteriorated. As shown in figure 1.13 (b), double digit-drops on the transatlantic route were recorded during the second quarter of 2020.

Owing to diminishing trade volumes as factory output in manufacturing regions slowed down and consumers reduced discretionary spending on non-essential items in Europe and North America, carriers cut capacity by introducing blank sailing, idling capacity and re-routing via the Cape of Good Hope to pare down costs while taking advantage of lower fuel prices (see chapters 2 and 4). This makes it possible to avoid the cost of transiting the Suez Canal ($600,000 and more for a one-way trip for ultralarge container ships) and absorbing excess capacity by extending sailing times. Re-routing vessels could

Table 1.12 International maritime trade development forecasts, 2020–2021
(Percentage change)

Forecasting entity	Annual growth (percentage)	Years	Source
UNCTAD	-4.1	2020	International Monetary Fund world GDP growth forecast
UNCTAD	4.8	2021	International Monetary Fund world GDP growth forecast
Clarksons Research Services	-4.0	2020	*Seaborne Trade Monitor,* October 2020
Clarksons Research Services	4.7	2021	*Seaborne Trade Monitor,* October 2020

Source: UNCTAD calculations, based on own analysis and forecasts published by the indicated institutions and data providers.

Figure 1.13 Containerized trade growth on main East–West routes
(a) in million 20-foot equivalent units;
(b) percentage change, first quarter 2019–first quarter 2020, second quarter 2019–second quarter 2020

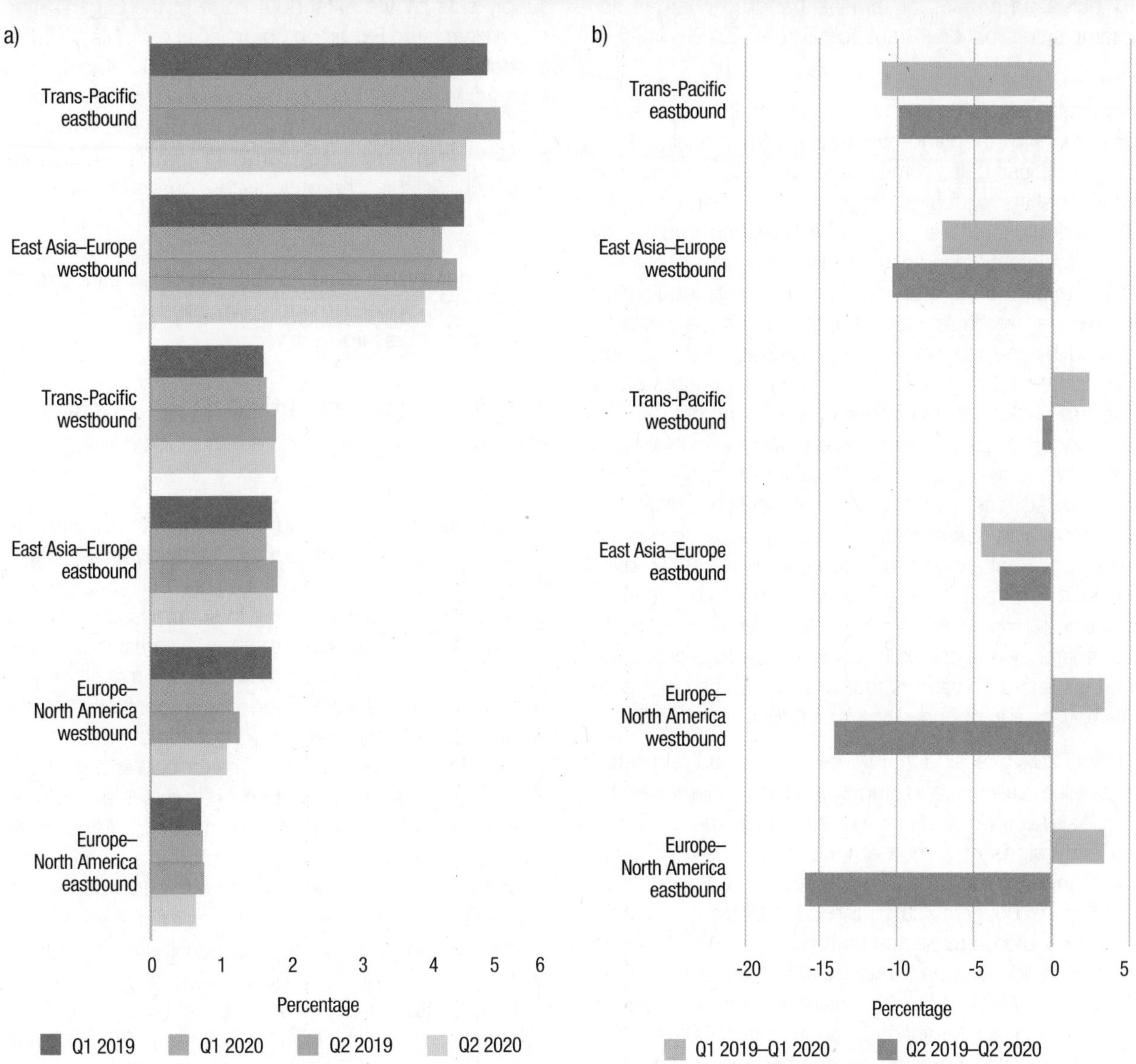

Source: UNCTAD calculations, based on MDS Transmodal, 2020b, World Cargo Database, 19 August.
Abbreviation: Q, quarter.

imply over $10 million in lost charges for the Suez Canal Authority. While a rebate scheme was announced in early May 2020, it failed to curtail the longer journeys via the Cape of Good Hope (DHL, 2020).

Blank sailing and service cancellations announced by the carriers without the usual notice periods affect service reliability and the ability of shippers to plan their supply chains. Deploying larger vessels means that any missed port calls caused by blank sailing has a greater impact on available capacity (JOC.com, 2020e). In June 2020, many ports reported that blank sailing had resulted in mega-sized vessels calling less often but when they did, the large volumes created peaks and operational challenges. These operational hurdles affect ports (ship-to-shore operations and yard activity), as well as landside distribution (Notteboom and Pallis, 2020).

Since container vessels move on a scheduled rotation, the cancellation of a sailing from the first port in the rotation cascades down to all the other ports served by that carrier in that rotation. Some smaller ports are particularly hard hit by multiple cancellations from different services. Ship capacity into and out of the ports of Manila and Odessa, the Russian Federation, for example, was reduced by 25 per cent in May 2020, that of the ports of Beirut and Visakhapatnam, India by 20 per cent, and larger ports such as Hamburg, Germany and Rotterdam, the Netherlands, by 10 per cent. Trans-shipment ports such as Colombo and Djibouti are also affected by such reductions, 13 per cent and 11 per cent, respectively (Clipper Data, 2020). In this context, it is argued that blank sailing could increase the bargaining power of carriers compared with terminals and canals, owing to increased arrears for terminal-handling charges, for example (International Transport Forum, 2020).

Shippers also contribute to the disruption by cancelling bookings without prior notice to carriers, thereby making any planning to optimize vessel capacity difficult. At the port level, less traffic sometime results in the cancellation of work shifts without advance notice to inland carriers. The operational challenges are combined and amplified by growing detention and demurrage charges for exceeding free storage time and the late return of equipment to marine terminals (see chapter 2). The experience shared by the Northern Corridor Transit and Transport Coordinating Authority in Eastern Africa highlights some of these challenges in the case of a cross-border corridor and underscores the need for effective trade-facilitation measures (see chapter 4). Pressure on warehousing capacity, such as shipments of non-essential merchandise idled, are also reported (JOC.com, 2020e). Rebalancing of empty containers is another challenge, as empties were in shortage in Europe, while they stagnated at ports in China (JOC.com, 2020f). Information sharing, transparency and communication are key to avoiding the hurdles and inefficiencies that arise while responding to disruptions (Lloyd's Loading List, 2020c).

In April 2020, reports that some carriers had reinstated cancelled sailings and announced rate increases for the Asia–Europe route were met with some optimism as early signs of a recovery. However, others argued that sailings had been reinstated in part because carriers had overestimated the fall in demand and that activity could be explained by a clearing of the backlog that had accumulated when China was in lockdown (JOC.com, 2020g). In all likelihood, the announced extension of blank sailings through August 2020 points to the expected pressure on demand and recovery in maritime trade volumes. Blank sailings could give some indication about trends in demand. (Drewry, 2020e). While a decline in the number of blank sailings could be one of the earliest signs that global trade may be picking up (Clipper Data, 2020), conclusions should not be drawn quickly. Blank sailings alone do not provide the full picture and should be assessed against scheduled supply capacities and other relevant indicators.

3. Oil and gas trade declines with restrictions in travel and transport in 2020

The pandemic has had a significant impact on trade in oil and gas. Global oil demand fell with the freezing of large parts of the global economy, restrictions on travel and transport, and cuts in industrial activity and refinery output. Together, these factors have depressed demand, as volumes of both crude oil and refined petroleum products have declined. Supply-side factors are another consideration. A surplus in oil production has practically filled all oil inventories, with many vessels being used as floating storage (see chapter 2). The implementation of supply cuts by the extended group of the Organization of the Petroleum Exporting countries in early May 2020 is expected to reduce the availability of crude oil. Disruptions in oil infrastructure in Libya, alongside declining outputs in the Islamic Republic of Iran and the Bolivarian Republic of Venezuela, are also curtailing growth (Clarksons Research, 2020j). The outlook for liquefied natural gas shipping is also affected by the pandemic. Disruptions in early 2020 depressed import demand in China during the first quarter. With the global outbreak of the pandemic in March 2020, global demand for liquefied natural gas also came under pressure.

4. Dry bulk trade affected by decline in industrial and automotive sector activities

Reductions in mining and industrial activity had an impact on dry bulk trade but to a relatively lesser extent than containerized trade. Global dry bulk trade came under pressure in 2020, owing to suppressed economic activity and demand. Nonetheless, a partial recovery in Brazilian iron ore exports and the rebuilding of stockpiles in China should support iron ore trade flows after a decline in 2019, the first in two decades. Trade in coal is projected to shrink, due to weaker power demand in many regions, and lower oil and gas prices are making coal power generation less competitive. Minor bulk trade commodities, such as steel products, cement and scrap metal, which are associated with construction and steel manufacturing, generally suffer from a weakening of the economy. The steel and aluminium sectors, on which the automotive industry depends, collapsed, and the automotive sector was hit hard (Baltic and International Maritime Council (BIMCO), 2020). Trade in minor bulk commodities is expected to deteriorate in 2020, although some of the stimulus measures that concentrate on infrastructure and housing investment may boost demand for such commodities. Overall, assuming commitments set out in phase 1 of the trade agreement between China and the United States are implemented, grain shipments from the latter are likely to pick up. Generally, food-based agricultural commodities are less exposed to a decline in economic output.

5. Shrinking port volumes in 2020 and need for more storage space

According to a baseline scenario provided by Drewry, global port container throughput is expected to contract by 7.3 per cent in 2020. The contraction could amplify and reach 12 per cent if the negative scenario is upheld. As shown by the quarterly trends depicted in figure 1.14, global container port volumes collapsed in the second quarter of 2020 at the height of the

Figure 1.14 World port-handling forecast, 2019–2021
(Million 20-foot equivalent units and percentage change)

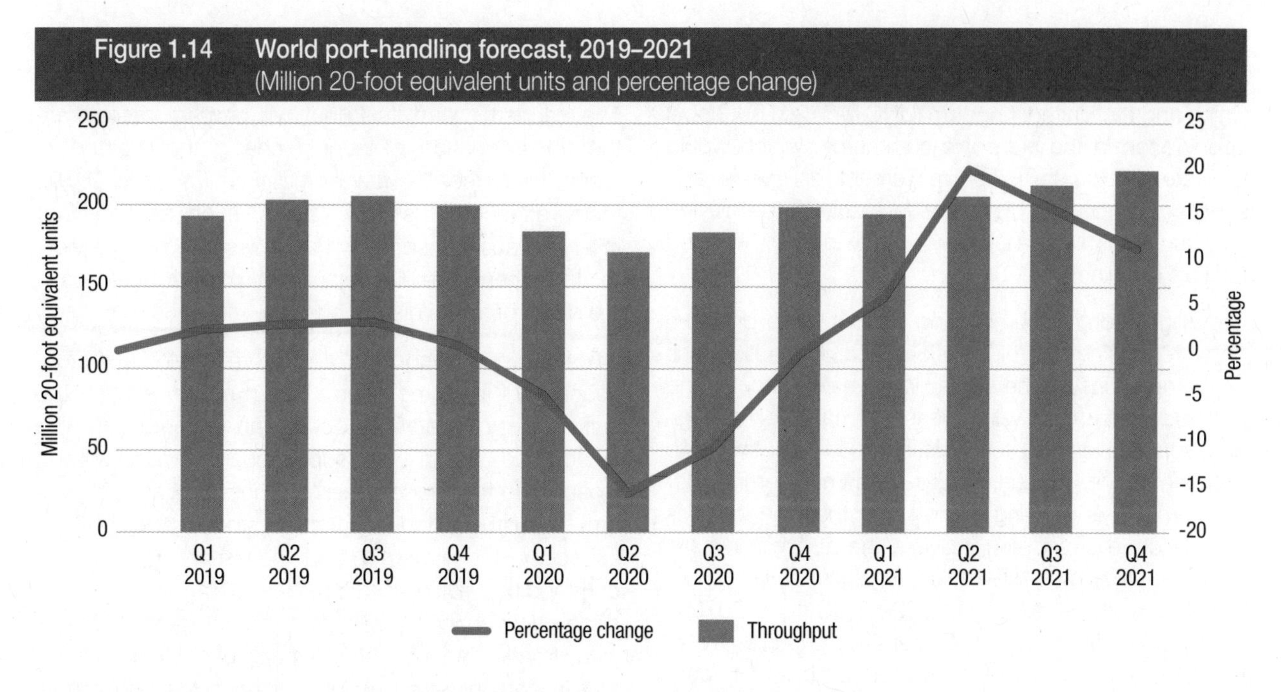

Source: Drewry baseline forecast; Drewry, 2020e, 2020 Container Forecaster Update, quarter 1, May.
Abbreviation: Q, quarter.

pandemic. Port volumes in 2021 will vary, depending on the scenario. Projected figures range between another contraction of 3 per cent and a jump of more than 10 per cent (Drewry, 2020f). The range of scenarios shows how unpredictable and volatile the short-term outlook can be.

Several ports reported an increase in port and terminal utilization due to a rise in imported essential goods, such as grains (rice, wheat). Other ports reported that traders began storing liquid bulk commodities in anticipation of future commodity price developments. Another situation faced by ports relates to the automotive industry, as many new cars were not collected, due to a collapse in sales, which resulted in the overcrowding of relevant storage areas. Storage space has also been used in cases where transit container shipping programmes have been suspended. For example, the Mediterranean Shipping Company applied the suspension of transit while using some of the world's leading trans-shipment hubs (Bremerhaven, Germany; Busan, the Republic of Korea; King Abdullah port, Saudi Arabia; Lomé; Rodman port, PSA Panama International Terminal, Panama; and Asyaport, Tekirdağ, Turkey). As reported in the experience shared by the Mediterranean Shipping Company, this allowed shippers to benefit from advance yard storage and start moving goods early in anticipation of a resumption in demand (see chapter 4).

Unlike shipping lines, which could mitigate the effect of volume reductions through, for example, blank sailings, service suspensions or capacity cuts, ports have no mitigation tools at their disposal and are likely to focus increasingly on costs. Developments in production and supply-chain-design choices are of relevance to ports. As noted above, the disruptions brought by the pandemic are likely to hasten a shift away from single country-centric sourcing. However, and as previously noted, while there may well be a shift away from China as a supplier, its supply chains have from some angles proved more resilient throughout the pandemic experience, compared with other locations.

Container ports will have an important role to play in servicing the migrating trade. The new locations will need to prepare for the potential growth in volumes. For example, Cambodia and Indonesia are said to have shortfalls in port capacity, that is, to handle more traffic and larger vessels. In Viet Nam, the major beneficiary of recent changes in container trade patterns, port capacity is considered suitable, although the country may need to invest in deepwater berths capable of handling larger vessels and direct calls. Closing the infrastructure gap in the region is estimated to require over $12 billion in investment (Drewry, 2020g).

6. Shifts in consumption and shipping patterns with the rise of e-commerce likely to continue

The pandemic revealed how e-commerce can be an important instrument to sustain consumption during crises. The pandemic and the lockdown may have boosted e-commerce uptake, which may continue as consumption patterns evolve. The potential for growth is significant. UNCTAD puts global e-commerce sales in 2018 at $25.6 trillion, up 8 per cent over 2017. In 2018, the estimated e-commerce sales value, which includes business-to-business and business-to-consumer sales, was equivalent to 30 per cent of global GDP. The United States continued to dominate the overall e-commerce market and remained among the top three countries in business-to-consumer e-commerce sales, namely China and the United Kingdom (UNCTAD, 2018). Global cooperation in the area of e-commerce, which would facilitate the cross-border movement of goods and services, narrow the digital divide and level the playing field for small businesses, will have to be enhanced (Lloyd's Loading List, 2020d).

Growing e-commerce shipping will put more pressure on warehousing and distribution capacity, as business will want to ensure the availability of safety stocks and buffers. In turn, this will increase demand for storage and space. Demand for logistics space continues to outpace supply in Asia, where consumer demand for e-commerce is growing much faster than the logistics infrastructure supporting it. More than $4 billion have been poured into Asia-based logistics development funds since the beginning of 2020 (JOC.com, 2020h). Demand for distribution centres and warehouses is also expected to increase, given the changes brought about by COVID-19-induced disruptions. For example, supply chains were re-appraised, inventories were increased and the geographical diversification and decentralization of supply chains pursued.

D. SUMMARY AND POLICY CONSIDERATIONS

The recovery is likely to vary with differences in the disruption caused by the pandemic, countries' levels of development and capacity to support economic growth, while providing social safety nets. International support and cooperation will be of paramount importance for developing countries, especially the least developed countries and small island developing States. Trade is a key component of recovery, and the maritime transport industry, which carries much of it, has a major role to play.

1. Maritime transport remains pivotal in an interdependent world

The COVID-19 outbreak revealed the high levels of global interdependency and is setting in motion new trends that will reshape maritime transport and trade. The sector is at a pivotal moment, as it needs to face the immediate concerns raised by the pandemic. However, longer-term considerations are also necessary: potential shifts in supply-chain design, globalization patterns, consumption and spending habits and, in general, a growing focus on risk assessment and vulnerability reduction. Further, the sector will need to continue mitigating the impact of inward-looking policies on trade and protectionism and to carry forward the sustainability and low-carbon agenda.

Various trends are likely to unfold and affect maritime transport and trade. In the post-COVID-19 pandemic world, there will probably be an element of shortened supply chains (near shoring, reshoring) and redundancy (maintaining excess inventory) (Flock Freight, 2020). The pandemic and its fallout will probably accelerate the transformation of supply chains that started in recent years (see *Review of Maritime Transport 2019*). Many aspects of supply chains, such as sourcing, inventory and transport, will be reassessed with a view to strengthening resilience and optimizing robustness in the event of future disruptions.

Investing in warehousing and storage, and therefore space, will become more important to ensure the sufficiency of safety stocks and inventories. The established just-in-time supply-chain model will be reassessed to include considerations such as resilience and robustness, for example, stocks and buffers, especially for strategic and necessary goods and commodities. Diversification in sourcing, routing and distribution channels will grow in importance. Moving away from a single country to multiple-location sourcing that is not only focused on cutting costs and delays but also on risk management and resilience will further evolve (JOC.com, 2020i). While the pandemic has brought into focus the notion of self-sufficiency, which is often equated with reshoring or near shoring, this approach is also not without vulnerabilities in case of localized disruptions. Decisions to uproot supply chains depend on more than labour costs and could be difficult to readily achieve.

2. Aftermath of the pandemic: Some potential implications

The pandemic will have a lasting impact on maritime transport and trade. The following five key trends in maritime transport and trade will be part of the pandemic's legacy:

- An accelerated shift in globalization patterns and supply-chain designs. While outright de-globalization may not occur because of the complexity and costs involved in uprooting and reshuffling highly integrated supply chains, the slower wave of globalization that started during the post-2008 financial crisis may decelerate further and the regionalization of trade is likely to gain momentum.
- A swifter uptake of technology and digitalization, with technology increasingly permeating supply chains and their distribution networks, including transport and logistics. Adopting technological solutions and keeping abreast of the most recent advances in the field will become a requisite, no longer an option. The pandemic and its disruptions have shown that first movers in terms of technological uptake are better able to weather the storm, for example, e-commerce and online platforms, blockchain solutions and information technology-enabled third-party logistics.

- Continued shifts in consumer spending and behaviour and evolving tastes that may change production and transport requirements. Examples include a further rise in online shopping in the post-COVID-19 era and a requirement for more customized goods. These trends are likely to emphasize the last-mile transport leg and promote shorter supply chains though the use of three-dimensional printing and robotics. These trends will trigger more demand for warehousing and space for stocks.
- Heightened importance of new criteria and metrics such as risk assessment and management on relevant policy agendas and industry's business plans and strategies. Risk assessments are likely to integrate considerations such as global interlinkages and interdependencies, including those underpinned by intertwined supply chains and financial channels.
- Adjustments in maritime transport to allow adaptation and change in line with the changing operating landscape. Industry stakeholders will probably continue to tap new business opportunities. Authorities at international maritime passages such as the Panama Canal are already assessing options on how to ensure preparedness in case of the reconfiguration of supply chains prompted by the pandemic (JOC.com, 2020j). The tapping of new business opportunities is a trend that had started before the pandemic. For instance, some shipping lines such as Maersk and port operators such as DP World, have been taking greater interest in business opportunities that may lie further down the supply chain through inland logistics. The aim is to be closer to shippers and emerge as reliable end-to-end logistics service providers (Riviera Maritime Media, 2019).

3. Priority action areas in preparation for a post-COVID-19 pandemic world

There are several priority action areas that can help address the ongoing challenges affecting the maritime transport and trade of developing countries, as well new challenges arising from the pandemic and its fallout. These are as follows:

- Fostering economic recovery. It is necessary to support economies on their path to recovery, especially developing countries that are more fiscally constrained, and to help them respond to the multiple shocks triggered by the crisis. Existing pledges and support packages are falling short of expectations. UNCTAD has called for a massive liquidity injection through extraordinary special drawing rights tailored to developing country needs and for re-scheduling and restructuring their external debt. Further, UNCTAD proposes that a $500 billion Marshall Plan be instituted for health care in developing countries to support their medical and social response to the pandemic.
- Allowing trade to support growth and development effectively. Trade tensions, protectionism and export restrictions, particularly for essential goods in times of a crisis, entail economic and social costs. These should be limited, to the extent possible. Further, non-tariff measures and other trade barriers should be addressed, including by stepping up trade-facilitation measures and customs automation.
- Helping reshape globalization for sustainability and resilience. It will be important to carefully assess all options regarding changes in supply-chain design to ensure the best economic, social and environmental outcomes, in line with the Sustainable Development Goals and the 2030 Agenda for Sustainable Development. For example, a shortening of supply chains through re-shoring or near shoring may reduce transport costs and fuel consumption, but it does not necessarily future-proof supply chains against disruptions that could occur anywhere, whatever the location. Multiple-sourcing approaches could prove more effective in resilience-building than concentrating all production in one location, whether at home or abroad. Strategies should aim to find ways in which unsustainable globalization patterns can be mitigated to generate more value to a wider range of economies.
- Strengthening international cooperation. The pandemic is a litmus test not only for globalization but for global solidarity as well (United Nations, 2020b). Addressing the impacts of the pandemic on global supply chains will require strengthened and coordinated global cooperation and action.
- Assisting shipping and ports in preparing for and adapting to the supply chains of the future. Maritime transport will need to adapt and ensure that it is prepared to support changes in supply chains that promote greater resilience and robustness. Shipping and ports will need to reassess business strategies and investment plans, including in terms of port capacity, shipping network configuration, vessels and capacity deployment. For example, investment in vessel capacity should take into account the shortening of some supply chains (for example, in critical and essential goods such as pharmaceuticals) and further regionalization in trade flows. Port and logistics capacity in countries receiving new businesses that have moved out of China should be upgraded and expanded as needed. More importantly, a key lesson drawn from

the pandemic experience is that cooperation, information sharing and the use of technology to support transport and coordinated action are crucial.

- Promoting resilience-building, including through investment in risk assessment and preparedness. It will be necessary to expand the visibility of supply chains through, among others, control towers and tools that allow for supply-chain disruptions to be predicted and analysed (Aylor et al., 2020). Plans should provide for how to respond to crises, as well as how to ensure business continuity through a set actions and protocols to be followed at different stages of a crisis (Knizek, 2020). For shipping, this may mean establishing priority lanes for handling critical cargo (for example, food, medicine or medical equipment) or limiting restrictions that affect labour such as crew changes and leave. Lessons learned from the pandemic should serve as guidance for informing preparedness and future-proofing maritime transport to allow for more resilient supply chains (see chapters 2, 4 and 5). Relevant actions could also include collecting and sharing information on potential concentration and bottlenecks, developing stress tests for essential supply chains and fostering an enabling regulatory framework that ensures greater certainty (OECD, 2020b). For example, following the 2008–2009 financial crisis, Governments developed stress tests for specific supply chains. These tests could be carried out in the context of policies related to the creation of strategic stockpiles to correctly assess the inventories and buffer stocks needed to prevent shortages in the future.

- Getting the priorities right and avoiding short-sighted policies. While the pandemic has been an overriding theme throughout 2020 and probably for years or decades to come, other important and potentially disruptive global issues should not be overlooked. For example, climate change is at risk of being pushed to the back burner, given the need to address the immediate concerns raised by the pandemic. Momentum on current efforts to address carbon emissions from shipping and the ongoing energy transition away from fossil fuels should be maintained. Governments could potentially direct the stimulus packages to support recovery while promoting other priorities at the same time, including climate-change mitigation and adaptation. Thus, policies adopted with a view to preparing for a world beyond the pandemic should support further progress in the shipping industry's transition to greening and sustainability. In particular, sustainability concerns such as the connectivity of small island developing States and progress made by the least developed countries towards the realization of Sustainable Development Goal 8.1 are ever more important in building their resilience to cope with future disruptions.

- Enabling greater uptake of technology while minding the digital divide. This means promoting efforts to accelerate the digital transformation to improve and build the resilience of supply chains and the supporting transportation networks. Digitalization efforts should enable enhanced efficiencies and productivity in transport, such as smart ports and shipping, but should also help countries to tap e-commerce capabilities and transport facilitation benefits that boost trade. Developing countries will need support to minimize the divide and ensure that they can also exploit the advantages of digitalization to build their resilience. For maritime transport to play its role in linking global economies and supply chains, it should leverage the crisis by investing in technology and adopting solutions that meet the needs of the supply chains of the future while supporting resilience-building efforts (Egloff, 2020).

REFERENCES

Aylor B, Gilbert M and Knizek C (2020). Responding to the coronavirus's impact on supply chains. Boston Consulting Group. 9 March.

BIMCO (2020). Dry bulk shipping: No quick recovery for the dry bulk market as COVID-19 digs deeper. 26 May.

Bloomberg (2020a). Japan to fund firms to shift production out of China. 8 April.

Bloomberg (2020b). India steps up effort to grab China's title of the world's factory. 4 June.

British Petroleum (2020). *BP Statistical Review of World Energy 2020*. London. June.

Clarksons Research (2006). *China Intelligence Monthly*. Volume 1. No. 1. June.

Clarksons Research (2020a). *Shipping Review and Outlook*. Spring.

Clarksons Research (2020b). *China Intelligence Monthly*. Volume 15. No. 6. June.

Clarksons Research (2020c). *Seaborne Trade Monitor*. Volume 7. No. 6. June.

Clarksons Research (2020d). *Seaborne Trade Monitor*. Volume 7. No. 4. April.

Clarksons Research (2020e). *Oil and Tanker Trades Outlook*. Volume 25. No. 1.

Clarksons Research (2020f). *Dry Bulk Trade Outlook*. Volume 26. No. 6. June.

Clarkson Research (2020g). *Dry Bulk Trade Outlook*. Volume 26. No. 1. January.

Clarksons Research (2020h). *Seaborne Trade Monitor*. Volume 7. No.1. January.

Clarksons Research (2020i). *Seaborne Trade Monitor*. Volume 7. No. 3. March.

Clarksons Research (2020j). Oil and Tanker Trades Outlook, Volume 25. No. 4. April.

Clipper Data (2020). Shipping lines slash May container capacity, extend cuts well into June. 7 May.

DHL (2020). Ocean freight market update. March.

Drewry (2019). *Container Trade Forecaster*. Quarter 4.

Drewry (2020a). Guinea set to supply iron ore from 2026. 21 May.

Drewry (2020b). *Ports and Terminals Insight*. Quarter 1.

Drewry (2020c). *Container Trade Forecaster*. Quarter 1.

Drewry (2020d). *Shipping Insight*. May.

Drewry (2020e). *2020 Container Forecaster Update*. Quarter 1. May.

Drewry (2020f). *Container Trade Forecaster*. Quarter 2. June.

Drewry (2020g). *Ports and Terminals Insight*. Quarter 2.

Economic and Social Commission for Asia and the Pacific (2013). *Building Resilience to Natural Disasters and Major Economic Crises*. Bangkok.

Egloff C (2020). Hit hard by COVID-19, transportation and logistics companies must adapt to keep supplies moving. Boston Consulting Group. 4 April.

Flock Freight (2020). How to build a pandemic-proof global supply chain. White Paper. 5 December.

Haven T and Van Der Marel E (2018). Servicification of manufacturing and boosting productivity through services sector reform in Turkey. Policy Research Working Paper 8643. World Bank Group.

International Energy Agency (2020). *Global Energy Review 2020: The Impacts of the COVID-19 Crisis on Global Energy Demand and CO2 [Carbon-dioxide] Emissions*. April. France.

International Monetary Fund (2020a). *World Economic Outlook:The Great Lockdown*. Washington, D.C. Available at www.imf.org/en/Publications/WEO/Issues/2020/04/14/weo-april-2020.

International Monetary Fund (2020b). *World Economic Outlook update: A Long and Difficult Ascent*. Washington, D.C. October.

International Transport Forum (2020). COVID-19 transport brief: Global container shipping and the coronavirus crisis. 29 April.

JOC.com (2020a). US [United States] accelerates import sourcing shift away from China. 15 May.

JOC.com (2020b). US [United States]–China trade war accelerates market share losses for West Coast ports. 4 May.

JOC.com (2020c). Importers rushing back to China for inventory replenishment. 2 July.

JOC.com (2020d). "Great Dispersal" to upend intermodal rail supply chains. 11 May.

JOC.com (2020e). Carriers see rise in booking cancellations, no-shows. 28 May.

JOC.com (2020f). Container cargo imbalance from COVID-19 deepens. 1 May.

JOC.com (2020g). Vessel port call frequency improving in Europe, falling in North America. 8 June.

JOC.com (2020h). COVID-19 sparks fresh investment in Asia warehouse capacity. 19 June.

JOC.com (2020i). Pandemic pushes US [United States] shippers to rethink sourcing, tech spend. 23 June.

JOC.com (2020j). Panama Canal preps for reduced container growth scenario. 7 May.

Kahneman D (2011). *Thinking Fast and Slow*. Penguin Books. London.

Knizek C (2020). Stabilizing supply chains in response to COVID-19. Boston Consulting Group. 26 March.

Lloyd's List (2020). *One Hundred Ports 2020*. Maritime Intelligence.1 May.

Lloyd's Loading List (2020a). UK outlines new customs and border arrangements for 2021. 16 June.

Lloyd's Loading List (2020b). US [United States] mulls paying firms to pull supply chains from China. 19 May.

Lloyd's Loading List (2020c). Current logistics challenges a 'perfect storm' for shippers. 20 April.

Lloyd's Loading List (2020d). Mixed picture globally for e-commerce logistics. 8 June.

MDS Transmodal (2020a). Will the coronavirus pandemic end globalization? 9 June.

MDS Transmodal (2020b). World Cargo Database. 19 August.

New York Times (2018). Congress approves $1.3 trillion spending bill, averting a shutdown. 22 March.

Notteboom T and Pallis T (2020). International Association of Ports and Harbours–World Ports Sustainability Programme Port Economic Impact Barometer. 6 July.

OECD (2020a). *OECD Economic Outlook: Volume 2020*. Issue 1: Preliminary version. No. 107. OECD Publishing. Paris.

OECD (2020b). COVID-19 and global value chains: Policy options to build more resilient production networks. 3 June.

Piattelli-Palmarini M (1994). *Inevitable Illusions: How Mistakes of Reason Rule Our Minds*. John Wiley and Sons. New York.

Port Economics (2020). Changing demand for maritime trades. Port Report No. 4. May. Available at www.porteconomics.eu.

Procurement Bulletin (2020). Understanding the "China, plus one" strategy. Available at www.procurementbulletin.com/understanding-the-china-plus-one-strategy/.

Riviera Maritime Media (2019). Maersk moves forward with goal to be "global integrator of container logistics". 21 February.

Rodrigue J-P (2020). *The Geography of Transport Systems*. Fifth edition. Routledge. New York.

Seara J, Denia L and Krueger F (2020). COVID-19 recovery scenarios for fashion and luxury brands. Boston Consulting Group. 25 March.

The Indian Express (2020). India's imports of palm oil: Dynamics of the trade with Malaysia. 29 January.

The Loadstar (2019). Maersk's inland terminal network to be integrated into Logistics and Service. 16 May.

United Nations (2020a). *World Economic Situation and Prospects*. Department of Economic and Social Affairs. New York.

United Nations (2020b). *World Economic Situation and Prospects as of Mid-2020*. Department of Economic and Social Affairs. New York.

UNCTAD (2018). *UNCTAD Estimates of Global E-commerce 2018*. Technical Notes on ICT for Development. No. 15. TN/UNCTAD/ICT4D/15.

UNCTAD (2020a). *Trade and Development Report 2020: From Global Pandemic to Prosperity for All – Avoiding Another Lost Decade*. (United Nations publication. Sales No. E.20.II.D.30. Geneva).

UNCTAD (2020b). Global Trade Update. June. UNCTAD/DITC/INF/2020/2.

UNCTAD (2020c). *Key Statistics and Trends in International Trade 2019: International Trade Slump* (United Nations publication. Sales No. E.20.II.D.8. Geneva).

UNCTAD (2020d). Global Trade Impact of the Coronavirus (COVID-19) Epidemic. Trade and Development Report Update. 4 March. UNCTAD/DITC/INF/2020/1.

UNCTAD (2020e). Textile and garment supply chains in times of COVID-19: Challenges for developing countries. UNCTAD Transport and Trade Facilitation Newsletter No. 86.

Whitten R and Ben-Moussa S (2020). A trade war on two fronts: U.S. [United States] considers more tariffs on European goods. Shepard Mullin. Global Trade Law Blog. Available at www.globaltradelawblog.com/2020/07/02/tariffs-european-goods-ustr/.

World Bank (2020). *World Economic Prospects: June 2020*. Washington, D.C.

World Steel Association (2019). Worldsteel Short Range Outlook October 2019. 14 October.

World Steel Association (2020). *2020 World Steel in Figures*. Brussels.

WTO (2020). Trade statistics and outlook: Trade falls steeply in first half of 2020. Press release No. 858. 23 June.

The present chapter focuses on key developments related to the supply of maritime transport during this past year. It also assesses the early impact of the COVID-19 pandemic on the supply of maritime transport services and industries and discusses the responses, lessons learned and possible implications of the pandemic in terms of forces shaping supply and the industry's long-term goal of decarbonization.

The pandemic has had a significant impact on the shipping industry. On the one hand, lockdowns and factory closures gradually affected demand for maritime transport, due to reduced cargo volumes (see chapter 1). On the other hand, safety measures applied to contain the spread of the virus, such as lockdowns and travel restrictions, affected the movement of maritime transport workers and procedural changes introduced in ports, and induced operational disruptions in the supply of maritime transport. These prompted changes in shipping operations and requests for government support in the sector. They made the industry reflect on ways to enhance resilience of the sector to future shocks.

This chapter reviews world fleet developments such as annual fleet growth, changes to the structure and age of the fleet. It considers selected segments of the maritime supply chain, such as shipbuilding, ship recycling, ship ownership, ship registration and the maritime workforce, emphasizing the impacts of the pandemic on maritime transport and marine manufacturing industries and on the supply of shipping services.

It also examines the impact of the pandemic on the container, dry bulk and tanker freight markets; government responses to support shipping; and industry prospects, in particular with regard to accelerated digitalization and the prioritization of environmental sustainability. Lastly, it explores the impact of the pandemic on the supply of port-related infrastructure and services, explaining how technology-based solutions relating to trade facilitation, automation and digitalization could support increased resilience to future shocks.

MARITIME TRANSPORT SERVICES AND INFRASTRUCTURE SUPPLY

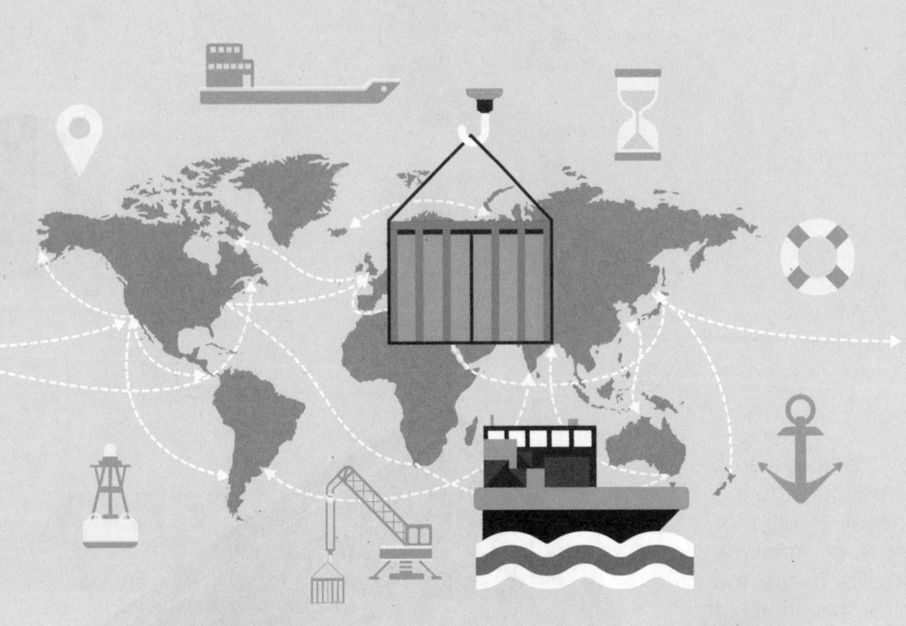

In early 2020, the world fleet totalled **98,140 ships** of 100 gross tons and above **2,061,944** dwt of capacity

Ship types representing largest proportion of global fleet value

Bulk carriers	**19.6%**
Oil tankers	**17.3%**
Offshore vessels	**17.1%**

FREIGHT RATES

Container ships

Blank sailing and other capacity-management measures applied to adapt supply capacity to reduced demand for seaborne trade and allow freight rates to remain strong

- Third quarter of 2020: capacity-reduction measures pursued in container shipping, although demand picking up, keeping freight rates on the rise
- Sustaining these measures for a long period during recovery may lead to dysfunctionalities in the sector, undermining performance of shippers and global supply chains

Tankers

Lockdown repercussions, geopolitical events, oil price fluctuations and increased use of vessels for floating storage led to higher freight rates in March–April 2020

Dry bulk carriers

Oversupplied market and shock of negative demand from China, owing to the outbreak of the pandemic, pulled down dry bulk freight rates

FLEET CHARACTERISTICS

Average fleet age

Age	Ship type
9.28 years	*Bulk carriers*
9.91 years	*Container ships*
10.38 years	*Tankers*
19.46 years	*General cargo ships*

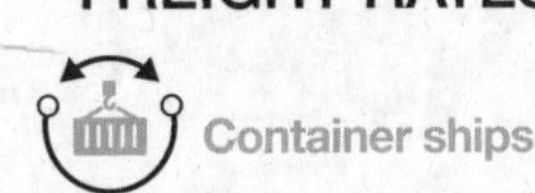

Average vessel sizes

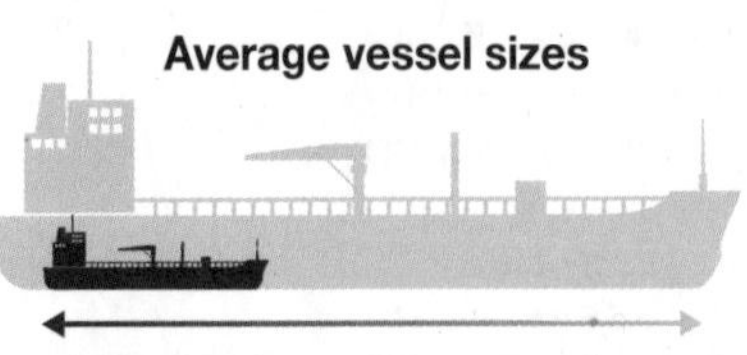

Ships built **20 years ago**

Ships built **in the last 4 years**

*Oil tankers **9 times** bigger*
*Container ships **4 times** bigger*
*General cargo ships **3 times** bigger*
*Bulk carriers **twice** as big*

FLEET GROWTH

Newbuildings
65,911,000 gross tons delivered in 2019 *of which:*

Bulk carriers	**34.5%**
Oil tankers	**30.2%**
Container ships	**16.5%**

Decreasing global volumes of **ship recycling** tonnage *(thousand gross tons)*

2016	**29,135**
2017	**23,138**
2018	**19,003**
2019	**12,218**

- COVID-19 crisis: container shipping industry sustained earnings as carriers applied discipline and strict capacity management
- 2008–2009 financial crisis: freight rates reached dramatic lows, as carriers sought to gain greater market shares through scale and capacity expansion, leading to great losses in container shipping trade

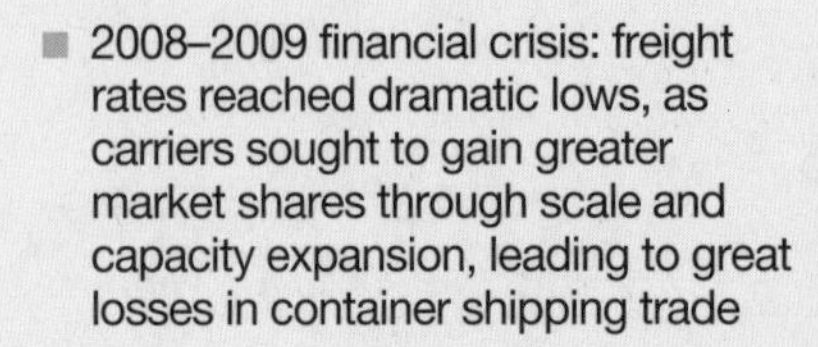

A. WORLD FLEET AND MARITIME WORKFORCE

1. Structure and age of fleet and vessel sizes

In early 2020, the total world fleet amounted to 98,140 ships of 100 gross tons and above, equivalent to 2,061,944,484 dwt of capacity. In the 12 months prior to 1 January 2020, the global commercial shipping fleet grew by 4.1 per cent (table 2.1), registering the highest growth since 2014, but still below levels observed during the 2004–2012 period. The market segment that achieved the highest growth was that of gas carriers, followed by that of oil tankers, bulk carriers and container ships. Gas carriers remained the most dynamic segment, recording the highest growth throughout the 2015–2020 period. In 2019–2020, growth in the oil tankers segment was the highest observed since 2015. In comparison, for the first time in many years, the slowest-growing segment was not that of general cargo ships, but of offshore vessels, where tonnage declined year on year (figure 2.1).

At the start of 2020, the average age of the global fleet was 21.29 years in terms of number of ships, and 10.76 years in terms of carrying capacity in dwt (table 2.2). In terms of dead-weight tonnage, bulk carriers are the youngest vessels, with an average age of 9.28 years, followed by container ships (9.91 years) and oil tankers (10.38 years). On average, general cargo ships are the oldest vessel type (19.46 years). Box 2.1 explains why the age of the fleet matters for decarbonization and provides an example illustrating the case of the Pacific islands.

The highest average vessel sizes are found within the youngest fleet segments (zero to four years). Among this group, oil tankers have the highest average size, followed by bulk carriers and container ships (figure 2.2). In terms of country groupings, developed and developing countries record higher average sizes fleets aged zero to nine years, whereas for countries with economies in transition, the highest average sizes are found in vessels that are between 10 and 19 years old.

Over the past 20 years, vessel sizes have been increasing to optimize costs through economies of scale (see chapter 3). Average bulker and container ship sizes have grown significantly since the 1990s – the average size of container ships has more than doubled since 1996.

The distribution of average sizes across vessel types (figure 2.2) suggests that the average capacity of vessels built in the last four years is much greater than those built 20 years ago. For example, compared with vessels built 20 years ago, the average capacity of oil tankers is nine times greater; of container ships, four times greater; of general cargo ships, three times greater; and of bulk carriers, two times greater.

Table 2.1 World fleet by principal vessel type, 2019–2020
(Thousand dead-weight tons and percentage)

Principal types	2019	2020	Percentage change 2020 over 2019
Bulk carriers	846 418	879 330	3.9
	43 per cent	*43 per cent*	
Oil tankers	568 244	601 163	5.8
	29 per cent	*29 per cent*	
Container ships	266 087	274 856	3.3
	13 per cent	*13 per cent*	
Other types	226 568	232 012	2.4
	11 per cent	*11 per cent*	
Other vessels	80 262	79 862	-0.5
	4 per cent	*4 per cent*	
Gas carriers	69 081	73 586	6.5
	3 per cent	*4 per cent*	
Chemical tankers	46 157	47 474	2.9
	2 per cent	*2 per cent*	
Ferries and passenger ships	7 096	7 289	2.7
	0 per cent	*0 per cent*	
Other/ not available	23 972	23 802	-0.7
	1 per cent	*1 per cent*	
General cargo ships	74 192	74 583	0.5
	4 per cent	*4 per cent*	
World total	**1 981 510**	**2 061 944**	**4.1**

Source: UNCTAD calculations, based on data from Clarksons Research.

Notes: Propelled seagoing merchant vessels of 100 tons and above; beginning-of-year figures.

2. Ship ownership and registration

Ship ownership

Greece, Japan, and China remain the top three ship-owning countries in terms of cargo-carrying capacity (table 2.3), representing 40.3 per cent of the world's tonnage and 30 per cent of the value of the global fleet (table 2.4). The list of the top 35 ship-owning countries in terms of cargo-carrying capacity has remained stable since 2016. In the 12 months prior to 1 January 2020, countries recording the highest increases in carrying capacity compared with the previous year included Nigeria (up 17.2 per cent), the United Arab Emirates (up 5 per cent) and the United Kingdom (up 11.9 per cent). By contrast, Germany, Saudi Arabia and Malaysia lost ground (minus 6.2 per cent, 3.6 per cent and 3.4 per cent, respectively).

Figure 2.1 Growth of world fleet by principal vessel type, 2014–2020
(Dead-weight tonnage and percentage change)

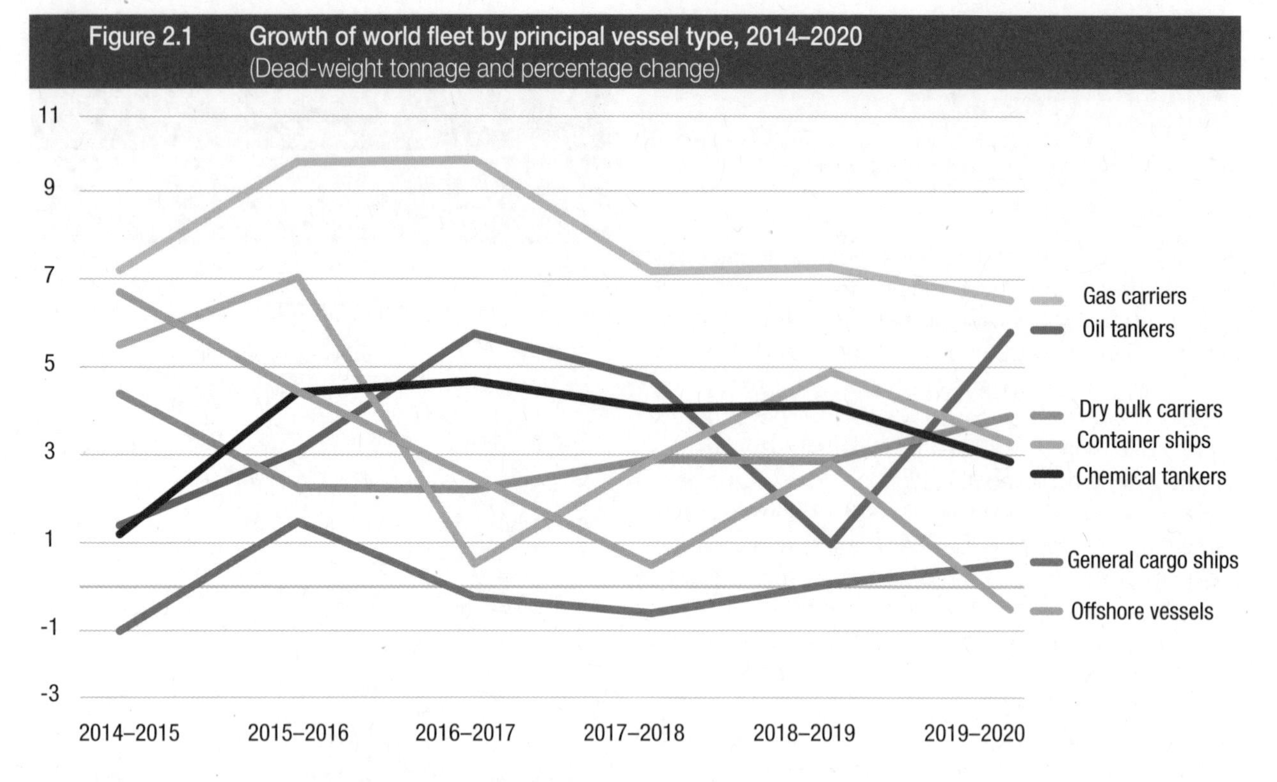

Source: UNCTAD, *Review of Maritime Transport*, various issues.

The value of the fleet is linked to expectations of revenue and performance of shipping markets (Hellenic Shipping News Worldwide, 2020a; Marine Insight, 2019) and hence to return on investment, an important consideration from the perspective of owners. The value of the fleet can also be linked to the transport and logistics value chain and to the level of sophistication of the fleet, that is, the embedded digital technology making it possible to improve efficiency, safety, equipment maintenance and operational processes (Riviera Maritime Media, 2020). At the beginning of 2020, the main ship types representing the highest proportion of the value in the global fleet were bulk carriers, oil tankers and offshore vessels (table 2.4).

The top three ship-owning economies (Greece, Japan and China) represent a higher share of the global carrying capacity than of the global value of the fleet (figure 2.3), unlike the fourth- and fifth-ranked countries (United States and Norway, respectively). The characteristics and composition of commercial fleets explain the contrast between the two percentage shares. In some countries, this is linked to high-value non-cargo ships. For instance, the highest proportion of the value of the fleet of Norway, the United Kingdom, the Netherlands and Brazil comes from offshore vessels, whereas in the case of the United States, Switzerland and Italy, it comes from cruise ships.

Ship registration

Panama, Liberia and the Marshall Islands remain the three leading flags of registration, in terms of carrying capacity (table 2.5) and of value of the fleet registered (table 2.6). As of 1 January 2020, they represented 42 per cent of the carrying capacity and 33.6 per cent of the value of the fleet. The flags of the Islamic Republic of Iran, Taiwan Province of China and Thailand registered the highest increases in terms of dead-weight tonnage. Ships under the flag of the Islamic Republic of Iran more than tripled their growth compared with 2019. The three registries that saw the level of tonnage decrease in the 12 months preceding 1 January 2020 were the United Kingdom, Bermuda and the Isle of Man.

The quadrupling of the number of ships flying under the flag of the Islamic Republic of Iran derives from increased pressure exerted by sanctions, which led several registries, including those of Liberia, Panama, Sierra Leone and Togo (Reuters, 2019a), to de-flag vessels associated with trade from that country (Lloyd's List, 2020a). The most recent guidance to the maritime industry, issued in May 2020 by the Office of Foreign Assets Control of the United States Department of the Treasury, was an important milestone. The guidance expanded the compliance responsibility for fleet control and monitoring to actors beyond shipowners and operators, including flag registries, port operators, freight forwarders, classification societies and financial institutions (Lexology, 2020; The Maritime Executive, 2020a).

Between 1 January 2019 and 1 January 2020, the registries from the United Kingdom and some of the international registries categorized as crown dependencies and overseas territories – Gibraltar and the Isle of Man – witnessed a reduction. Tonnage registered under the flag of the United Kingdom

Table 2.2 Age distribution of world merchant fleet by vessel type, 2019–2020
(Percentage and dead-weight tonnage)

Country grouping		Years					Average age	Average age
		0–4	5–9	10–14	15–19	More than 20	2020	2020
World								
Bulk carriers	Percentage of total ships	20.22	42.17	18.70	8.99	9.93	10.18	9.69
	Percentage of dead-weight tonnage	23.30	44.86	16.73	8.22	6.89	9.28	8.87
	Average vessel size (dead-weight tonnage)	84 714	78 169	65 767	67 246	50 973		
Container ships	Percentage of total ships	15.60	20.39	32.79	14.67	16.55	12.72	12.29
	Percentage of dead-weight tonnage	24.41	29.14	28.19	11.74	6.53	9.91	9.43
	Average vessel size (dead-weight tonnage)	80 070	73 137	43 993	40 934	20 186		
General cargo ships	Percentage of total ships	4.64	12.34	15.67	7.99	59.36	26.93	26.30
	Percentage of dead-weight tonnage	8.52	23.16	19.76	9.88	38.69	19.46	18.89
	Average vessel size (dead-weight tonnage)	7 933	8 029	5 455	5 902	2 772		
Oil tankers	Percentage of total ships	14.45	18.95	20.19	11.11	35.32	19.12	18.77
	Percentage of dead-weight tonnage	24.73	24.99	26.57	17.52	6.20	10.38	10.11
	Average vessel size (dead-weight tonnage)	93 311	72 952	71 391	86 251	9 924		
Other	Percentage of total ships	11.21	18.05	15.53	8.28	46.93	23.18	22.70
	Percentage of dead-weight tonnage	21.56	16.94	22.22	10.57	28.71	15.59	15.42
	Average vessel size (dead-weight tonnage)	11 613	6 267	8 682	8 034	4 304		
All ships	Percentage of total ships	11.64	20.11	17.42	8.98	41.85	21.29	20.83
	Percentage of dead-weight tonnage	23.14	33.04	21.85	11.72	10.25	10.76	10.43
	Average vessel size (dead-weight tonnage)	47 901	40 986	30 290	32 742	6 661		
Developing economies (all ships)								
	Percentage of total ships	11.26	21.72	17.31	8.49	41.21	20.38	19.90
	Percentage of dead-weight tonnage	21.75	33.21	18.22	11.62	15.21	11.56	11.15
	Average vessel size (dead-weight tonnage)	37 438	32 440	20 900	27 950	7 544		
Developed economies (all ships)								
	Percentage of total ships	13.33	20.35	19.82	10.67	35.84	19.95	19.54
	Percentage of dead-weight tonnage	24.52	33.42	24.42	11.68	5.97	9.96	9.71
	Average vessel size (dead-weight tonnage)	61 465	52 885	40 792	38 294	7 305		
Countries with economies in transition (all ships)								
	Percentage of total ships	6.38	8.19	8.63	4.34	72.47	30.33	29.82
	Percentage of dead-weight tonnage	8.94	20.19	27.46	15.58	27.83	16.99	16.39
	Average vessel size (dead-weight tonnage)	12 644	18 987	25 905	25 880	2 724		

Source: Clarksons Research.

Note: Propelled seagoing vessels of 100 gross tons and above; beginning-of-year figures.

Box 2.1 Reducing carbon dioxide emissions: The case of the Pacific islands

The average age of a vessel can be an indirect indication of its environmental performance. In most cases, younger vessels are more fuel-efficient and less polluting because of technological advances. Bringing down the carbon footprint of shipping is not only a function of the age of the fleet (which could be associated with the introduction of technical improvements) but could also be a function of operational measures, such as speed optimization, or of shifting to alternative fuels. Other factors that also come into play are maintenance schemes or fleet-renewal trends linked to scrapping patterns and financial incentives (either to scrap or to order newbuildings).

Recent studies were conducted in the Pacific to assess different carbon dioxide reduction pathways, as several of the islands in the region have launched regional and national initiatives to develop low-carbon coastal maritime transport. The age of the fleet was an important consideration to inform decision-making related to maritime transport strategies and objectives. According to recent estimates by the Pacific Community, 41 per cent of the vessels from Fiji, Kiribati, the Marshall Islands, Samoa, Solomon Islands and Vanuatu are less than 20 years old; 20 per cent, between 20 and 30 years old ; and 38 per cent, more than 30 years old. There is a large proportion of older vessels because many of them were donated or bought second-hand. These vessels have low carrying capacity (less than 5,000 tons) and entail economic costs due to increasing maintenance and survey costs.

Although newbuildings would result in an 80–90 per cent improvement in operational efficiency, they would require significant investment to enable fleet replacement to meet the emission-reduction targets set in regional and national decarbonization strategies, highlighting the need for financing.

To abate emissions in the existing fleet, the Pacific islands are retrofitting vessels with wind propulsion and using wind and solar as auxiliary power supply. Such retrofits were found to be more suitable to the characteristics, financial capabilities, level of technological uptake and maritime heritage of the Pacific fleet than other options being considered in other countries, such as shifting to some alternative fuels and the use of onshore electrification. The studies found a potential to scale up such retrofits but acknowledged that retrofits could not achieve the same degree of savings and emission reduction as newbuilds.

Sources: Government of Fiji, 2018; Micronesian Centre for Sustainable Transport, 2019a, 2019b, 2020.

declined by 29.8 per cent, that of the Isle of Man by 13.5 per cent and Gibraltar, by 7.4 per cent. These developments could be linked to geopolitical tensions

Figure 2.2 Average vessel size and age distribution, selected vessel types, 2020
(Dead-weight tons)

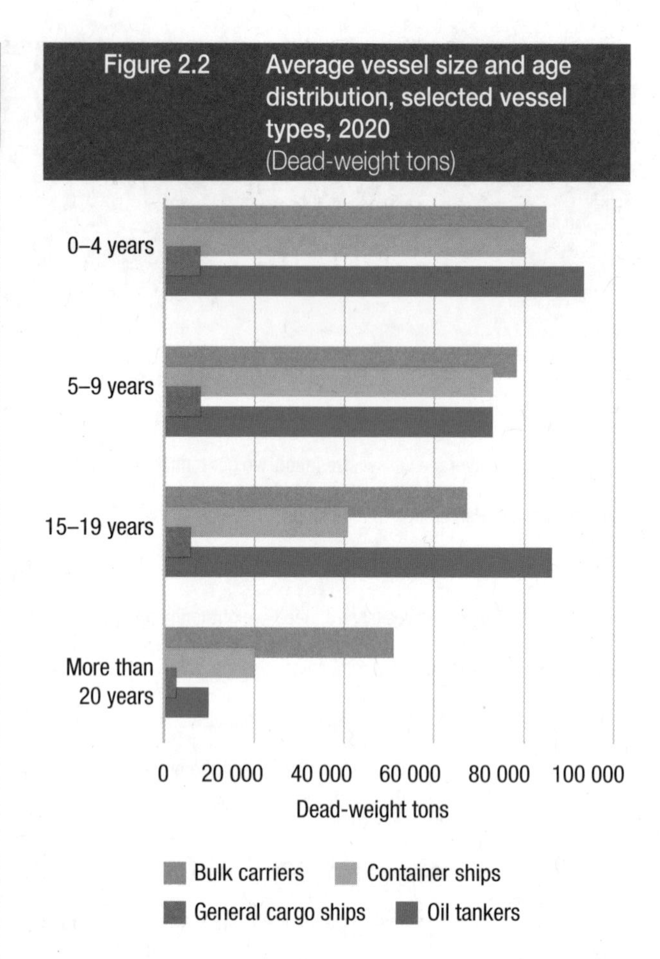

Source: UNCTAD calculations, based on data from Clarksons Research.

Note: Propelled seagoing vessels of 100 gross tons and above; beginning-of-year figures.

with the Islamic Republic of Iran, which led to changes in ship registration (United Kingdom Department for Transport, 2020) but also to uncertainty related to the Brexit process (Lloyd's List, 2019a; Reuters, 2019b).

Plans for improving the competitiveness and attractiveness of the United Kingdom registry, particularly for low or zero-emission technology vessels and, in the long-term, for autonomous and semi-autonomous ships, include digitalization initiatives. These are aimed at reinforcing paperless maritime governance and e-registration and enhancing the quality of service through new standards and practices pertaining to inspections, certifications and business facilitation (United Kingdom Department for Transport, 2019).

3. Shipbuilding, new orders and ship recycling

Shipbuilding

China, the Republic of Korea and Japan maintained their traditional leadership in shipbuilding, representing 92.5 per cent of the newbuilding deliveries in 2019

Table 2.3 Ownership of world fleet, ranked by carrying capacity in dead-weight tons, 2020

	Country or territory of ownership	Number of vessels			Dead-weight tonnage				
		National flag	Foreign flag	Total	National flag	Foreign flag	Total	Foreign flag as a percentage of total	Total as a percentage of total
1	Greece	671	3 977	4 648	60 827 479	303 026 753	363 854 232	83.28	17.77
2	Japan	909	3 001	3 910	36 805 225	196 329 652	233 134 877	84.21	11.38
3	China	4 569	2 300	6 869	99 484 023	128 892 849	228 376 872	56.44	11.15
4	Singapore	1 493	1 368	2 861	74 754 209	62 545 517	137 299 726	45.55	6.70
5	Hong Kong, China	883	807	1 690	72 505 185	28 452 208	100 957 393	28.18	4.93
6	Germany	205	2 299	2 504	8 340 596	81 062 481	89 403 077	90.67	4.37
7	Republic of Korea	778	837	1 615	14 402 899	66 179 736	80 582 635	82.13	3.93
8	Norway	383	1 660	2 043	1 884 535	62 051 275	63 935 810	97.05	3.12
9	Bermuda	13	529	542	324 902	60 088 969	60 413 871	99.46	2.95
10	United States	799	1 131	1 930	10 237 585	46 979 245	57 216 830	82.11	2.79
11	United Kingdom	317	1 027	1 344	6 835 508	46 355 337	53 190 845	87.15	2.60
12	Taiwan Province of China	140	850	990	6 636 271	44 255 009	50 891 280	86.96	2.48
13	Monaco		473	473		43 831 888	43 831 888	100.00	2.14
14	Denmark	25	921	946	31 435	42 683 049	42 714 484	99.93	2.09
15	Belgium	113	188	301	10 040 106	20 658 108	30 698 214	67.29	1.50
16	Turkey	449	1 079	1 528	6 656 989	21 433 413	28 090 402	76.30	1.37
17	Switzerland	26	401	427	1 113 387	25 365 225	26 478 612	95.80	1.29
18	India	859	183	1 042	16 800 490	9 035 433	25 835 923	34.97	1.26
19	Indonesia	2 132	76	2 208	22 301 493	1 604 369	23 905 862	6.71	1.17
20	Russian Federation	1 403	339	1 742	8 292 932	14 812 631	23 105 563	64.11	1.13
21	United Arab Emirates	118	852	970	480 283	20 271 823	20 752 106	97.69	1.01
22	Islamic Republic of Iran	238	8	246	18 245 935	353 441	18 599 376	1.90	0.91
23	Netherlands	700	492	1 192	5 584 365	12 437 918	18 022 283	69.01	0.88
24	Saudi Arabia	137	132	269	13 303 057	4 126 462	17 429 519	23.68	0.85
25	Italy	499	179	678	11 005 343	6 400 010	17 405 353	36.77	0.85
26	Brazil	302	94	396	4 963 496	8 984 821	13 948 317	64.42	0.68
27	France	106	333	439	898 897	12 448 289	13 347 186	93.27	0.65
28	Cyprus	141	165	306	4 958 311	6 659 094	11 617 405	57.32	0.57
29	Viet Nam	910	150	1 060	8 390 791	2 357 014	10 747 805	21.93	0.52
30	Canada	222	159	381	2 723 583	7 247 389	9 970 972	72.68	0.49
31	Malaysia	464	156	620	6 378 887	2 164 848	8 543 735	25.34	0.42
32	Oman	5	51	56	5 704	8 069 314	8 075 018	99.93	0.39
33	Qatar	59	67	126	1 056 669	6 054 422	7 111 091	85.14	0.35
34	Sweden	88	213	301	929 401	5 580 520	6 509 921	85.72	0.32
35	Nigeria	182	74	256	3 227 668	3 031 686	6 259 354	48.43	0.31
	Subtotal, top 35 shipowners	**20 338**	**26 571**	**46 909**	**540 427 639**	**1 411 830 198**	**1 952 257 837**	**72.32**	**95.33**
	Rest of world and unknown	*3 037*	*3 015*	*6 052*	*36 513 130*	*59 204 480*	*95 717 610*	*61.85*	*4.67*
	World total	**23 375**	**29 586**	**52 961**	**576 940 769**	**1 471 034 678**	**2 047 975 447**	**71.8**	**100.0**

Source: UNCTAD calculations, based on data from Clarksons Research.

Notes: Propelled seagoing vessels of 1,000 gross tons and above, as at 1 January 2020. For the purposes of this table, second and international registries are recorded as foreign or international registries, whereby, for example, ships belonging to owners in the United Kingdom registered in Gibraltar or on the Isle of Man are recorded as being under a foreign or an international flag. In addition, ships belonging to owners in Denmark and registered in the Danish International Ship Register account for 45 per cent of the Denmark-owned fleet in dead-weight tonnage, and ships belonging to owners in Norway registered in the Norwegian International Ship Register account for 27.4 per cent of the Norway-owned fleet in dead-weight tonnage. For a complete listing of nationally owned fleets, see http://stats.unctad.org/fleetownership.

Table 2.4 Top 25 ship-owning economies, as at 1 January 2020
(Million dollars)

	Country or territory	Bulk carriers	Oil tankers	Offshore vessels	Ferries and passenger ships	Container ships	Gas carriers	General cargo ships	Chemical tankers	Other /not available	Total
1	Greece	34 426	37 873	187	2 404	7 936	12 238	189	1 064	468	96 785
2	Japan	34 027	9 981	4 713	3 030	11 805	15 173	3 482	4 937	9 150	96 298
3	China	30 108	13 278	10 189	5 089	17 243	4 267	5 244	3 126	3 008	91 553
4	United States	3 352	6 308	20 392	52 130	1 190	1 458	1 122	1 971	732	88 655
5	Norway	4 213	6 217	23 156	3 088	1 852	7 847	950	2 423	3 002	52 748
6	Singapore	12 860	13 975	5 189	25	6 845	4 428	1 043	4 695	566	49 626
7	Germany	5 857	2 121	630	9 630	17 211	1 966	3 429	791	360	41 996
8	United Kingdom	3 760	4 106	13 226	4 575	4 592	5 318	920	1 457	2 581	40 535
9	Hong Kong, China	10 209	7 239	601	2 723	10 082	1 173	898	282	1 027	34 234
10	Bermuda	4 826	5 895	5 779		2 079	8 431		375	62	27 447
11	Republic of Korea	7 319	5 999	264	366	2 400	4 914	710	1 595	2 816	26 383
12	Denmark	1 412	4 008	2 373	999	10 642	2 014	752	971	111	23 282
13	Switzerland	813	821	3 244	10 243	7 337	225	236	213	9	23 142
14	Netherlands	747	535	13 457	619	386	753	3 411	1 228	1 938	23 076
15	Italy	1 162	2 319	2 655	8 944	4	305	2 068	553	504	18 515
16	Brazil	145	1 029	15 345	69	298	131	35	84	1	17 138
17	Monaco	3 292	7 232		32	997	3 712		32	30	15 327
18	Taiwan Province of China	7 057	1 668	37	79	4 088	396	632	156	105	14 219
19	France	374	130	5 393	1 813	4 174	521	179	141	224	12 949
20	Turkey	3 208	1 433	691	346	1 290	145	1 892	1 121	42	10 168
21	Russian Federation	246	3 966	1 456	74	72	1 489	1 227	633	849	10 014
22	Malaysia	166	239	6 409	14	73	1 897	138	142	166	9 245
23	Belgium	1 515	4 070	88		262	1 221	811	167	529	8 663
24	Indonesia	838	2 091	849	1 942	790	517	1 105	348	47	8 528
25	United Arab Emirates	1 530	2 300	3 051	59	216	473	75	584	72	8 359
	Other	13 157	19 676	23 857	12 120	3 135	15 552	8 345	4 169	3 317	103 328
	World total	**186 622**	**164 511**	**163 232**	**120 413**	**116 998**	**96 568**	**38 894**	**33 258**	**31 718**	**952 213**

Source: UNCTAD calculations, based on data from Clarksons Research, as at 1 January 2020 (estimated current value).
Note: Value is estimated for all commercial ships of 1,000 gross tons and above.

(table 2.7). Each country specializes in different shipping segments. China is the leading builder of bulk carriers (56.2 per cent), offshore vessels (58 per cent) and general cargo ships (34.6 per cent); the Republic of Korea, of gas carriers (62.8 per cent), oil tankers (59.4 per cent) and container ships (41.7 per cent); and Japan, chemical tankers (54.1 per cent).

Compared with 2019, the market share of the Republic of Korea increased by 7.7 percentage points, whereas that of China decreased by 5.1 percentage points. Bulk carrier and oil tanker newbuildings registered the largest increases (7.8 and 5.2 percentage points, respectively) whereas container ships and gas carriers registered the greatest decreases (-2 and -3.2 percentage points, respectively).

New orders

In early 2020, the world order book had declined with respect to dry bulk carriers, oil carriers, container ships and general cargo ships (figure 2.4). Orders for three of these shipping segments have been shrinking since 2017 (except for dry bulk carriers, which increased in 2019). Widening disparity between newbuilding prices and earnings, geopolitical instability, persistent financing challenges and broad uncertainty over fuel and technology choices explain this trend (Barry Rogliano Salles, 2020).

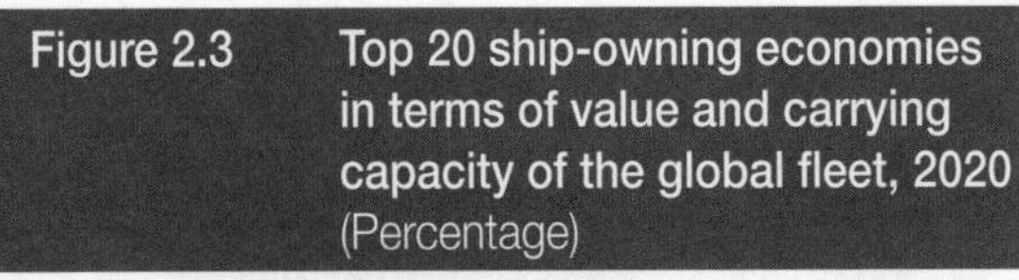
Figure 2.3 Top 20 ship-owning economies in terms of value and carrying capacity of the global fleet, 2020 (Percentage)

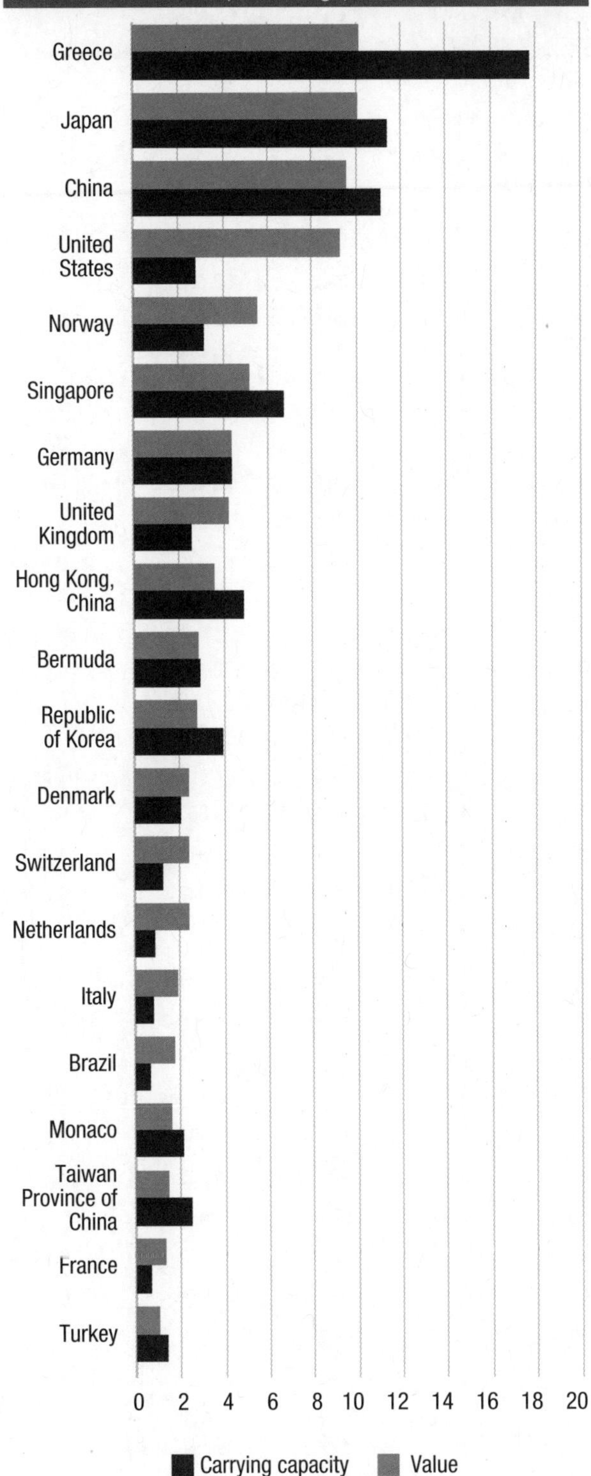

Source: UNCTAD calculations, based on data from Clarksons Research.

Note: Value is estimated for all commercial ships of 1,000 gross tons and above.

Ship recycling

Bangladesh remains the country with the largest global share of recycled tonnage, accounting for more than half of the ships recycled in 2019. Together with India and Turkey, these three countries represented 90.3 per cent of the ship recycling activity in 2019. The same year, bulk carriers constituted most of the recycled tonnage (about one third), followed by container ships and oil tankers (table 2.8). Since 2016, global volumes of recycled tonnage have been on the wane. Volumes fell to 29,135 thousand gross tons in 2016, 23,138 thousand gross tons in 2017, 19,003 thousand gross tons in 2018 and 12,218 thousand gross tons in 2019. Steel price developments in scrapping destinations and expectations concerning the evolution of freight rates are factors underpinning these trends (Hellenic Shipping News Worldwide, 2019).

The only country among the top five scrapping destinations that increased its ship-recycling volumes in 2019 was Turkey (figure 2.5), linked reportedly to certification of Turkish shipyards by the European Union, enabling them to be on the list of approved facilities for the recycling of ships flying European Union flags (Hellenic Shipping News Worldwide, 2018). In 2019, Turkey also ratified the Hong Kong [China] International Convention for the Safe and Environmentally Sound Recycling of Ships, 2009 of IMO. Among the other countries, the reduction in the share of Pakistan was most significant, motivated by adverse conditions related to taxation and exchange rates (The Maritime Executive, 2019). In 2019, bulk carriers increased their percentage share in global recycling volumes by 172 per cent; container ships, by 145 per cent; and offshore vessels, by 88 per cent. By contrast, oil tankers and gas carriers registered significant decreases of 71 and 55 per cent, respectively.

Impacts of the coronavirus disease pandemic, responses and prospects: Labour shortages affect newbuilding and ship recycling and weak investor sentiment affects ordering

The pandemic led to reductions and delays in newbuilding delivery and to a standstill in ship recycling. This can be attributed to lockdown-induced labour shortages in the shipbuilding and ship recycling industries. In addition, other measures implemented to reduce the spread of the pandemic, such as travel restrictions, made it impossible for owners to arrange visits or obtain a crew for final delivery. Port closures also affected tonnage arrival into scrapping destinations on the Indian subcontinent (Hellenic Shipping News Worldwide, 2020b).

The pandemic also had a significant impact on the manufacturing segments of the maritime supply chain. In February 2020, deliveries from China fell to their lowest level in 15 years, with only four ships delivered. As lockdowns were gradually lifted, industrial activity resumed. China was reported to have returned to 50 per cent of its 2019 output average in March 2020

Table 2.5 Leading flags of registration by dead-weight tonnage, 2020

	Flag of registration	Number of vessels	Share of world vessel total (percentage)	Dead-weight tonnage (thousand dead-weight tons)	Share of total world dead-weight tonnage (percentage)	Cumulated share of dead-weight tonnage (percentage)	Average vessel size (dead-weight tonnage)	Growth in dead-weight tonnage 2020 over 2019 (percentage)
1	Panama	7 886	8	328 950	16	16.0	41 713	-1.3
2	Liberia	3 716	4	274 786	13	29.3	73 947	13.0
3	Marshall Islands	3 683	4	261 806	13	42.0	71 085	6.5
4	Hong Kong, China	2 694	3	201 361	10	51.7	74 744	1.3
5	Singapore	3 420	3	140 333	7	58.5	41 033	8.3
6	Malta	2 207	2	115 879	6	64.2	52 505	4.7
7	China	6 192	6	100 086	5	69.0	16 164	3.0
8	Bahamas	1 381	1	77 869	4	72.8	56 386	0.1
9	Greece	1 294	1	68 632	3	76.1	53 039	-0.7
10	Japan	5 041	5	40 323	2	78.1	7 999	3.4
11	Cyprus	1 065	1	34 533	2	79.8	32 425	-0.1
12	Indonesia	10 137	10	25 574	1	81.0	2 523	6.9
13	Isle of Man	356	0	24 129	1	82.2	67 779	-13.5
14	Danish International Register	575	1	23 044	1	83.3	40 077	3.0
15	Norwegian International Register	647	1	20 780	1	84.3	32 118	4.8
16	Madeira	526	1	20 698	1	85.3	39 351	6.0
17	Islamic Republic of Iran	877	1	19 700	1	86.3	22 463	362.3
18	India	1 768	2	17 339	1	87.1	9 807	-0.2
19	Republic of Korea	1 889	2	14 942	1	87.8	7 910	14.9
20	Saudi Arabia	376	0	13 554	1	88.5	36 047	3.2
21	United States	3 650	4	11 985	1	89.1	3 284	0.6
22	United Kingdom	945	1	11 962	1	89.6	12 658	-29.8
23	Italy	1 310	1	11 953	1	90.2	9 124	-10.8
24	Belgium	203	0	10 349	1	90.7	50 980	-1.1
25	Malaysia	1 772	2	10 260	0	91.2	5 790	-0.4
26	Russian Federation	2 808	3	9 797	0	91.7	3 489	6.9
27	Viet Nam	1 909	2	9 123	0	92.1	4 779	7.7
28	Germany	606	1	8 468	0	92.5	13 974	-0.9
29	Bermuda	138	0	7 662	0	92.9	55 525	-18.9
30	Turkey	1 216	1	6 993	0	93.3	5 751	-6.5
31	Netherlands	1 200	1	6 982	0	93.6	5 818	-1.4
32	Taiwan Province of China	407	0	6 739	0	93.9	16 557	16.0
33	Antigua and Barbuda	727	1	6 657	0	94.2	9 157	-11.1
34	Thailand	840	1	6 642	0	94.6	7 907	15.7
35	Cayman Islands	163	0	6 636	0	94.9	40 713	-1.1
	Top 35 total	**73 624**	**75**	**1 956 529**	**95**	**94.9**	**26 575**	
	World total	**98 140**	**100**	**2 061 944**	**100**	**100.00**	**21 010**	**4.1**

Source: UNCTAD calculations, based on data from Clarksons Research.

Notes: Propelled seagoing merchant vessels of 100 gross tons and above, as at 1 January 2020. For a complete listing of countries, see http://stats.unctad.org/fleet.

Table 2.6 Leading flags of registration, ranked by value of principal vessel type, 2020
(Dollars)

	Flag of registration	Bulk carriers	Oil tankers	Offshore vessels	Ferries and passenger ships	Container ships	Gas carriers	General cargo ships	Chemical tankers	Other/not applicable	Total
1	Panama	40 369	13 462	17 612	12 037	17 035	10 632	3 899	5 306	7 412	127 765
2	Marshall Islands	27 870	29 606	17 257	1 284	6 150	15 110	515	4 511	2 207	104 511
3	Liberia	23 729	22 944	12 662	150	17 217	5 756	1 010	2 590	1 488	87 544
4	Bahamas	4 950	7 759	23 781	31 330	606	13 295	73	106	2 566	84 466
5	Hong Kong, China	23 280	11 360	289	42	21 030	5 987	1 607	1 878	120	65 592
6	Malta	9 418	11 192	4 758	15 420	12 173	4 929	1 681	1 793	873	62 236
7	Singapore	12 226	14 540	8 748		11 673	7 473	1 066	3 541	1 458	60 725
8	China	14 910	7 012	7 914	4 412	3 456	678	2 880	1 451	2 887	45 599
9	Greece	2 831	10 710	1	1 561	272	5 587	47	77	90	21 176
10	Italy	671	1 064	501	14 235	77	244	2 106	388	504	19 791
	Subtotal top 10	**160 253**	**129 650**	**93 521**	**80 469**	**89 689**	**69 692**	**14 883**	**21 642**	**19 606**	**679 405**
	Other	***26 370***	***34 861***	***69 711***	***39 944***	***27 309***	***26 876***	***24 011***	***11 615***	***12 112***	***272 808***
	World total	**186 622**	**164 511**	**163 232**	**120 413**	**116 998**	**96 568**	**38 894**	**33 258**	**31 718**	**952 213**

Source: UNCTAD calculations, based on data from Clarksons Research, as at 1 January 2019 (estimated current value).
Note: Value is estimated for all commercial ships of 1,000 gross tons and above.

Table 2.7 Deliveries of newbuildings by major vessel types and countries of construction, 2019
(Thousand gross tons)

Vessel type	China	Republic of Korea	Japan	Philippines	Rest of world	Total	Percentage
Bulk carriers	12 773	1 010	7 942	652	338	**22 716**	34.5
Oil tankers	4 200	11 827	2 811	128	946	**9 912**	30.2
Container ships	3 712	4 545	2 521	19	94	**10 891**	16.5
Gas carriers	420	3 888	1 881		1	**6 189**	9.4
Ferries and passenger ships	214	3	59	3	1 903	**2 182**	3.3
General cargo ships	452	202	267		387	**1 307**	2.0
Offshore vessels	651	135	4		332	**1 121**	1.7
Chemical tankers	368	49	574		71	**1 063**	1.6
Other	285	12	182	0	50	**530**	0.8
Total	**23 074**	**21 670**	**16 242**	**802**	**4 122**	**65 911**	**100.0**
Percentage	35.0	32.9	24.6	1.2	6.3	**100.0**	

Source: UNCTAD calculations, based on data from Clarksons Research.
Notes: Propelled seagoing merchant vessels of 100 gross tons and above. For more data on other shipbuilding countries, see http://stats.unctad.org/shipbuilding.

and to 60 per cent in May. However, by May 2020 global shipbuilding output in dwt was down 14 per cent year over year (Clarksons Research, 2020a). In March, when the pandemic erupted in the Europe and the United States, lockdowns in Bangladesh, India and Pakistan gradually halted ship recycling (Vessels Value, 2020). In June 2020, Indian recycling yards were reported to be operating at just 30 to 40 per cent of full capacity (Clarksons Research, 2020b).

The COVID-19 pandemic has brought widespread uncertainty related to economic performance in 2020 and 2021 (see chapter 1). As a result, strategic investment decisions had to be reconsidered, for instance, newbuilding ordering and repairs were postponed. Ordering contracts were down 53 per cent year over year in July 2020 (Clarksons Research, 2020c). In addition, many companies decided to delay scrubber installation because of the impact of the pandemic

Figure 2.4 World tonnage on order, 2000–2020
(Thousand dead-weight tons)

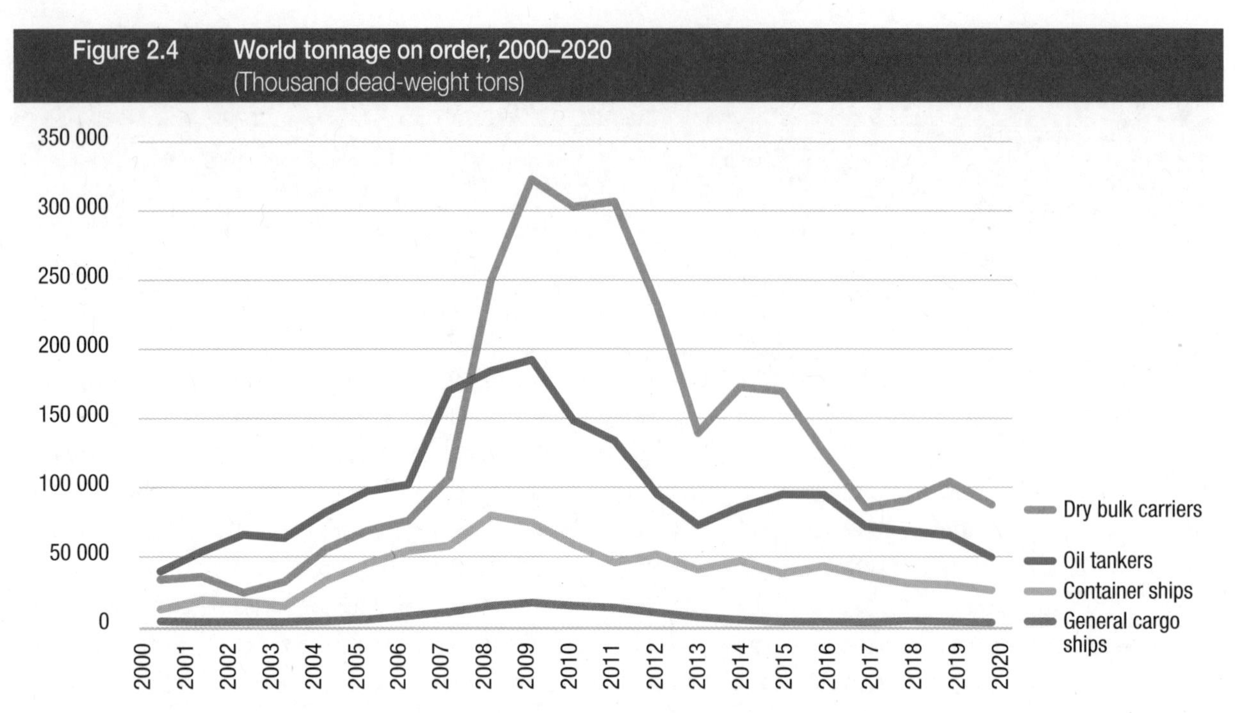

Source: UNCTAD calculations, based on data from Clarksons Research.
Notes: Propelled seagoing merchant vessels of 100 gross tons and above; beginning-of-year figures.

Table 2.8 Reported tonnage sold for ship recycling by major vessel type and country of ship recycling, 2019
(Thousand gross tons)

Vessel type	Bangladesh	China	India	Pakistan	Turkey	Rest of world	World total	Percentage
Bulk carriers	3 426	238	582	132	161	32	**4 570**	37.4
Chemical tankers	64	4	125	7	3	9	**211**	1.7
Container ships	1 015	24	964	12	10	86	**2 111**	17.3
Ferries and passenger ships	71	2	46	27	76	5	**226**	1.8
General cargo ships	140	62	150	12	174	36	**575**	4.7
Liquefied gas carriers	169		70		30	9	**279**	2.3
Offshore vessels	326	4	543	9	435	197	**1 514**	12.4
Oil tankers	1 271	14	387	56	119	153	**1 999**	16.4
Other	200	35	384	13	87	12	**732**	6.0
Total	**6 682**	**383**	**6 682**	**267**	**1 095**	**540**	**12 218**	**100.0**
Percentage	54.7	3.1	26.6	2.2	9.0	4.4	**100.0**	

Source: Clarksons Research.
Notes: Propelled seagoing vessels of 100 gross tons and above. Estimates for all countries available at http://stats.unctad.org/shipscrapping.

on financial cash flow (Clarksons Research, 2020d; *Manifold Times*, 2020). This is also linked to fuel price dynamics since January 2020, namely the narrowing of the price differential between high and low sulphur fuel, which increased the time to recover the investment cost of installing scrubbers (IHS Markit, 2020; *Seatrade Maritime News*, 2020a).

Before the pandemic, the shipbuilding sector had already been facing a challenging environment of fierce competition and declining orders. Increased consolidation and government finance helped to cope with this situation (UNCTAD, 2019a). Seeking to minimize costs and losses and restructuring their businesses to improve balance sheets, the world's

Figure 2.5 Reported tonnage sold for ship recycling by major vessel type and country of ship recycling, 2017–2019
(Thousand gross tons and percentage change)

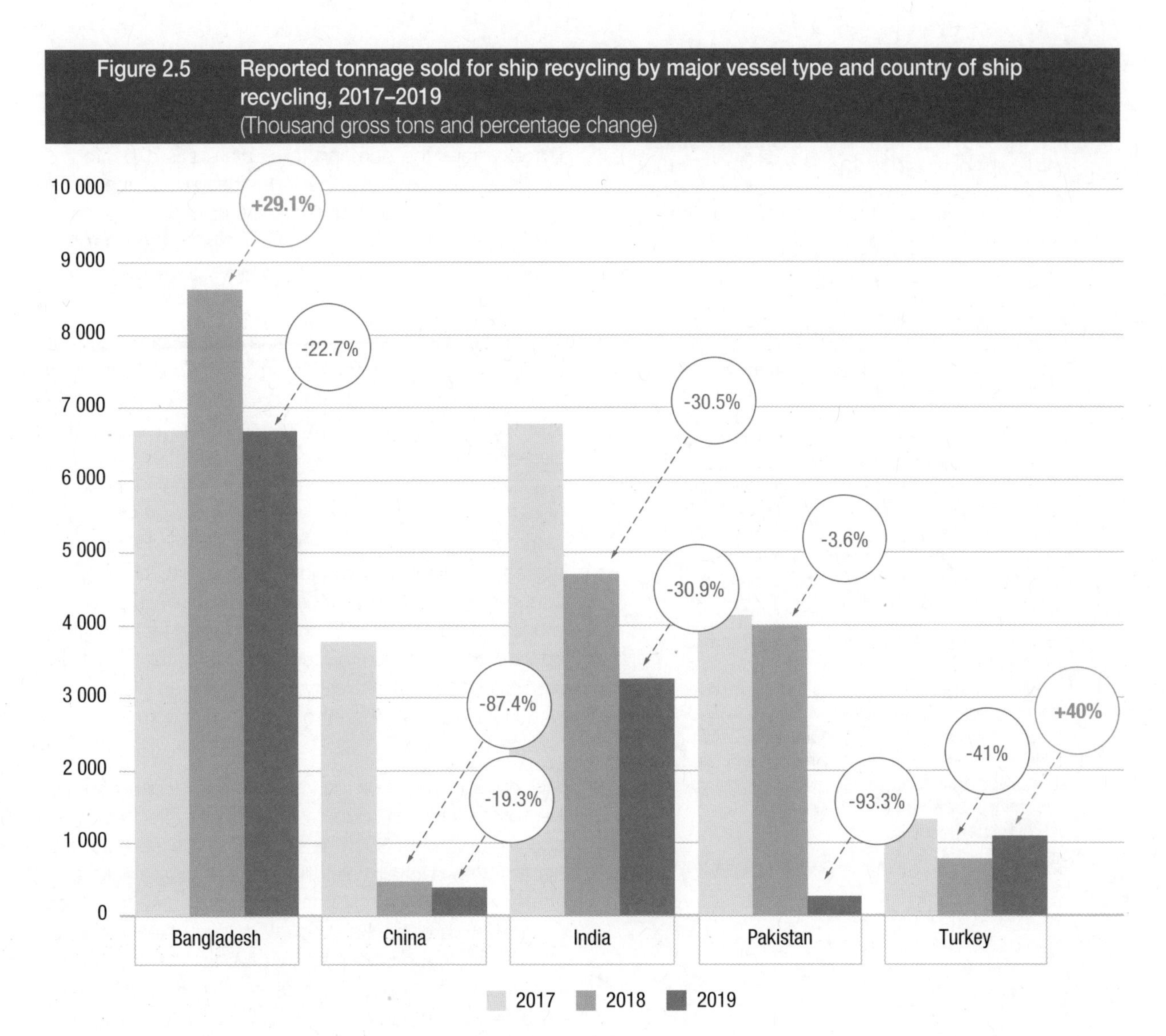

Source: UNCTAD calculations, based on data from Clarksons Research.

Notes: Propelled seagoing vessels of 100 gross tons and above. Estimates for all countries available at http://stats.unctad.org/shipscrapping.

largest shipbuilder (Hyundai Heavy Industries Company of the Republic of Korea) signed in March 2020 a formal agreement with the State-run Korea [Republic of] Development Bank to buy Daewoo Shipbuilding and Marine Engineering. The merger will be completed upon approval by antitrust authorities in China, the European Union, Japan, Kazakhstan, the Republic of Korea and Singapore (*The Korea Times*, 2020). The European Union and Japan have voiced concerns about the potential of this merger to lead to an uneven trading playing field (WTO, 2020) and reduced competition in shipbuilding markets of large container ships, oil tankers, liquefied natural gas carriers and liquefied petroleum gas carriers (European Commission, 2019).

Against this background, the pandemic further accentuated challenges, reducing demand and affecting orders, production and delivery. Box 2.2 describes some of these challenges, from perspective of the European Union.

The slowdown in shipbuilding contributes to lower fleet growth. Fewer newbuilding deliveries during the April–September 2020 period could result in relatively lower fleet growth, bringing it to about 1.6 per cent for 2020 (Clarksons Research, 2020e). The extent to which this will improve supply–demand balance in 2021 will depend on how demand and economic activity will recover and on developments in ship recycling.

In comparison, ship recycling offers more positive prospects. In June 2020, container ship recycling volumes were nearly as high as levels reported from January to May 2020 (Hellenic Shipping News Worldwide, 2020c). By the end of that month, ship-recycling activity had partially recovered in the bulk carriers segment. In this segment, scrapped volumes for the first half of 2020 exceeded levels for the full year 2019 (Clarksons Research, 2020e). Ship recycling is expected to increase, as the shipping industry copes with idling fleets and plans to scrap older vessels (more than 15 years old) that are not fuel efficient (Lloyd's List, 2020b).

Box 2.2 Shipbuilding at a crossroads in the European Union

In the face of production halts, temporary layoffs and liquidity issues stemming from the COVID-19 pandemic, the European shipbuilding and maritime equipment manufacturing industries have sought additional support – beyond horizontal industrial policies and financial support – calling for sector-specific support measures.

By doing so, they aim to preserve the economic contribution of the sector but, more importantly, to prevent potential dependence on Asian foreign suppliers for maritime technology, a strategic element to generate value in the maritime supply chain. The European Shipbuilding and Maritime Equipment Association estimated that this scenario could mean losing about €120 billion of added value created by the maritime technology sector; 1 million jobs in maritime technology companies and Europe's innovation and technological global leadership in complex ship types.

Concerns also relate to the role played by the shipbuilding and equipment industries in achieving longer-term goals such as promoting technological development and innovation to ensure carbon neutral shipping by 2050, as foreseen in the European Green Deal. In this sense, losing European shipyards could mean becoming dependent on Asian nations to achieve such goals.

Sources: Safety4sea, 2020; SWZ|Maritime, 2020; World Maritime News, 2020.

4. Seafarers and the maritime workforce

Emerging challenges for the maritime workforce as a result of the changing nature of work due to technological change

Historically, innovation and technology have played a crucial role in increasing the economic efficiency of the shipping industry. More recently, they have also become drivers and enablers of improved environmental performance of this sector. From a social perspective, technological advances and automation represent both opportunities and challenges for the shipping industry. Many emerging technologies in the maritime industry aim to improve safety and efficiency on board. Technological change also entails challenges. Disruptions in the labour market are expected because the sets of skills in demand and work routines will change.

According to a recent report by the International Transport Workers' Federation (2019),[3] forecast scenarios suggest that, although technology has the potential to reduce labour requirements, expanding international trade will counterbalance this reduction. For example, the demand for seafarers is expected to continue mounting up to 2040, albeit not at the same rate.[4] In some cases, a decrease in jobs in transport is offset by an increase in other parts of the transport system. Thus, more transport workers will be needed in the future.

The impact of technology and automation on the global maritime workforce, from 2020 until 2040, will vary, depending on the skills and tasks performed and workers' demographic groups. Low and middle-skilled jobs (that is to say, support activities for deep-sea transport workers such as cargo handlers in ports, dockers, crane operators, and maintenance and repair workers) and ageing or higher-wage workforces face a greater risk of redundancy. By contrast, high-skilled occupations, such as ship captains and officers, are less prone to automation, with automation and technological applications being introduced to assist them in their work. Younger and lower-wage workforces are likely to witness a delay in the introduction of automation and new technologies.

The impact on labour markets will also depend on the level of readiness of countries to adopt new technologies and automation. Such readiness is defined as the capability to capitalize on the future, mitigate risks and challenges, and be resilient and agile in responding to unknown future shocks. A country's level of readiness for automation is measured against five factors: innovation and technology, infrastructure quality, regulation and governance, human capital and skills, and business and investment. According to the above-mentioned report, there is a readiness gap in the maritime sector between developed and developing countries. A higher level of readiness is observed in Australia, East Asia, Europe and the United States, whereas countries in Africa and South America are positioned at the other end, due to insufficient technological advancement and investment, as well as to regulation and infrastructure gaps and weaknesses in terms of business models.

This means that most developing countries will witness a slower adoption rate of technology and automation, although low and middle-skilled jobs in industrialized countries face a more substantial risk of disappearing due to automation probability. This is likely to be accompanied by lower capital investments and research and development expenditures, leading to smaller productivity increases and the risk of falling behind in terms of maritime sector capabilities and competitiveness.

In all likelihood, the future of work in the maritime sector will look very different from what it is today, and there will be less jobs onboard ships and more

[3] The report analyses several modes of transport and explores readiness based on 17 country case studies (Australia, Brazil, China, Denmark, France, Ghana, Japan, Nigeria, Norway, Panama, Peru, the Philippines, the Republic of Korea, South Africa, Sweden, Turkey and the United States).

[4] For complete statistics on the supply of seafarers, see http://stats.unctad.org/seafarersupply.

onshore jobs, requiring a more adaptable workforce. Re-skilling and retraining will be crucial in preparing workers for the transformations that will arise as result of advanced technologies and automation. However, most countries have not elaborated long-term plans for automation in the maritime sector (International Transport Workers' Federation, 2019).

To support the successful transition of workers, the report of the Federation recommends the following actions:

- Raising awareness of the implications of further introduction of automation and technology into transport systems.
- Facilitating dialogues between stakeholders in global transport for a better understanding of the different positions of all parties concerned.
- Establishing national strategies and policies to address the ramifications of further automation and technology in transport.
- Supporting developing countries in dealing with the effects of introducing more automation and technology in transport.
- Identifying essential skills needed to work effectively in a world of advanced automation and technology in transport, implementing them in education and training.

Impacts, responses and prospects in relation to the COVID-19 pandemic: Sailors stranded at sea

Each month about 150,000 seafarers need to be changed over to and from the ships they operate to ensure compliance with international maritime regulations for ensuring safety, crew health and welfare, and the prevention of fatigue. The pandemic has led to restrictions in the cross-border movement of persons, closures of consulates affecting visa processing, port closures, disembarkation restrictions and lack of air services, which have impaired the ability to repatriate or resupply crews.

To mobilize action towards addressing this problem, several international organizations, maritime industry and labour organizations approached the relevant authorities and issued guidance documents to facilitate crew changes and repatriation of seafarers while, at the same time, taking steps to minimize the risk of contagion of the coronavirus disease (see chapter 5 for a detailed description of guidance documents).

In May 2020, some Governments started allowing crew changes at port under strict protocols. Despite all efforts, crew changes advanced slowly. In June, many seafarers were working beyond their contractual terms, could not disembark or be replaced. In mid-June 2020, IMO estimated that as many as 300,000 seafarers each month required international flights to enable crew changeovers. About half of them needed to be repatriated home by aircraft, while the other half needed to join ships. Additionally, about 70,000 cruise ship staff were waiting for repatriation (IMO, 2020).

Countries have faced several challenges at the local level to enact crew changes. These include difficulty to engage through a systematic approach the wide range of domestic agencies that need to be involved in the process. Countries have also faced difficulties related to the lack of infrastructure or of protective equipment and to unclear procedures on how to mitigate risks, while enabling the logistics of crew change amid restrictions and lockdown protocols and shortages of staff involved in the process (Lloyd's List, 2020c).

The pandemic has brought visibility to seafarers with the recognition that they provide an essential service because they ensure trade in essential goods, such as medical supplies and food, and they keep supply chains running. However, the slow pace of concrete actions highlights the challenges of balancing the safety and well-being of workers with operational continuity, which raises the question as to whether practices and procedures regarding crew changeover, disease management, health care and welfare need to evolve to enhance support for seafarers.

Further, the pandemic has provided an opportunity to raise awareness of the importance of gender in the maritime sector, including seafarers. Today, women represent only 2 per cent of the world's 1.2 million seafarers; 94 per cent of women seafarers are working in the cruise industry (www.imo.org/en/OurWork/TechnicalCooperation/Pages/WomenInMaritime.aspx#.) It is important to move forward and promote a safe and attractive sector that supports greater engagement for women (see box 2.3).

B. SHIPPING COMPANIES, EARNINGS AND REVENUES AND OPERATIONS DURING AND BEYOND THE PANDEMIC CRISIS

1. Impact of the pandemic on freight rates and earnings

This section describes the impact of the COVID-19 pandemic and relevant developments in maritime freight markets, namely containerized trade, dry bulk and tankers, during the first half of 2020. With the coronavirus taking a toll on the global economy and seaborne trade in early 2020, freight rates in shipping were strongly affected and continued to be determined by the way supply capacity was handled. This was the case of the container ships segment, which practised blank sailing and applied other capacity-management measures to adapt supply capacity to reduced demand for seaborne trade and allow freight rates to remain strong. Tanker freight rates were also affected not

Box 2.3 Promoting diversity and inclusion in the maritime sector

On 27 January 2020, the Women's International Shipping and Trading Association and IMO signed a memorandum of understanding under which they agreed to enhance technical cooperation activities in the maritime field to build opportunities for diversity and inclusion, professional development and skill competency.

In particular, the parties agreed to the following:

- To look for opportunities to partner on maritime issues, which could include organizing workshops or speaking on panels at annual conferences or other events held by the parties, with a focus on panel diversity.
- To promote greater engagement for women in maritime occupations, among their members, the broader ocean business community, ocean stakeholders and the public.
- To develop and participate in relevant training, workshops, among other business related to their areas of mutual interest.
- To support the implementation of IMO Assembly resolution 1147(31) of 4 December 2019 on preserving the legacy of the world maritime theme for 2019 and achieving a barrier-free working environment for women in the maritime sector.

UNCTAD has also been collaborating with the Association and is currently discussing further collaboration in terms data collection and dedicated capacity-building activities.

Sources: See also: www.imo.org/en/OurWork/TechnicalCooperation/Pages/WomenInMaritime.aspx#; Women's International Shipping and Trading Association, 2020.

only by repercussions of the lockdowns relating to the pandemic, but also by geopolitical events, oil price fluctuations and the increased use of vessels for storage floating, which led to a rise in freight rates, mainly in March–April 2020. Dry bulk freight rates, pulled down by an oversupplied market, were further affected by the shock of negative demand, namely from China, owing to the outbreak of the coronavirus disease.

Container freight rates and earnings: Strong freight rates despite abrupt drop in seaborne trade

The container segment of the shipping industry was already struggling with an oversupplied market and slow demand growth before the pandemic, which had kept the level of container freight rates generally low over the past few years. As the pandemic brought economies to a halt and took a toll on trade, this industry segment experienced a major setback. The start of 2020 had witnessed some recovery in demand and freight rates before the pandemic but with the outbreak of the pandemic, prospects for demand not only decreased, but fleet development was affected as well. With lockdowns having come into force in March 2020, reducing demand for containerized goods, shipping companies engaged in strategies to manage supply capacity and reduce costs to cope and to keep freight rates from falling.

As shown in table 2.9, 2020 began with better freight rates compared with average rates in 2019 for most routes, driven mainly by the surcharge applied by carriers to compensate for higher bunker costs and reduced supply capacity due to scrubber retrofits in compliance with IMO 2020 sulphur cap regulations. With the spread of the coronavirus pandemic in early 2020, which led to a sudden drop in demand for seaborne transport, carriers applied strategies such as increased blank sailing and idling of vessels, and re-routing (MDS Transmodal, 2020) as a way of adjusting supply to low demand (see also chapter 1). This allowed freight rates to remain stable at a time of lower demand for ocean shipping. Although blank sailings, accompanied by low oil bunker prices, helped shipping lines to manage supply capacity and reduce costs, blank sailings still cost carriers about 40 per cent of the operating cost of a vessel (Drewry, 2020a) and have an impact on revenue due to capacity withdrawals.

From the perspective of shippers, these strategies meant severe space limitations to transport goods and delays in delivery dates, which had an impact on supply chains and the proper functioning of ports.

With regard to idling, 11 per cent of the container fleet was estimated to be idle during the first half of 2020. The vessel types showing a higher proportion of idle fleet – between 7 and 9 per cent – included containers, tankers and car carriers (Clarksons Research, 2020c). Those showing the highest increases in the idle fleet compared with January 2020 were car carriers – which more than tripled – liquefied natural gas carriers and liquefied petroleum gas carriers.

With regard to the charter market, declining demand and an increase in idling and blank sailings applied by carriers to reduce supply it after capacity had a negative impact on all segments of container charter rates, particularly the larger vessels within that segment. The ConTex charter rate decreased to an average of 368 points during the first six months of 2020, compared with an annual average of 407 points in 2019 (figure 2.6). However, rates did not reach the low level witnessed in 2016, when earnings for most segments fell beneath operating costs due to an oversupplied market. Some improvements were witnessed in July 2020, as the volume of activity picked up slightly, namely with regard to large and medium-sized vessels. It remains unclear whether these improvements will persist.

During the third quarter of 2020 container ships continued extending capacity-reduction programmes,

Table 2.9 Container freight market rates, 2010–2020

Freight market	2010	2011	2012	2013	2014	2015	2016	2017	2018	2019	2020 (average January–April)	January 2020	February 2020	March 2020	April 2020	2019 (April January–April)	Percentage change
Trans-Pacific	(Dollars per 40-foot equivalent unit)																
Shanghai–United States West Coast	2 308	1 667	2 287	2 033	1 970	1 506	1 272	1 485	1 736	1 525	1 521	1 572	1 395	1 509	1 608	1 711	-11.10
Percentage change	68.22	-27.77	37.19	-11.11	-3.1	-23.6	-15.5	16.7	16.9	-12.2							
Shanghai– United States East Coast	3 499	3 008	3 416	3 290	3 720	3 182	2 094	2 457	2 806	2 634	2 775.5	2 898	2 714	2 784	2 706	2 807.25	-1.13
Percentage change	47.82	-14.03	13.56	-3.69	13.07	-14.5	-34.2	17.3	14.2	-6.1							
Far East–Europe	(Dollars per 20-foot equivalent unit)																
Shanghai–Northern Europe	1 789	881	1 353	1 084	1 161	629	690	876	822	760	85.5	1 040	829	805	740	814.5	4.79
Percentage change	28.24	-50.75	53.58	-19.88	7.10	-45.8	9.7	27.0	-6.2	-7.5							
Shanghai–Mediterranean	1 739	973	1 336	1 151	1 253	739	684	817	797	811	976.75	1 181	979	898	849	841.5	16.07
Percentage change	24.48	-44.05	37.31	-13.85	8.9	-41,0	-7.4	19.4	-2.4	1.8							
North–South	(Dollars per 20-foot equivalent unit)																
Shanghai–South America (Santos)	2 236	1 483	1 771	1 380	1 103	455	1 647	2 679	1 703	1 673	1 551	2 069	1 714	1 426	995	1 387	11.82
Percentage change	-7.95	-33.68	19.4	-22.08	-20.1	-58.7	262.0	62.7	-36.4	-1.8							
Shanghai–Australia/New Zealand (Melbourne)	1 189	772	925	818	678	492	526	677	827	596	884.75	944	868	815	912	441	100.62
Percentage change	-20.73	-35.07	19.82	-11.6	-17.1	-27.4	6.9	28.7	22.2	-27.9							
Shanghai–West Africa (Lagos)	2 305	1 908	2 092	1 927	1 838	1 449	1 181	1 770	1 920	2 474	2 857.75	2 856	2 930	2 891	2 754	2 603.5	9.77
Percentage change	2.58	-17.22	9.64	-7.89	-4.6	-21.2	-18.5	49.9	8.5	28.9							
Shanghai–South Africa (Durban)	1 481	991	1047	805	760	693	584	1 155	888	802	986.5	1 120	1 032	969	825	753.75	30.88
Percentage change	-0.94	-33.09	5.65	-23.11	-5.6	-8.8	-15.7	97.8	-23.1	-9.7							
Intra-Asian	(Dollars per 20-foot equivalent unit)																
Shanghai–South-East Asia (Singapore)	318	210	256	231	233	187	70	148	146	138	193.25	189	187	201	196	149.75	29.05
Percentage change		-33.96	21.90	-9.77	0.9	-19.7	-62.6	111.4	-1.4	-5.5							
Shanghai–East Japan	316	337	345	346	273	146	185	215	223	233	239.75	241	236	240	242	228.5	4.92
Percentage change		6.65	2.37	0.29	-21.1	-46.5	26.7	16.2	3.7	4.5							
Shanghai–West Japan								215	223	229	227	226	221	227	234	227.5	-0.22
Percentage change								..	3.7	2.7							
Shanghai–Republic of Korea	193	198	183	197	187	160	104	141	163	128	119	120	118	118	120	144.25	-17.50
Percentage change		2.59	-7.6	7.65	-5.1	-14.4	-35.0	35.6	15.6	-215							
Shanghai–Mediterranean Gulf/Red Sea	922	838	981	771	820	525	399	618	463	735	983.25	1 161	1 034	997	741	704.25	39.62
Percentage change		-9.11	17.06	-21.41	6.4	-36.0	-24.0	54.9	-25.1	58.7							

Source: Shanghai containerized freight index; Shanghai Sh pping Exchange; and data from Clarksons Research, *Container Intelligence Monthly*, various issues.

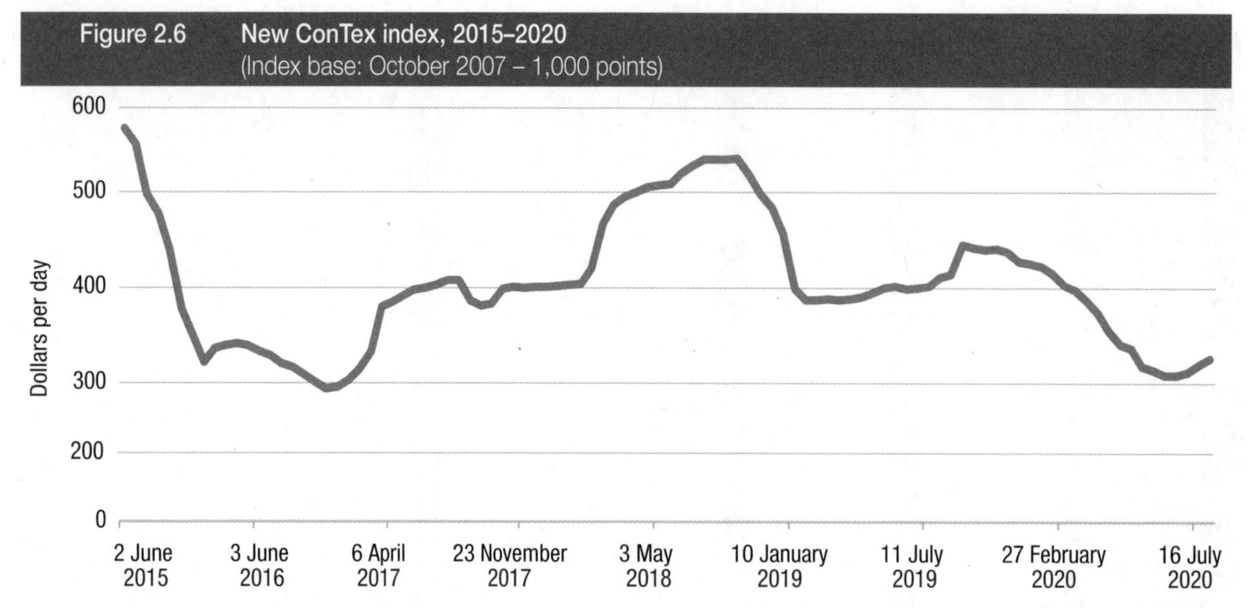

Source: UNCTAD calculations, based on data from the New ConTex index produced by the Hamburg Shipbrokers Association (www.vhss.de).

Notes: The New ConTex index is based on assessments of current day charter rates of six selected container ship types, which are representative of their size categories: Types 1,100 TEUs and 1,700 TEUs (charter period of one year); Types 2,500, 2,700, 3,500 and 4,250 TEUs (charter period of two years).

although demand was picking up, keeping freight rates on the rise. This may be a signal that shipping lines are expecting a slow recovery from the effects of the crisis caused by the pandemic. However, the persisting application of reduced capacity measures appears to be causing severe problems. For example, carriers are offering sailings with delays of two to three weeks, and containers (empty and filled) are building up at ports because sailings are not taking place as scheduled. Filled containers are arriving at ports booked for a particular sailing but have to wait for a longer period of time until the arrival of the next vessel, resulting in port delays (Hellenic Shipping News Worldwide, 2020d).

The situation is exacerbated when vessels are being given only a limited window at ports due to labour shortages (as is the case in India, where the pandemic was still spreading in July 2020).

Another example is empty containers piling up in ports. Ports in the United Kingdom, for example, reported being overwhelmed with empty containers stacking up and causing congestion in limited port storage yards (Hellenic Shipping News Worldwide, 2020d) (See also Box 2.6).

Tankers freight rates and earnings: Sharp freight rate fluctuations and surge in demand for tankers to be used as floating storage

Lockdowns induced by the pandemic, geopolitical events and oil price fluctuations had an impact on developments in the oil tanker freight market, maintaining freight rates high during the first quarter of 2020. During this period, the freight rates market experienced highly volatile trends, despite a weak market balance due to an oversupplied fleet market and low demand.

In March and April 2020, tanker rates rose sharply, as demand for these vessels increased, despite global demand for crude oil and petroleum products falling dramatically due to the pandemic (see chapter 1). This is explained by the hiring of many vessels as floating storage, following the lack of agreement within the Organization of the Petroleum Exporting Countries and its wider group regarding further production cuts that had led to a temporary increase in output from Saudi Arabia at a time when there was no such need on the consumption side (see chapter 1). The oil market was in a state of super contango where front-month prices were much lower than they would be in future months, making the storage of oil for future sales profitable. Traders rushed to charter large tankers for floating storage so they could sell the oil at higher prices later, thus reducing the availability of vessels in the market and triggering a sharp rise in tanker rates.

As shown in table 2.10, time-charter equivalent earnings also picked up in all tanker segments during March and April 2020, with huge peaks in the very large crude carrier segment. A case in point is the Arabian Gulf–Japan single voyage route. This route saw a surge from an average 48 Worldscale points in February to an average 137 Worldscale points in March and 174 Worldscale points in April 2020. This worked out to an average daily time-charter equivalent of $124,000 in March and $170,900 in April, spiking by almost 10 times compared with average earnings in February 2020.

Table 2.10 Crude oil and product tanker spot rates and time-charter equivalent earnings
(Worldscale and dollars per day)

			2020						2019
		2020	January	February	March	April	May	June	December
Crude oil tankers									
Very large crude carriers	Arabian Gulf–Japan	Worldscale	100	48	137	174	66	57	105
		Dollars per day	63 500	16 500	124 000	170 900	51 700	38 800	87 800
		Change in earnings (percentage)	-28	-74	652	38	-70	-25	
	Arabian Gulf–China	Worldscale	94	44	125	159	60	52	109
		Dollars per day	70 000	18 300	128 200	176 000	53 800	40 600	83 400
		Change in earnings (percentage)	-16	-74	601	37	-69	-25	
	Arabian Gulf–north-western Europe	Worldscale	127	33	127	104	38	106	61
		Dollars per day	63 200	20 900	205 600	169 200	169 400	167 000	66 100
		Change in earnings (percentage)	-4	-67	884	-18	0	-1	
Suezmax crude tankers	West Africa–north-western Europe	Worldscale	136	82	126	146	82	49	
		Dollars per day	54 800	26 400	59 700	77 400	37 600	14 400	57 800
		Change in earnings (percentage)	-5,19	-51,82	126,14	29,65	-51,42	-61,70	
	West Africa–Caribbean/east coast of North America	Worldscale	103	79	121	141	78	54	
		Dollars per day	35 900	24 800	59 600	76 800	36 200	18 200	41 500
		Change in earnings (percentage)	-13	-31	140	29	-53	-50	
	Black Sea–Mediterranean	Worldscale	147	90	134	151	86	54	
		Dollars per day	62 900	24 700	65 700	82 700	33 400	6 200	61 200
		Change in earnings (percentage)		-60,73	165,99	25,88	-59,61	-81,44	
Aframax crude tankers	Mediterranean–Mediterranean	Worldscale	149	81	143	157	107	63	193
		Dollars per day	34 200	5 700	42 000	50 800	26 500	3 400	55 400
		Change in earnings (percentage)	-38	-83	637	21	-48	-87	
	North-western Europe–North-western Europe	Worldscale	147	118	136	170	109	74	209
		Dollars per day	41 500	25 200	42 900	69 100	28 300	2 200	83 200
		Change in earnings (percentage)	-50	-39	70	61	-59	-92	
	Caribbean–east coast of North America	Worldscale	324	169	161	155	122	68	225
		Dollars per day	91 600	36 900	39 700	41 300	28 000	5 300	53 800
		Change in earnings (percentage)	70	-60	8	4	-32	-81	
	South-East Asia–east coast of Australia	Worldscale	151	99	121	156	132	73	178
		Dollars per day	30 100	15 000	31 000	50 500	39 400	12 900	44 300
		Change in earnings (percentage)	-32	-50	107	63	-22	-67	
Product tankers									
Medium-range tankers 1	Baltic–United Kingdom or continental Europe	Worldscale	190	195	187	247	160	103	205
		Dollars per day	18 400	21 400	22 800	36 400	19 300	6 900	22 300
		Change in earnings (percentage)	-17	16	7	60	-47	-64	
Medium-range tankers 2	United States Gulf-north-western Europe	Worldscale	161	97	120	150	108	76	122
		Dollars per day	16 100	5 200	13 600	22 100	13 000	5 200	10 700
		Change in earnings (percentage)	50	-68	162	63	-41	-60	
Long-range tankers 1	Arabian Gulf-Japan	Worldscale	127	100	153	304	254	82	157
		Dollars per day	12 300	9 900	28 600	70 400	56 700	10 800	23 000
		Change in earnings (percentage)	-47	-20	189	146	-19	-81	
Long-range tankers 2	Arabian Gulf-Japan	Worldscale	121	93	155	319	263	87	156
		Dollars per day	15 800	11 600	40 400	102 200	81 400	17 000	31 600
		Change in earnings (percentage)	-50	-27	248	153	-20	-79	

Source: UNCTAD calculations, based on *Drewry Shipping Insight*, various issues.

As noted in table 2.10, the product tanker market also witnessed a surge in earnings supported by increased floating storage demand, particularly for large vessels. However, after peaking in March–April, freight rates and vessel earnings in both segments declined sharply in May, as about a third of total vessels locked in floating storage returned to active trade, inflating supply. The tonnage locked in floating storage dropped from about 45 million dwt at the end of April to 30 million dwt at the end of May (Drewry, 2020b). The number of very large crude carriers storing crude oil dropped sharply from 83 vessels to 56 vessels over this period. This, nevertheless, remains a historically high number.

Tanker rates in the crude oil and product tankers market continued to decrease in June 2020, although many countries were easing up the lockdowns measures. Demand for oil remained significantly lower in the second quarter of 2020 compared with 2019. At the same time, continued cuts in output by the Organization of the Petroleum Exporting Countries and its wider group led to a return of vessels locked in floating storage, increasing supply capacity.

With regard to the outlook, freight rates might remain low, as the tanker market fundamentals appear highly uncertain. Recession projections in the global economy would obviously reduce the demand for oil and oil products. Oil price development and geopolitics will also have an impact. Consequently, tanker supply will remain high for some time. The management of vessel order books and recycling will therefore be crucial to improve market imbalances and reduce freight volatility.

Dry bulk freight rates and earnings: Weakened fundamentals due to the COVID-19 pandemic and increased freight rate volatility

During the first six months of 2020, the market for dry bulk freight rates continued to be shaped by imbalances in supply and demand, which was aggravated by the impact of the pandemic and resulted in high fluctuations, namely among larger vessels during this period. As discussed earlier, overcapacity was already affecting the dry bulk market, as supply growth had been outstripping demand for many years. This was further exacerbated by the negative demand shock caused by the pandemic, which added downward pressure on shipping freight rates.

At the beginning of 2020, dry bulk shipping industry freight rates and earnings were severely affected, namely the Capesize market. This was mainly due to the combination of a drop in seasonal dry bulk demand and the outbreak of the coronavirus disease in China, which imports the majority of globally shipped dry bulk cargo volumes, including iron ore, coal, and major grains and oilseeds. The outbreak of the pandemic in early 2020 disrupted industrial activities in China, which resulted in reduced demand for dry bulk vessels, particularly for Capesize vessels that carry industrial raw materials to China. At the same time, low exports of iron and ores out of Brazil (see chapter 1) added pressure to dry bulk volumes, further exacerbating freight rate volatility and leading to unprecedented low and negative levels in Capsize market freight rates. The Baltic Exchange Capesize index became negative in February and March, dropping to -243 and -221 points because of a sudden massive drop in globally shipped dry bulk cargo volumes due to the shutdown in China (figure 2.7). In June 2020, the index increased to high levels of 2,267 points boosted by a higher demand for iron ore in China following the easing of the COVID-19-related restrictions.

Although freight rates for smaller vessel sizes did not experience such a decline, they remained highly volatile and very low. Demand for Panamax and Supramax vessels, mainly used for global shipping of grain and oil seeds, was higher, as trade volumes remained relatively stronger (see chapter 1).

Time-charter rates across all segments were also affected by the pandemic that weakened market fundamentals, already plagued by an oversupply of vessels. In June 2020, the average of one-year time-charter rates for Capesize bulk carriers was $11,050 per day, $9,785 per day for Panamax bulk carriers, $8,513 per day for Handysize bulk carriers and $8,150 per day for Supramax bulk carriers (figure 2.8).

Sector recovery will depend on global economic growth. However, with the prospect of global recession and uncertainties concerning the impact of the pandemic across developed and developing economies, the development of freight rates remains uncertain. A key feature is development in China, which would be the biggest driver for the recovery of the dry bulk industry. At the same time, overcapacity remains a threat to industry market fundamentals and an increase in the market arising from additional supply could offset any growth in demand.

2. Government-backed financial support for the shipping industry in times of pandemic: The case of the container segment

With the abrupt and significant drop in seaborne trade and uncertainties about the future caused by the pandemic, the financial viability of the container segment of the shipping industry was at risk, having already been confronted with freight rates volatility and low profits for more than a decade. Financial support by Governments to ensure the proper functioning of maritime transport services became a global necessity. Unlike the airline industry, such financial assistance was not a common practice in the shipping industry, except in Asia (namely East-Asian and South-Asian countries such as China, the Republic of Korea, Singapore and Taiwan Province of China) where the sector could rely on bailout funds or financial relief from Governments (Drewry, 2020b).

Figure 2.7 Baltic Exchange dry index, 2017–2020

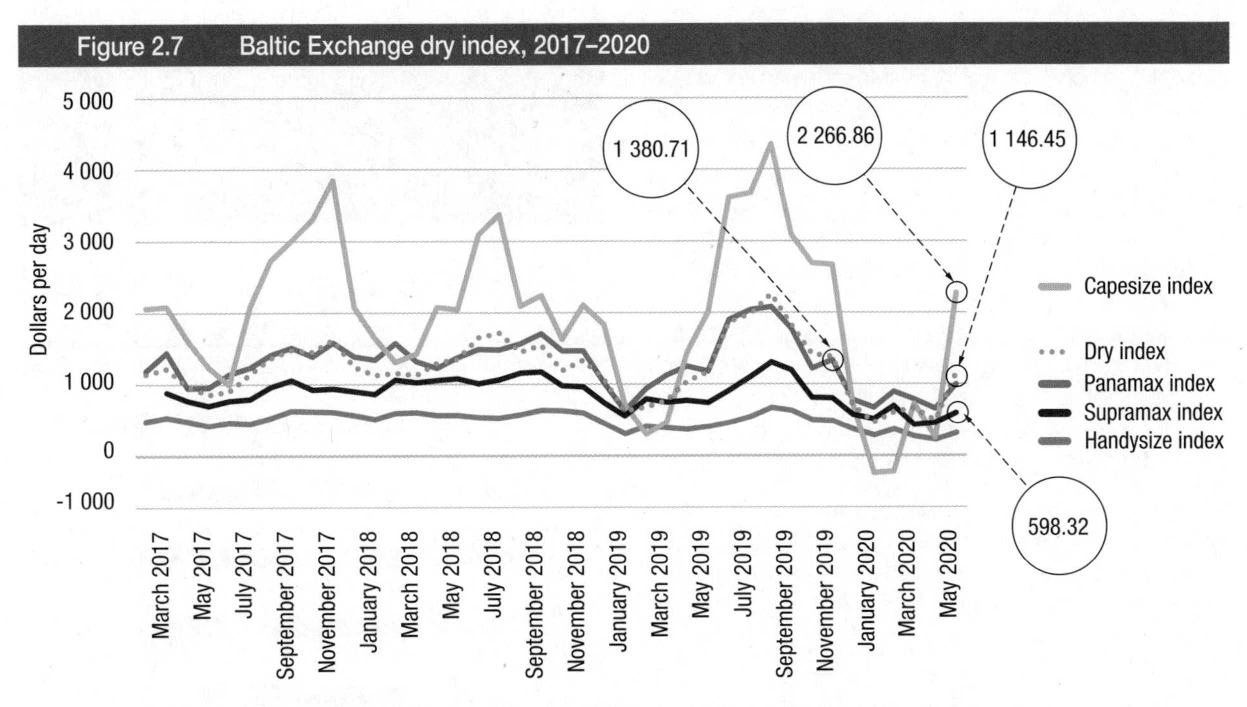

Source: Baltic Exchange; data derived from Clarksons Research, Shipping Intelligence Network Time Series.

Notes: Panamax index: basis – 82,500-dwt vessel from start 2020, 74,000-dwt vessel prior. Handysize index: basis – 38,200-dwt vessel from start 2020, 28,000-dwt vessel prior.

However, government intervention and support are not always well perceived by the industry, as it disrupts its equilibrium and impedes market reform.

Nonetheless, given the pandemic crisis and growing uncertainties on when and how demand will recover, several carriers applied for State-backed financial support in various regions, including Europe. For example, in May 2020, CMA CGM secured $1.14 billion (€1.05 billion) of State-guaranteed syndicated loans from the Government of France (JOC.com, 2020a) to strengthen the company's cash position to confront uncertainties in the global economy resulting from the pandemic. In addition, the Republic of Korea launched a $33 billion rescue fund to protect seven of its mainstay sectors (Hellenic Shipping News Worldwide, 2020e), including the

Figure 2.8 One-year time-charter rates for bulk carriers, 2015–2020
(Dollars per day)

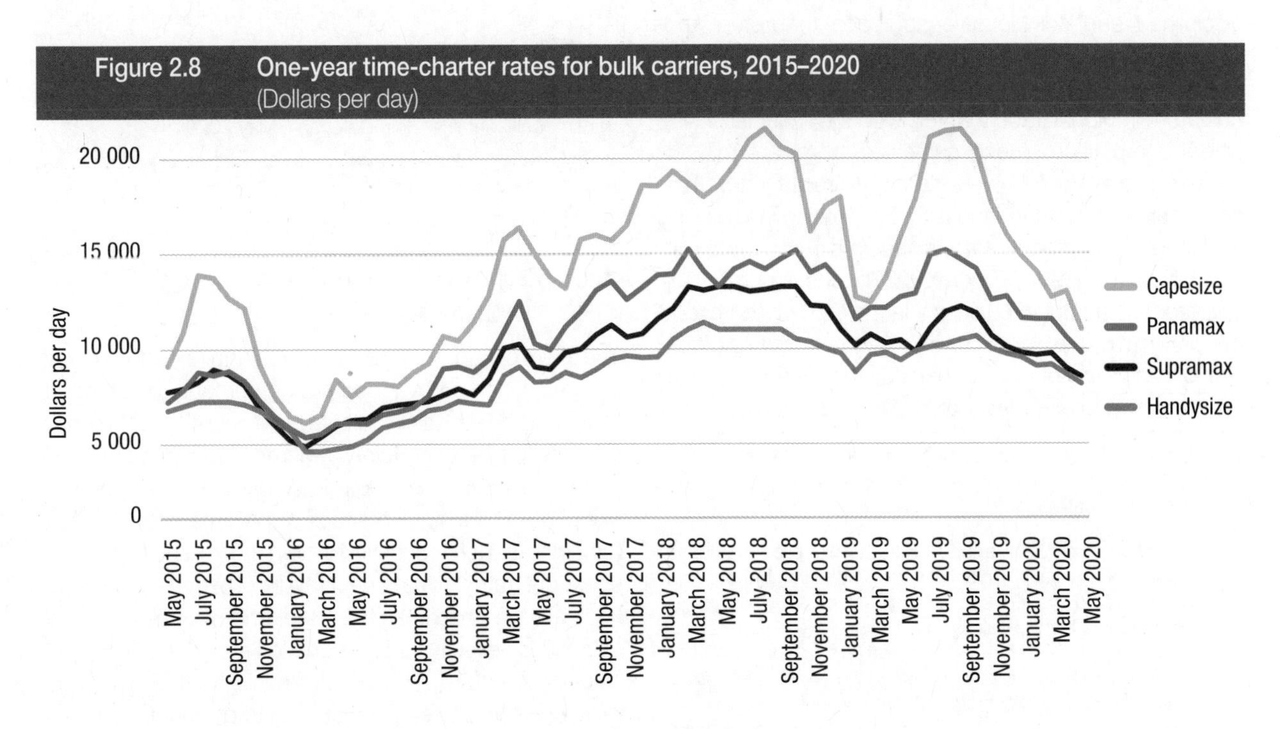

Source: Baltic Exchange; data derived from Clarksons Research, Shipping Intelligence Network Time Series

Note: Long-run historical series.

shipping and shipbuilding sectors, which were allocated about $1 billion[5], of which HMM, formerly known as Hyundai Merchant Marine, received about $400 million (Pulse, 2020).[6] Evergreen and Yang Ming Marine Transport Corporation will receive State-backed loans totalling about $568 million as part of the plan of Taiwan Province of China to alleviate the financial pressure facing the local shipping sector (Lloyd's List, 2020d). Under the plan, the Government has pledged to provide guarantees for at least 80 per cent of the approved loans plus subsidies for interest, which would allow local shipping companies and ports to have access to additional financing. The four above-mentioned carriers are among the world's top 10 deep-sea container shipping lines (figure 2.9).

Moreover, in addition to industry involvement in recovery, reliable governmental policies and support for new sustainable business models are fundamental to building the resilience of the sector.

Figure 2.9 Top 10 deep-sea container shipping lines, ranked by deployed capacity and market share, May 2020
(20-foot equivalent units and percentage)

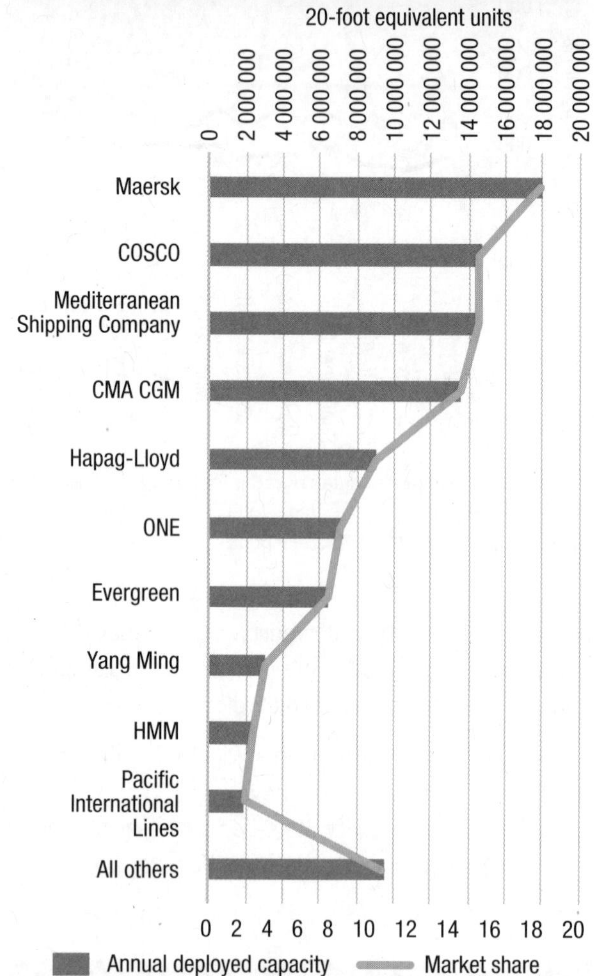

Source: UNCTAD calculations, based on data from MDS Transmodal Container Ship Databank, May 2020.

Note: Data refer to fully cellular container ship tonnage and do not include intraregional services.

3. Industry prospects in times of pandemic and beyond: Supply discipline and collaboration, accelerated digitalization and prioritization of environmental sustainability

Disciplinary and collaborative approach to the container ship segment in the face of the pandemic

With regard to the measures applied during the pandemic crisis and how the container ship segment of the industry handled the crisis compared with the financial crisis in 2009, the industry has taken a more disciplined and collaborative approach to protect the industry and ensure its long-term recovery and viability. There have been some lessons learned from the downturn in global trade that followed the financial crisis, where competition among carriers to dominate market through scale. Vessels were sailing at freight rates that could barely cover operational costs, resulting in losses in the container segment of the shipping industry of about $20 billion in 2009 (JOC.com, 2020b) and a number of operators going out of business. In the current context of the pandemic, the container ship segment did not look into gaining market share. Instead, it concentrated on maintaining a positive level of freight rates by managing capacity supply in line with demand while reducing costs and ensuring sector viability.

The effect of the pandemic crisis on container shipping was obvious, reflected by a decreasing demand for seaborne trade and a reduction in fleet deployment. In an effort to address future uncertainty regarding the prospects for demand growth (see box 2.5), carriers may continue exercising flexibility in managing maritime networks and matching supply capacity to demand to support freight cost and rates. It is true that freight rates should be kept at level that ensures the economic viability of the sector. However, if supply-reduction measures applied by shipping lines are sustained for a long period during the recovery in volumes, this may lead to dysfunctionalities in the sector, including ports, undermining performance of shippers and global supply chains.

[5] Other industries include airlines, automotive manufacturing, machinery manufacturing, power generation and telecommunications.

[6] In addition, the State agency Korea [Republic of] Ocean Business Corporation planned to buy 100 billion won worth of subordinated bonds from shippers by accepting the shippers' loan-to-value ratio of up to a maximum of 95 per cent from the current average of 60 to 80 per cent. The agency will also directly buy 100 billion won worth of debts of small- and mid-sized shippers (https://pulsenews.co.kr/view.php?year=2020andno=423920 and www.seatrade-maritime.com/finance-insurance/south-korea-pledges-1bn-support-ailing-shipping-sector).

Box 2.4 Policies to support shipping for a sustainable recovery beyond the pandemic crisis

The global shipping industry will be at the forefront of the recovery as a vital enabler of smooth functioning of international supply chains. As countries turned to consider economic stimulus packages to promote recovery, many of them asked themselves how they could leverage this support to build economies that could drive sustainable economic prosperity. Such a reflection requires going beyond short-term priorities (job creation and boosting economic activity) and thinking about long-term objectives.

Long-term objectives refer to support for growth potential, resilience to future shocks and a sustainable growth trajectory, including decarbonization. An important consideration in this respect is climate-proofing infrastructure investments to avoid future disruption to transport operations. Following this line of thinking, several countries have considered strategic for diverse reasons to include some of these elements in policies related to their maritime transport strategies as part of their recovery plans beyond the pandemic crisis, as follows:

- To avoid having stranded assets (that is, assets that lose economic value well ahead of their anticipated useful life) and investing in declining technology by supporting investment in emerging technologies that can bring simultaneous economic and environmental benefits instead. For example, the British Ports Association proposed a plan to utilize ports and maritime industries to stimulate future growth, which involved a maritime green fund to invest in green equipment and vessels, and a study to identify barriers to increase the uptake of onshore electricity, which could bring financial savings to ports and contribute to reduce air pollution.
- To build resilience to future shocks, for instance by promoting digitalization. This is the case of an initiative launched by the Maritime and Port Authority of Singapore, Singapore Shipping Association and Infocomm Media Development Authority to support maritime companies in digital transformation, which includes support to formulate their digitalization road maps, guide execution and benefit from maritime digital platforms covering port clearances and services, trade documentation, and trade operations and financing.
- To develop new export markets, create domestic value chains, generate jobs and be prepared for a future without fossil fuels. An example of this is the national hydrogen strategy of Germany, aimed at promoting use of this alternative fuel across several industries, including shipping. It offers market incentives to make green hydrogen competitive and investments of at least €9 billion of onshore electricity, which could bring financial savings to ports and contribute to reduce air pollution.

Sources: Chambers, 2020; Elgie and McNally, 2020; Greenport, 2020; Hammer and Hallegatte, 2020; *Seatrade Maritime News*, 2020b.

Box 2.5 The changing landscape of international production, the COVID-19 pandemic, resilience-building and maritime transport fleet deployment

International production patterns have been changing since the financial crisis of 2008–2009. The slowdown in overall trade and in global value chain trade is linked to a shift in the trade and investment policy environment, which is trending towards greater interventionism, rising protectionism and a shift to regional and bilateral frameworks. Other drivers for changes in the landscape of international production include technological advancements and sustainability trends. UNCTAD analysis suggests that changes are taking place in the degree of fragmentation and length of value chains and in the geographical spread of value added, pointing towards shorter value chains and more concentrated value added.

The COVID-19 crisis brought to the spotlight the exposure of international production to systemic risks, particularly from the perspective of securing continuity of supply. As such, building resilience in the supply chain can translate into diversifying sources of inputs. Thus, the crisis accentuated pre-existing trends related to changes in the length and fragmentation of value chains. Depending on the starting configuration of different industries, possible trajectories that the system of international production could follow include reshoring, diversification, regionalization and replication.

Although it may be too early to fully grasp supply-chain redesign patterns in a post-pandemic recovery scenario, it is inevitable that the shipping industry will be fundamentally affected, regardless of the specific trajectories that different industries follow. For instance, a reshoring trajectory, leading to shorter and less fragmented value chains, could have an impact on deep-sea cargo volumes and the capacity to generate economies of scale through mega-sized vessels, which also provide less flexibility than smaller ships to adapt to sharp fluctuations between supply and demand. On the other hand, a regionalization trajectory, leading to short physical supply chains that are not less fragmented, could increase the attractiveness of short sea networks between countries, opening up opportunities for regional cooperation and cabotage services:

Sources: Sánchez, forthcoming; The Loadstar, 2020a; UNCTAD, 2020a.

Accelerated digitalization and prioritization of environmental sustainability

The current context has accentuated the industry trend towards digitalization. Companies have leveraged digitalization to adapt to the new circumstances, increasingly favouring online tools to simplify processes and cut costs. For example, in June 2020, the Mediterranean Shipping Company introduced the instant-quote tool to provide easy access to its rates for ocean shipping, to make its customers' supply chain easier to manage and improve end-to-end efficiency (Port Technology, 2020).

Companies have also sought to improve data accessibility and transparency, to adapt to evolving consumer expectations in an environment characterized by supply-chain disruption, remote working and increased engagement through business-to-consumer e-commerce. For instance, in mid-April 2020, Maersk's online application, which features cargo release, the calculation of fees and online payment for immediate release functionalities, registered an 85 per cent increase in transactions as customers started ordering more remotely and sought to track cargo more efficiently (Maersk, 2020a).

The current context has also accelerated the interest for data-driven services to support decision-making and the emergence of new services and business opportunities. For example, Cubex Global is a digital marketplace built on collaborative blockchain principles, which enables the buying and selling of cubic metres of container space, enabling capacity management through a digital platform. The platform promises gains in operational efficiency ranging between 25 and 40 per cent in less than container load state and 100 per cent in full container load state and empties (Khalid and Tariq, 2020). In conclusion, collaborative innovation, accelerated though digital solutions to cope with the impacts of the pandemic and respond to changing consumer needs, is likely to remain in the long term, confirming the need to embark on digital transformation and customer-centric service development.

The long-term goal of shipping decarbonization is linked to the Initial IMO Strategy on reduction of greenhouse gas emissions from ships, which is aimed at cutting annual emissions by at least 50 per cent by 2050 and the carbon intensity of emissions by 40 per cent by 2030 and 70 per cent by 2050, compared with 2008 levels. Maintaining the commitment to reach this goal will require significant resources and investment.

Notwithstanding the impacts of the pandemic, this long-term goal remains a priority for the industry (Shell International, 2020). This is due to the increased awareness that technical progress to improve sustainability of operations can help unlock savings and generate new commercial opportunities and that there is a need to adapt to a changing regulatory environment as a result of the Initial IMO Strategy.

During the first semester of 2020, several companies announced that they were maintaining, and even initiating, investment plans related to developing carbon-neutral fuels and new technologies, and setting new ambitious company targets to reduce carbon-dioxide emission (Maersk, 2020b; S and P Global, 2020).

C. PORT SERVICES AND INFRASTRUCTURE SUPPLY

Ports play an essential role in facilitating the movements of goods across supply chains. They are a key node in the transport system as gateways connecting countries through maritime transport networks, and maritime transport with domestic and regional markets though multimodal transport connections in the hinterland.

Past editions of the *Review of Maritime Transport* discussed the heightened pressure ports had experienced in recent years, in view of larger and more powerful alliances seeking to raise network efficiency. This led ports to enhance productivity to adapt space, infrastructure and equipment to increased vessel size and competitive pressure among ports seeking to attract investment and diversify sources of income to other activities. Like other maritime transport activities, this sector is subject to pressure to incorporate sustainable criteria in port development and to a wave of horizontal and vertical consolidation, affecting mainly container terminals.

1. Vertical integration between shipping companies, terminal operators and inland logistics intensifies

From 2010–2020, container shipping companies sought to expand their services offer to include shipping, terminal operations and inland logistics to reduce exposure to volatile freight rates and generate alternative revenue streams providing end-to-end logistic solutions.

Table 2.11 identifies the 21 main global players that control 80 per cent of global terminal operations. Several of these companies are part of or are closely linked to shipping lines (APM Terminals/Maersk; Terminal Investment Limited/Mediterranean Shipping Company; Mitsui Osaka Shosen Kaisha Lines; Yang Ming Marine Transport Corporation; HMM and COSCO).

Similarly, terminal operators are engaging in vertical integration by taking greater control of inland logistics and aiming to provide integrated service offerings and generate more value. Examples of these developments in 2020 include Maersk's acquisition of a customs brokerage firm and a warehousing and distribution services provider (JOC.com, 2020c), CMA CGM's partnership with an online platform that links couriers to online retailers (Lloyd's List, 2019b) and DP World's acquisitions in the global feeder network,

Table 2.11 Top 21 global terminal operators, throughput and capacity, 2019
(Million 20-foot equivalent units)

Ranking	Operator	Throughput				Capacity	
		Total port handling (million TEUs)	Share of world throughput (percentage)	Growth or decline (million TEUs)	Growth or decline (percentage)	Total capacity (million TEUs)	Growth or decline (percentage)
1	COSCO	109.8	13.7	4.0	3.8	141.6	8.9
2	PSA International	84.8	10.6	4.8	5.9	117.0	3.9
3	APM Terminals	84.2	10.5	5.5	7.0	107.6	7.9
4	Hutchison Ports	82.6	10.3	0.1	0.1	113.0	0.9
5	DP World	69.4	8.7	-0.6	-0.9	91.0	1.5
6	Terminal Investment Limited	50.8	6.3	3.1	6.4	72.8	16.8
7	China Merchants Ports	35.6	4.4	1.1	3.1	44.2	3.1
8	CMA CGM	26.1	3.3	0.5	2.0	43.1	12.3
9	SSA Marine	13.0	1.6	0.4	3.3	20.5	1.4
10	ICTSI	11.8	1.5	2.0	20.9	20.0	11.7
11	Eurogate	11.7	1.5	-1.9	-14.2	20.6	-9.1
12	Evergreen	10.1	1.3	-0.3	-3.0	17.0	-0.9
13	Hyundai	9.5	1.2	2.0	25.8	12.1	-2.1
14	NYK Lines (Nippon Yusen Kabushiki Kaisha)	8.2	1.0	-2.4	-22.4	22.5	-5.3
15	MOL (Mitsui Osaka Shosen Kaisha Lines)	7.8	1.0	0.5	6.7	10.7	6.6
16	HHLA ((Hamburger Hafen und Logistik)	7.7	1.0	0.2	3.2	10.5	1.5
17	Yildirim/Yilport	6.1	0.8	-0.3	-4.4	11.9	16.8
18	Bollore	6.0	0.7	0.7	12.7	9.8	4.5
19	Yang Ming Marine Transport Corporation	4.3	0.5	0.0	-1.1	8.4	0.0
20	SAAM Puertos (Sudamericana Agencia Aéreas y Marítimas)	3.1	0.4	0.0	-0.3	5.6	8.2
21	"K" Line (Kawasaki Kisen Kaisha)	3.1	0.4	-0.2	-4.6	5.7	0.0
	Global operators total	**645.8**		**19.1**	**3.1**	**905.6**	**5.2**

Source: Drewry, 2019, Global *Container Terminal Operators Annual Review and Forecast: Annual Report 2020/21*.

Note: World throughput refers to data estimated by Drewry, not to container port throughput data reported in table 1.11 of chapter 1 of this report.

as well and freight forwarding services providers (The Loadstar, 2020b).

A recent study of a representative group of ports in Latin American and Caribbean countries (Argentina, the Bahamas, Brazil, Colombia, Jamaica, Mexico, Panama and Peru) suggests that a significant proportion of container volumes in the region (see table 2.12) is handled at port terminals controlled by shipping companies that are part of the three major alliances (2M, Ocean Alliance and THE Alliance) (Sánchez, forthcoming).

From the perspective of port development, terminal investments by shipping lines can have a positive impact. For example, these investments can make it possible to secure more capital investment to upgrade port facilities to serve ever-larger vessels, increase efficiency and service reliability, and reduce costs and operating times (Zhu et al., 2019). Yet, increased vertical integration between shipping and port services could also discourage other lines from calling at ports, limit choices available to shippers and influence approaches to terminal concessions (UNCTAD, 2018).

2. Impact of the pandemic and responses thereto

Worker shortages at ports and port closures resulting from the pandemic affected the ability of ports and terminal operators to complete vessel-related operations in a timely fashion and to provide key services associated with the port–hinterland interface. This situation led to interrupted cargo movement in and out of ports, inducing port congestion, additional costs for shippers and container shortages. Reduced port calls (see chapter 3) also caused a decline in port stock prices and revenues. To mitigate the impact of congestion and the economic impacts on carriers and

Table 2.12 Share of integrated port terminals in container volumes handled, selected countries of Latin America and the Caribbean
(Percentage)

Country	Ports	Share of integrated terminals in these ports (percentage)	Share of integrated terminals in country total throughput (percentage)
Argentina	Buenos Aires	67.7	56.8
Bahamas	Freeport	100.0	89.8
Brazil	Itapoa, Itajaí, Paranaguá, Pecém, Rio de Janeiro, Santos	67.2	48.6
Colombia	Buenaventura, Cartagena	11.1	10.3
Jamaica	Kingston	81.9	81.9
Mexico	Lázaro Cardenas, Progreso	72.9	15.1
Panama	Balboa, Cristobal, Rodman	10.8	10.7
Peru	Callao	41.2	34.6
	Total	**37.3**	**32.36**

Source: Sánchez, forthcoming, Latin America: Concerns about the evolution of shipping markets in the post-pandemic era.

shippers, many ports cut or deferred fees and charges, which further accentuated their diminishing revenues, increasing debt and insolvency risks. Box 2.6 expands the discussion to consider the case of ports in India.

Ports have been central in keeping supply chains open and allowing maritime trade to continue. They became the first line of defence in stopping the spread of the pandemic and protecting essential staff in their daily tasks, while letting goods flow. To respond to this challenge, ports had to introduce significant changes in procedures and operations. To help them in this endeavour, a large set of documentation was collected from port members of the UNCTAD TrainForTrade Port Management Programme and other relevant entities to help build generic guidelines and share best practices (box 2.7). Further, a crisis protocol for port entities was drawn up outlining immediate response measures, based on four colour-coded levels of intervention ranging from green, yellow and orange to red, indicating worst case scenarios with confirmed COVID-19 cases in the port area.

3. Prospects and lessons learned: Building supply-chain resilience from the perspective of supply of port services and infrastructure

Trade facilitation: Remote documentary processes to ensure continuity of cross-border trade

During the COVID-19 crisis, the role of information and communications technologies (ICTs) in promoting trade facilitation has become increasingly prominent. Digital trade facilitation commonly refers to making full use of ICTs and going paperless for all stages of the cross-border trade process. Digital trade facilitation means higher efficiency, more convenience and cost savings for cross-border trade operations, and it also means that the entire process can be completed with significantly less – or even without – in-person physical contact and interaction. It proved crucial during the COVID-19 crisis for ensuring the continuity of cross-border trade, while reducing direct physical contact among people through remote operations.

International agreements enabled the mainstreaming of digital trade facilitation. For example, the IMO Convention on Facilitation of International Maritime Traffic, 1965 requires national Governments to facilitate electronic information exchange between ship and ports, recommending the use of maritime single windows. Several initiatives are seeking to transpose physical documentation of maritime cargo to digital working methods (see chapter 5). Another international legal instrument, the WTO Agreement on Trade Facilitation, makes several references to ICT tools as a means to make cross-border trade regulations more transparent and predictable and to expedite the movement, release and clearance of goods.

During the COVID-19 crisis, several developing countries launched or expanded initiatives to allow traders to present documents remotely and enable border officials to undertake remote verification and clearance processes in a more transparent manner. For example, in Morocco, the National Single Window of Foreign Trade (Portnet) shifted to 100 per cent online tools allowing the completion of import-export formalities and access to related

Box 2.6 Challenges faced by ports in India as a result of the COVID-19 pandemic

Attempting to minimize the spread of the pandemic, India implemented lockdown measures from 24 March 2020, which led to acute workforce shortages in its ports. This was due to widespread migrant labour in many of the country's industrial and port hubs: workers returned to their home towns after the announcement of lockdown, sometimes despite offers of additional remuneration and facilities.

Labour shortages had an impact on the emptying of import containers, reducing daily outward moves. Shortages of drivers severely restricted the movement of cargo out of the ports until June 2020, affecting inland logistics.

Worker shortages also had an impact on the ability of ports to undertake cargo-clearance activities. Customs clearance procedures were also affected by other operational issues such as the decision on 22 June 2020 to conduct 100 per cent physical verification of import consignments from China at ports.

Limited cargo movements in and out of ports led to port congestion. By end April 2020, 100,000 TEUs were reported to have remained uncollected from container freight stations near Jawaharlal Nehru port, and about 50,000 TEUs remained uncleared at Chennai port. In some instances, such as in the case of Hazira port, this situation forced ports to close their gates to imports and exports.

Uncleared cargo also blocked carriers' equipment. By mid-May 2020, Indian ports reported a 50–60 per cent shortage in cargo containers for export. As a result, carriers began imposing an equipment imbalance surcharge, citing additional inventory repositioning costs. For instance, the Mediterranean Shipping Company was reported to be asking for $300 per container on cargo shipped from the ports of Jawaharlal Nehru, Mundra and Hazira to ports in eastern and southern Africa. Different media sources suggest an increase of freight of containers in India of between 25 and 32 per cent.

Authorities in India introduced several measures aimed at coping with these challenges. These include an allotment of additional land for storage to accommodate the needs of port users who faced issues related to cargo movement and a waiver of penalty charges to port users for delays due to late loading, unloading or evacuation of cargo. Other measures include deferment of payment of vessel-related charges by shipping lines, as well as waivers on some lease rentals and licence fees.

In view of labour scarcity and other factors beyond their control that affected the ability of ports to meet shippers' expectations, several ports in India declared force majeure as of end March 2020.

Sources: Grainmart News, 2020; Hellenic Shipping News Worldwide, 2020f; *Hindustan Times*, 2020; JOC.com, 2020d; Reuters, 2020; *Seatrade Maritime News*, 2020c; Standard Club, 2020; *The Economic Times*, 2020; The Loadstar, 2020c.

Box 2.7 Measures to protect staff working in port communities and to ensure continuity of port operations: Generic guidelines

Based on information from the Port Management Programme of UNCTAD and other entities, the following guidelines on protecting staff working in port communities and to ensure continuity of port operations were drawn up:

- Constantly promote and enforce preventive hygiene measures (handwashing).
- Limit physical interaction between onboard and onshore staff. Ship crew should communicate with quayside staff by radio or telephone.
- Respect physical distancing rules: stay two metres apart.
- Expand the use of digital documentation to limit human contact to the minimum
- Provide adequate and sufficient protective equipment to staff (face masks, gloves, hand sanitizers, protective eyewear).
- Increase the sanitation of surfaces that come in contact with hands.
- Establish a point of control in the perimeter of the port area to monitor temperature and related symptoms (automated temperature screening) and equip it with antibacterial solutions and sanitizers.
- Establish a waste disposal policy for suspicious cases.
- Fumigate and disinfect all passenger terminals and areas.
- Disinfect and monitor cargo.
- Set up a passenger information system for easy contact tracing and an isolated holding and testing area for port users displaying symptoms of the coronavirus disease.
- Institute a protocol for disembarking passengers and crew requiring immediate medical care in coordination with national health authorities
- Identify decontamination areas in port buildings.

Source: UNCTAD, 2020b.

governmental services 24 hours a day, 7 days a week (Morocco World News, 2020). Oman capitalized on electronic procedures that were put in place before the pandemic, which made possible the virtual clearance of officers in trade processes and online submission of cargo manifests 48 hours before vessel arrival and expanded e-services to exchange documents, payments and data (Global Alliance for Trade Facilitation, 2020).

Leveraging automation and digitalization to develop port resilience

The pandemic has brought to the fore the concept of building the resilience of supply chains. From the perspective of trade logistics, and more specifically of the supply of port services and infrastructure, this means improving risk management to developing capabilities to avoid severe threats to operators. Technology appears to hold the key to achieving these objectives.

Workforce shortages during the pandemic and resulting lockdowns severely disrupted maritime cargo operations and multimodal transport connections, highlighting the extent to which the movement of goods to keep supply chains running depends on human labour. From this perspective, increased automation could be a useful strategy to protect the workforce, ensure business continuity in port and terminal operation processes and vessel visits, and reduce processing times. Potential applications include remote piloting, alternative communications with ship navigation systems to assist increasingly autonomous ship navigation, automated cranes, automated rubber-tyre port vehicles and automated intermodal connections (The Maritime Executive, 2020b).

Digitalization can enhance port resilience by enabling better collaboration and decision-making. Port-call optimization is an example of how enhanced digital data exchange across actors involved in the port-call process can contribute to proper planning and predictable timings to achieve more efficient operations while offering opportunities for more environmentally sustainable transport, reducing emissions with just-in-time sailing (UNCTAD, 2019b).

In addition, digitalization can play a key role in diversifying business opportunities for ports, going beyond charging fees for the use of space, towards providing services that add value but do not lead to unnecessary costs. For example, digital solutions enabling shared warehouses with shared logistics assets and transport-capacity sharing could allow service providers to raise asset and capacity utilization rates and cut logistics costs (Economist Intelligence Unit, 2020; World Ports Sustainability Programme, 2020).

Leveraging digitalization to enhance port resilience will require increased investment in technological innovations and strengthened cybersecurity to protect digital infrastructure (see analysis of cyberrisks, chapter 5). As many ports are lagging behind in terms of electronic commerce and data exchange, it will be necessary to boost Internet capabilities and accessibility inside and outside port areas for port workers and users alike and engage in innovative training approaches to scale up the use of and maximize benefits from technological innovations. Advancing towards data standardization and interoperability to enable improved data sharing among different actors of the supply chain will also be necessary.

D. CONCLUSIONS AND POLICY CONSIDERATIONS

Past editions of the *Review of Maritime Transport* have identified low profitability – underpinned by oversupply – and more stringent environmental standards as the main drivers shaping the supply of maritime transport, leading to heightened pressure to increase cost-cutting efficiencies and improve sustainability in operations. Hence the growing size of vessels, the diversification of business activities combining the supply of maritime and land-side logistic services, and company partnerships to share assets, combine operations and improve fleet utilization. In this context, digitalization becomes an enabler of change, providing solutions to optimize costs and to improve efficiency and sustainability in operations.

Managing capacity to cope with oversupply

During 2019, fleets experienced the highest growth rate since 2014, with vessel sizes continuing to increase. At the beginning of 2020, the contraction of cargo volumes caused by the pandemic brought an additional challenge to structural market imbalance. To avoid low profitability and declining freight rates, carriers exercised more discipline to manage capacity and cut costs, particularly through blank sailings.

In an effort to address future uncertainty regarding the prospects for demand growth (see box 2.5), carriers may continue exercising flexibility in managing maritime networks and matching supply capacity to demand to support freight cost and rates. It is true that freight rates should be kept at a level ensuring the economic viability of the sector. However, if supply-reduction measures applied by shipping lines are sustained for a long period during the recovery in volumes, this may lead to dysfunctionalities in the sector, including ports, undermining performance of shippers and global supply chains.

Leveraging technology to cope with disruption

Workforce shortages during the pandemic and resulting lockdowns seriously disrupted manufacturing segments of the maritime supply chain and port services, highlighting the extent to which maritime transport supply and particularly, the movement of goods involved in keeping supply chain running depends on human labour. In this context, the pandemic gave new impetus to digitalization because it emerged as a vehicle to overcome an important challenge during the pandemic, that is, maintaining continuity in transport operations and trade processes while reducing the risk of contagion. Quick deployment of technological solutions made it possible to ensure continuity of business activities and government processes linked to cross-border trade and to respond to new consumer expectations in an environment characterized by supply-chain disruption, remote working and increased engagement through

business-to-consumer e-commerce for business operations.

Therefore, technological solutions featuring digital trade facilitation and digitalized processes at ports are likely to become an important element of a toolbox designed to build resilience to potential disruption that could have an impact on the performance of maritime transport in supply chains. The use of automation in maritime cargo operations and multimodal transport connections at ports could also become increasingly used to introduce improvements to ensure business continuity and workforce safety in case of disruptions, as well as to optimize efficiency. Expanding the supply of port services through digital technology and developing services that enable better collaboration across port actors and improved visibility across the supply chain could also contribute to enhancing resilience and diversifying business opportunities for ports.

Supply-chain redesign patterns can have an impact on future ship-deployment patterns

The pandemic has put a spotlight on the exposure of international production to systemic risks, particularly from the perspective of securing continuity of supply. Thus, the crisis has accentuated pre-existing trends related to changes in the length and fragmentation of value chains. Although it may be too early to fully grasp supply-chain redesign patterns in a post-pandemic recovery scenario, the shipping industry will be affected, regardless of the specific trajectories that different industries will follow, potentially influencing patterns in ship deployment.

Priority action areas in preparation for a post-COVID-19 world

The COVID-19 crisis has revealed the importance of maritime transport as an essential service ensuring the continuity of trade and supply of critical supplies and the global flow of goods during the pandemic. Ensuring the proper functioning of maritime transport services is a precondition for economic recovery. Policies that consider long-term objectives for the sector will be crucial to "build back better" in a future beyond the pandemic crisis. This means considering climate change as a global challenge that poses a threat of increased disruption to transport operations. It also means prioritizing investments that can bring simultaneous economic and environmental benefits, for example by expediting the adaptation of alternative fuels, as well as the use of wind and solar energy for ships. Reducing the carbon footprint of the fleet, either through fleet renewal or retrofits, represents a significant challenge (UNCTAD, 2020c). Given the characteristics of shipping markets and age of the fleet in many small island developing States and least developed countries, additional investment and capacity-building will be required.

To meet the challenges of post-pandemic recovery, including the need to acknowledge asymmetric capabilities across countries, the following priorities should be considered:

- Promote the use of technological tools, including through digital trade facilitation reforms, to enhance sectoral resilience to future disruptions in transport and supply-chain operations.
- Increase the accessibility of ICT tools.
- Develop data infrastructure capabilities.
- Build local capacities on ICT tools and solutions.
- Develop skills to work effectively in a world of advanced automation and technology.
- Mitigate cybersecurity risks.
- Make use of available international technical support for digital trade facilitation reforms.

In conclusion, it is also important to enhance collaboration across port States and among different actors within countries to improve crew-changeover processes and to ensure standards of procedure and risk-management protocols at the national level so as to achieve a better balance between the safety and well-being of workers and the imperatives of operational continuity.

REFERENCES

Barry Rogliano Salles Group (2020). Annual Review: Shipping and Shipbuilding Markets 2020.

Chambers S (2020). Germany paves way for green hydrogen future. Splash 24/7. 11 June.

Clarksons Research, *Container Intelligence Monthly*, various issues.

Clarksons Research (2020a). COVID-19: Monitoring the supply-side metrics. 1 May.

Clarksons Research (2020b). *Shipping Intelligence Weekly.* 26 June.

Clarksons Research (2020c). *Shipping Intelligence Weekly*. 31 July.

Clarksons Research (2020d). COVID-19: Ship repair impact.19 June.

Clarksons Research (2020e). COVID-19: Shipping market impact assessment. Update No. 6. 2 July.

Drewry (2020). *Global Container Terminal Operators Annual Review and Forecast 2020: Annual Report 2020/21*. London.

Drewry (2020a). *Maritime Financial Insight*. April.

Drewry (2020b). *Shipping Insight*. June.

Economist Intelligence Unit (2020). *The great unwinding: COVID-19 and the regionalization of global supply chains*. London.

Elgie S and McNally J (2020). 3 ingredients for smart stimulus. Blog. 30 April.

European Commission (2019). Mergers: Commission opens in-depth investigation into proposed acquisition of DSME [Daewoo Shipbuilding and Marine Engineering Company] by HHIH [Hyundai Heavy Industries Holdings]. Press release. 17 December.

Global Alliance for Trade Facilitation (2020). How COVID-19 is accelerating the digital transformation of trade at Oman's ports. 19 June.

Government of Fiji (2018). Fiji Low Emission Development Strategy 2018–2050. Ministry of Economy.

Grainmart News (2020). Shortage of labour and containers at ports leading to fall in India's exports. 29 May.

Greenport (2020). Green maritime fund proposed for UK [United Kingdom] recovery. 21 May.

Hammer S and Hallegatte S (2020). Planning for the economic recovery from COVID-19: A sustainability checklist for policymakers. World Bank Blog. 14 April.

Hellenic Shipping News Worldwide (2018). Turkish yards' EU [European Union] shipbreaking approval increases scrap possibilities. 10 December.

Hellenic Shipping News Worldwide (2019). Demolition market fired up. 19 December.

Hellenic Shipping News Worldwide (2020a). Hellas: Shipping fleet value reaches $100.5 billion. 11 May.

Hellenic Shipping News Worldwide (2020b). Ships' demolition market facing the doldrums. 7 April.

Hellenic Shipping News Worldwide (2020c) June container ship demolitions increase year-to-date volumes by 100. 2 July.

Hellenic Shipping News Worldwide (2020d). Shipping lines adopt desperate measures as pandemic disrupts operations. 28 July.

Hellenic Shipping News Worldwide (2020e). As coronavirus weighs on trade, South Korea [Republic of Korea] launches world's largest container ship. 27 April.

Hellenic Shipping News Worldwide (2020f). COVID-19 and its India impact: Issues, green shoots and way forward. 1 June.

Hindustan Times (2020). Shortage of workers, choked ports disrupt supply chains. 14 April.

IHS Markit (2020). Scrubber installation delays and maintenance incidents. Safety at Sea. 18 June.

International Transport Workers' Federation (2019). *Transport 2040: Automation, Technology, Employment – The Future of Work*. World Maritime University. London.

IMO (2019). Resolution 1147 of the IMO Assembly, thirty-first session. 4 December.

IMO (2020). FAQ on crew changes and repatriation of seafarers. IMO Media Centre. 16 June.

JOC.com (2020a). CMA CGM lands $1.14 billion state-backed loan. 13 May.

JOC.com (2020b). High debt levels leave container lines exposed. 6 April.

JOC.com (2020c). Maersk acquisition deepens North American warehouse reach. 19 February.

JOC.com (2020d). Labour scarcity at warehouses compounds Indian port flow woes. 20 May.

Reuters (2020). Singapore's PSA, container freight operators warn of congestion at Indian ports. 27 April.

Khalid W and Tariq S A U (2020). How to decarbonize shipping without spending billions. World Economic Forum Blog. 13 March.

Lexology (2020). U.S. [United States] sanctions compliance guidance released for the global maritime, energy and metals sectors. 3 June.

Lloyd's List (2019a). UK [United Kingdom] flag abandons growth targets as Brexit wipes out 30 of tonnage, 19 June.

Lloyd's List (2019b). CMA CGM takes stake in delivery firm. 29 October.

Lloyd's List (2020a). Flag registries drop [Islamic Republic of] Iran-linked ships amid sanctions scrutiny. 14 January.

Lloyd's List (2020b). Shipping's short-term pain won't stifle decarbonization. 25 June.

Lloyd's List (2020c). Shipping struggles to overcome political inertia as crew change crisis starts to bite. 10 June.

Lloyd's List (2020d). Evergreen and Yang Ming set to receive State-backed loans. 9 June.

Maersk (2020a). Maersk app sees record use amid COVID-19. Press release. 6 May.

Maersk (2020b). New research centre will lead the way for decarbonizing shipping. Press releases. 25 June.

Manifold Times (2020). Shipowners delay or postpone scrubber retrofits to minimize COVID-19 financial impact.

Marine Insight (2019). Greece #1 In 2019's Global Fleet Value Rankings – VesselsValue. 6 February.

MDS Transmodal (2020). More container service cancellations likely. 9 April.

Micronesian Centre for Sustainable Transport (2019a). Country profiles and fleet data. Technical Working Paper No. 1.

Micronesian Centre for Sustainable Transport (2019b). Potential shipping emissions abatement measures. Technical Working Paper No. 2.

Micronesian Centre for Sustainable Transport (2020). Pacific domestic shipping emissions abatement measures and technology transition pathways for selected ship types. Technical Working Paper No. 3.

Morocco World News. (2020). COVID-19: Portnet launches online services for import/export. 16 March.

Port Technology (2020). MSC [Mediterranean Shipping Company] launches online quotation tool in digitization drive. 1 July.

Pulse (2020). Korean govt to offer $1 bn aid to sea carriers, nearly $400 mn going to HMM. 23 April.

Safety4Sea (2020). European shipbuilding sector calls for urgent support due to COVID-19 crisis. 2 April.

Sánchez R (forthcoming). Latin America: Concerns about the evolution of shipping markets in the post-pandemic era.

Reuters (2019a). Flags of inconvenience: Noose tightens about Iranian shipping. 26 July.

Reuters (2019b). Shipping firms drop British flag as Brexit risks loom. 2 July.

Riviera Maritime Media (2020). Creating future-proof bulkers and tankers. 11 June.

Shell International (2020). Decarbonizing shipping: All hands on deck – Industry perspectives. Available at www.shell.com/energy-and-innovation/the-energy-future/decarbonising-shipping/.

S and P Global (2020). After IMO 2020, decarbonization in spotlight for shipping sector: Fuel for thought. 29 July.

Seatrade Maritime News (2020a). Bunker prices slip, high-low sulphur spread narrows to $56. 24 April.

Seatrade Maritime News (2020b). Singapore launches 'Playbook' to accelerate maritime's digital transformation. 22 June.

Seatrade Maritime News (2020c). Container freight rates for Indian subcontinent ports rise sharply. 18 March.

Standard Club (2020). Indian ports declare force majeure due to the COVID-19 outbreak. 27 March.

SWZ [Schip en Werf de Zee] |Maritime (2020). Pandemic may break European taboo on state aid for maritime industry. 11 June.

The Economic Times. (2020). All Chinese cargo being checked following nationwide risk alert. 25 June.

The Korea Times (2020). EU [European Union], KFTC delaying HHI–DSME merger process. 9 March.

The Loadstar (2020a). Diversify post-COVID, but 'shorter supply chains aren't a shortcut to more efficiency'.12 June.

The Loadstar (2020b). DP World to take a 60 stake in South Korean [Republic of Korea] forwarder Unico. 27 July.

The Loadstar (2020c). Uncleared import boxes clogging India's ports being used as 'free warehousing'. 24 April.

The Maritime Executive (2019). Developments in ship recycling in 2019. 15 December.

The Maritime Executive (2020a). U.S. [United States] sanctions against Iranian shipping interests enforced. 6 June.

The Maritime Executive. (2020b). Automation could aid port operations during pandemic. 9 April.

UNCTAD (2018). Market consolidation in container shipping: What next? Policy Brief No. 69.

UNCTAD (2019a). *Review of Maritime Transport 2019*. (United Nations publication. Sales No. E.19.II.D.20. Geneva).

UNCTAD (2019b). *Digitalizing the Port Call Process*. Transport and Trade Facilitation Series. No. 13. (United Nations publication. Geneva).

UNCTAD (2020a). *World Investment Report 2020: International Production Beyond the Pandemic* (United Nations publication. Sales No. E.20.II.D.23. Geneva).

UNCTAD (2020b). Port responsiveness in the fight against the 'invisible' threat: COVID-19. Technical Note. Available at https://tft.unctad.org/ports-covid-19/.

UNCTAD (2020c). Decarbonizing maritime transport: Estimating fleet renewal trends based on ship scrapping patterns. UNCTAD Transport and Trade Facilitation Newsletter No. 85.

United Kingdom Department for Transport (2019). Maritime 2050: Navigating the future.

United Kingdom Department for Transport (2020). Shipping Flag Statistics 2019. Statistical Release. 15 April.

Vessels Value (2020). Monthly market report. March.

Women's International Shipping and Trading Association (2020). WISTA [Women's International Shipping and Trading Association] International signs MOU [memorandum of understanding] with the International Maritime Organization, IMO. Available at https://wistainternational.com/news/wista-international-signs-mou-with-the-international-maritime-organisation-imo/.

World Maritime News (2020). European shipyards, equipment manufacturers call for EU [European Union] protection from COVID-19 crisis. Offshore Energy. 2 April.

World Ports Sustainability Programme (2020). Second World Ports COVID-19 survey: Some ports seeing significant changes in storage utilization at ports with some overcrowded car terminals. Available at https://sustainableworldports.org/second-world-ports-covid-19-survey-some-ports-seeing-significant-changes-in-storage-utilization-at-ports-with-some-overcrowded-car-terminals/.

WTO (2020). Japan initiates second WTO dispute complaint regarding Korean support for shipbuilders. Dispute settlement. 10 February.

This chapter looks at a series of performance indicators relating to the maritime transport sector. It provides an update on port activity, with a focus on the liner shipping connectivity index, the time ships spend in ports and data on the operation of container terminals. It also offers insights from the port performance scorecard of the TrainForTrade Port Management Programme of UNCTAD. Finally, the chapter presents novel metrics on greenhouse gas emissions from shipping in terms of flag, vessel type and other parameters.

The port data offer useful information on the determinants of port performance, including infrastructure investments, private sector participation and trade facilitation. The data also show the relevance – and the limits – of economies of scale as they apply to container shipping and port operations. Each of the different data sources is helpful in the analysis of complementary information:

- Section A uses automatic identification system data for the complete world fleet and port calls at the country level, with a high level of detail about the vessels and the time they spent in port in 2018, 2019 and early 2020.[7]
- Section B is devoted to data relating to container ships. It employs data on their shipping schedules and presents statistics on the network of the services and companies from 2006 to early 2020.[8] Unlike the automatic identification system data discussed in section A, the data in section B do not cover other vessel types.
- Section C utilizes data obtained from 10 of the world's largest shipping companies on container ports of call of these companies in 2019. The section provides a detailed analysis of the performance of container terminals for these ports.[9]
- Section D uses data from selected ports that are members of the TrainForTrade Port Management Programme, based on a detailed questionnaire elaborated by UNCTAD.[10]
- Section E makes use of automatic identification system data, coupled with information about vessel types and other ship characteristics, to discuss a key performance indicator for the shipping side of maritime transport, notably carbon-dioxide emissions. By doing so, it is possible to provide statistics on the annual carbon-dioxide emissions of the world fleet.[11]

It is reassuring that the statistics generated by different means from different sources are consistent in their main metrics, for example, as regards the relationships between vessel sizes, their position in the shipping network and economic development on the one hand, and performance indicators on the other.

7 Underlying data provided by MarineTraffic.

8 Underlying data provided by MDS Transmodal.

9 Underlying data provided by Journal of Commerce–IHS Markit.

10 Underlying data provided by the ports in annual surveys.

11 Underlying data provided by Marine Benchmark.

PERFORMANCE INDICATORS

Most connected port pairs

1. **Ningbo–Shanghai, China**
   52 liner shipping companies providing 154 direct services and a total deployed annualized capacity of 50.1 million TEUs
2. **Port Klang, Malaysia–Singapore**
   41 companies
3. **Busan, Republic of Korea–Shanghai, China**
   38 companies
4. **Shanghai–Qingdao, China**
   37 companies

Port calls in 2019

Recorded arrivals
4,362,737

Median time in port
0.966 day

Average age of vessels
18 years

Average size of vessels
14,980 gross tons

Maximum size of vessels
234,006 gross tons

Maximum container carrying capacity of vessels
23,756 TEUs

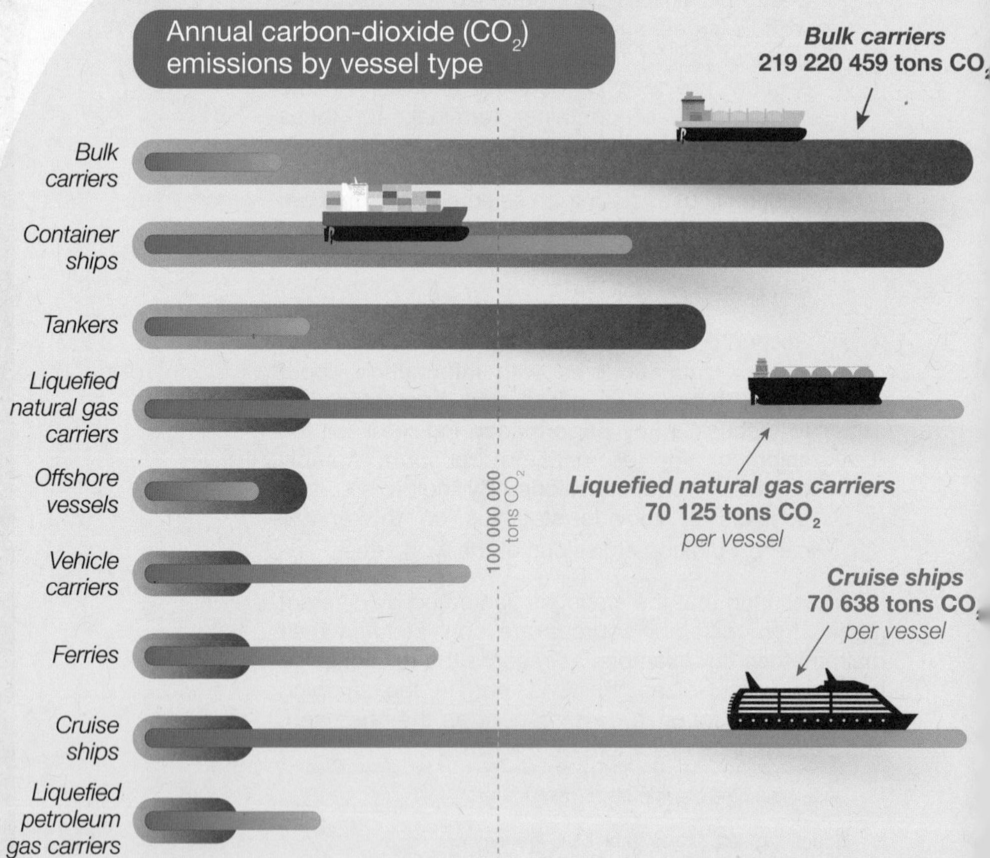

A. PORT CALLS AND TURNAROUND TIMES

1. Port calls increase and turnaround times improve

The global number of recorded commercial shipping port calls of ships of 1,000 gross tons and above rose by 6.07 per cent between 2018 and 2019 (figure 3.1). Ports further improved their overall efficiency, as the median time a ship spent in port decreased slightly by 0.41 per cent (table 3.1), from 0.970 days to 0.966 days.

The performance of seaports is an important determinant of trade costs and connectivity (Sánchez et al., 2003; UNCTAD, 2017a). The longer ships spend in port, the less time they have at sea to carry cargo for international trade. Longer times in port will lead to either higher speeds at sea and thus greater fuel consumption and carbon-dioxide emissions or the use of additional vessels to maintain the same frequency of services. This also results in longer transit times and higher inventory-holding costs. Neither of these outcomes is desirable for carriers or shippers. For ports, too, faster turnaround times are of interest, as they effectively increase their throughput capacity with the same fixed assets. Port efficiency and prompt turnarounds are therefore mutually rewarding.

A shorter time in port is a positive indicator of a port's efficiency and trade competitiveness, although there may also be good reasons for a ship to spend more time in a port, as it may bunker, purchase goods or services, or simply load and unload high volumes of goods for import and export. Benefiting from a data set provided by MarineTraffic, which draws on automatic identification system data emitted by the world's commercial fleet, this section provides an update on the time ships spent in port during calls in 2018 and 2019, including initial trends that can be observed during that period.[12]

In 2019, more than half (55 per cent) of recorded port calls worldwide were passenger ships, followed by tankers and other wet bulk carriers (12 per cent), container ships (1 I per cent) and general cargo break bulk ships (10 per cent) (table 3.2). Container ships had the fastest turnaround time, with a median of 0.69 days, an improvement of one per cent over 2018. Dry bulk carriers took the longest to load and unload – more than two days' median time. For all vessel types, 2019 recorded an increase in port calls and a slight decrease in the median turnaround time, as compared with 2018.

[12] UNCTAD calculations are based on data provided by MarineTraffic (www.marinetraffic.com). Aggregated figures are derived from the fusion of automatic identification system data with port-mapping intelligence by MarineTraffic, covering ships of 1,000 gross tons and above. Passenger ships and roll-on roll-off carriers are not included in the computation of turnaround times. Only arrivals have been taken into account to measure the number of port calls. Cases with less than 10 arrivals or 5 different vessels on a country level per commercial market as segmented are not included. The data will be updated every six months on the maritime statistics portal of UNCTAD (http://stats.unctad.org/maritime).

2. Turnaround times vary by vessel type

Container ships

The maximum vessel size of container ships in gross tons went up by 6.87 per cent between 2018 and 2019, while the increase in TEUs was even greater, at more than 10.94 per cent. The largest container ships are now de facto as big as the largest wet bulk carriers and bigger than the largest dry bulk carriers and cruise ships (table 3.2; see also chapter 2 for more details of the world fleet).

The countries with the most container ship port calls in 2019 (table 3.3, figure 3.2), were China (72,583), Japan (39,066) and the Republic of Korea (23,933). Among the top 25 countries in container port calls, only 4 recorded median turnaround times of more than one day, notably Australia, Indonesia, Viet Nam and the United States, while in Japan and Taiwan Province of China, a container ship spent a median time of less than half a day in port (table 3.3).

Section C discusses in more detail the possible determinants of why container ships may spend more time in port in some countries than in others. Most importantly, the time in port is associated with the number of containers that are loaded and unloaded during each port call.

Tankers and other liquid bulk carriers

With 44,633 port calls to its name, Japan continued to record the largest number of arrivals of tankers and other liquid bulk carriers in 2019, albeit slightly less (-0.55 per cent) than in 2018. It is followed by the Netherlands (41,042 arrivals), China (40,702) and Singapore (36,187). Together, these four countries account for 30.9 per cent of the world total for this vessel type, while the top 20 countries account for 74.6 per cent.

Japan (7.4 hours) and Germany (8.5 hours) represent the shortest median turnaround times, compared with India and the United States, whose tankers spent the longest time in port. There is a close relationship between vessel sizes and time spent in port, as smaller ships take less time to load or unload. Most countries among the top 20 receive ships of 300 000 dwt and above. The exceptions are Belgium, Hong Kong, China and the Russian Federation, where port depth and infrastructure do not accommodate vessels of this size.

Dry bulk carriers

The largest dry bulk carriers of 404,389 dwt are deployed for the transportation of iron ore from Brazil to China or to a distribution hub in Malaysia. With regard to port calls, China received by far the largest number of dry bulk carriers in 2019 (60,420 arrivals), followed by Japan (30,528 arrivals) and Australia (15,399 arrivals).

Figure 3.1 Port calls, all vessel types, 2019

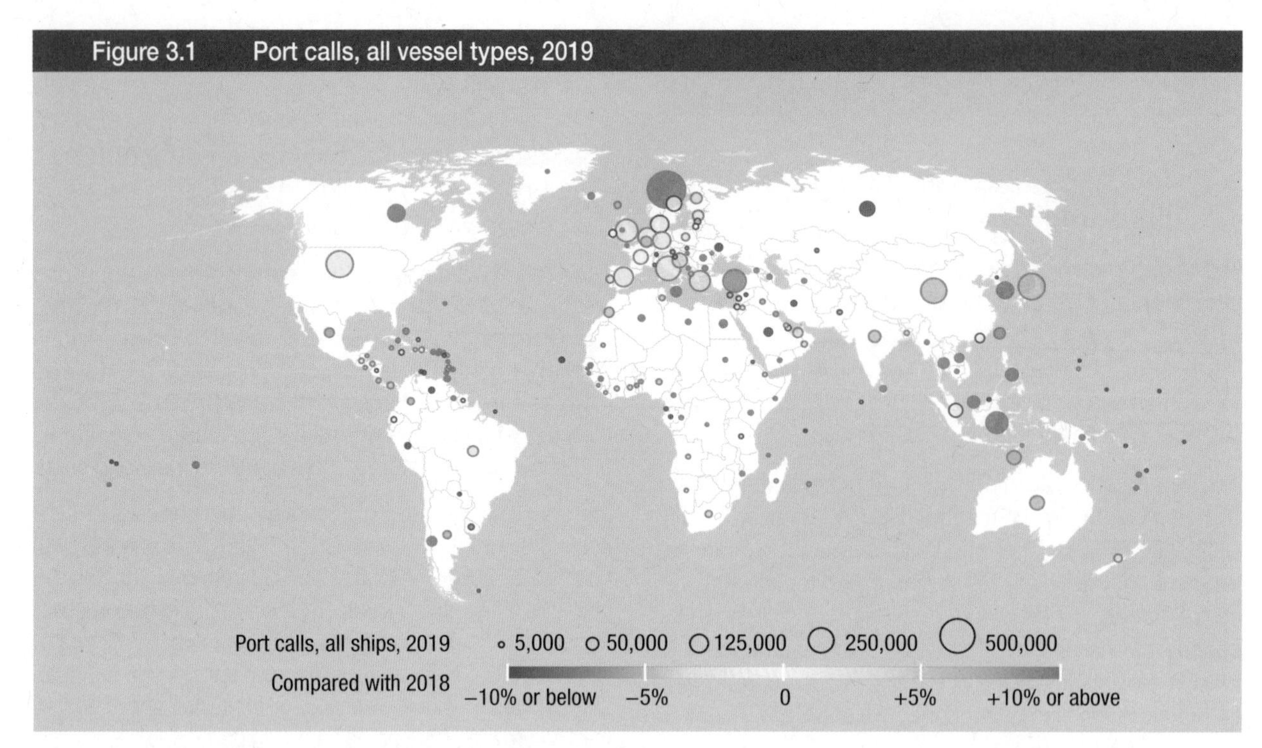

Source: UNCTAD calculations, based on data provided by MarineTraffic.
Notes: Ships of 1,000 gross tons and above. For data that include all countries, see http://stats.unctad.org/maritime.

Table 3.1 Recorded port calls and time in port, 2018 and 2019

Port calls	2018	2019	Change 2019 over 2018
Number of recorded arrivals	4 112 944	4 362 737	6.07
Median time in port (days)	0.970	0.966	- 0.41
Average age of vessels (years)	18	18	0.00
Average size of vessels (gross tons)	15 066	14 980	- 0.57
Maximum size of vessels (gross tons)	234 006	234 006	0.00
Maximum container-carrying capacity of vessels (20-foot equivalent units)	21 413	23 756	10.94
Total	**7.66**	**1.58**	**0.53**

Source: UNCTAD calculations, based on data provided by MarineTraffic (www.marinetraffic.com).

Roll-on roll-off carriers

Japan leads the world in roll-on roll-off ship arrivals, with 34,995 port calls in 2019. It is followed by the United Kingdom (16,465), the Netherlands (12,494), Spain (11,529) and Italy (9,465). This vessel type mainly includes ferries for coastal and inter-island transport, as well as car carriers. As an island economy and major automobile exporter, Japan is particularly dependent on roll-on roll-off shipping.

Passenger ships

In 2019, Norway accounted for the largest share of port calls (535,649) of passenger ships of 1,000 gross tons, followed by the United States (213,902) and Italy (194,992). The latter two are home ports to many cruise ships that are included in this category. In the Baltic and Mediterranean seas, as well as in countries with large archipelagos, such as Indonesia, Japan, Norway, the Philippines and Turkey, maritime passenger transport often replaces buses and trains as the most economical and environmentally friendly mode of public transport.

Liquefied natural gas carriers

The number of arrivals of liquefied natural gas carriers rose significantly between 2018 and 2019 (more than 15 per cent), in line with the growing demand for this source of energy and the corresponding fleet growth (table 2.1). The countries with the most port calls in this segment were Japan (1,901), Australia (1,179) and Qatar (1,043). Among the top 20 countries, ships spent

Table 3.2 Port calls and time in port by vessel type, 2019

Vessel type	Number of arrivals	Number of arrivals, change over 2018 (percentage)	Median time in port (days)	Median time in port (days), change over 2018 (percentage)	Average size of vessels (gross tons)	Average size of vessels, change over 2018 (percentage)	Average age of vessels	Maximum size of vessels (gross tons)
Container ships	474 553	4.52	0.69	-1.09	38 172	-0.90	13	232 618
Dry break bulk carriers	446 817	3.83	1.10	-0.71	5 476	0.70	20	91 784
Dry bulk carriers	277 872	7.06	2.01	-2.14	32 011	0.22	15	204 014
Liquefied natural gas carriers	12 222	15.12	1.11	-0.15	95 469	1.79	10	168 189
Liquefied petroleum gas carriers	55 227	11.89	1.01	-0.60	10 300	-3.40	14	59 226
Passenger ships	2 378 937	6.80	-		8 859	-0.77	21	228 081
Roll-on roll off carriers	190 907	1.80	-		25 277	-0.36	19	100 430
Wet bulk carriers	526 202	6.49	0.93	-0.56	15 702	1.02	14	234 006
All	**4 362 737**	**6.07**	**0.97**	**-0.41**	**14 980**	**-0.57**	**18**	**234 006**

Source: UNCTAD calculations, based on data provided by MarineTraffic (www.marinetraffic.com).
Note: Ships of 1,000 gross tons and above.

Table 3.3 Port calls and median time spent in port by container ships: Top 25 countries, 2019

Country	Number of arrivals	Median time in port (days)	Average age of vessels (years)	Average size of vessels (gross tons)	Maximum size of vessels (gross tons)	Maximum cargo-carrying capacity of vessels (TEU)
China	72 583	0.60	12	50 062	232 618	23 756
Japan	39 066	0.35	12	17 205	219 688	20 388
Republic of Korea	23 933	0.58	14	30 951	232 618	23 756
United States	19 574	1.03	13	59 336	194 250	19 462
Taiwan Province of China	16 733	0.44	14	29 571	219 775	20 388
Malaysia	16 459	0.75	14	41 499	232 618	23 756
Singapore	16 299	0.77	13	54 612	228 741	21 413
Spain	15 137	0.65	14	35 592	232 618	23 756
Indonesia	14 715	1.05	14	15 475	131 332	11 356
Hong Kong, China	12 355	0.53	14	39 826	228 741	21 413
Netherlands	12 155	0.80	13	32 385	232 618	23 756
Turkey	11 011	0.63	16	34 599	176 490	15 908
Viet Nam	10 041	1.03	16	18 459	175 688	16 000
Germany	9 543	0.74	13	42 018	232 618	23 756
United Kingdom	8 395	0.73	14	36 766	232 618	23 756
India	8 211	0.91	15	46 994	153 666	13 386
Italy	8 171	0.91	15	44 772	194 849	19 462
Thailand	8 130	0.68	17	22 653	154 000	14 220
Brazil	8 050	0.73	9	62 947	119 441	11 923
United Arab Emirates	7 082	0.94	15	47 830	219 277	21 200
Philippines	5 492	0.84	15	19 124	71 786	6 800
Belgium	5 190	1.00	14	52 967	232 618	23 756
France	4 468	0.75	13	56 344	219 277	20 776
Australia	4 400	1.18	12	48 715	109 712	9 971
Panama	4 347	0.63	11	45 162	150 000	14 000
World total	**474 553**	**0.69**	**13**	**38 172**	**232 618**	**23 756**

Source: UNCTAD calculations, based on data provided by MarineTraffic.
Notes: Ships of 1,000 gross tons and above. For data that include all countries, see http://stats.unctad.org/maritime.

Figure 3.2 Port calls by container ships, 2019

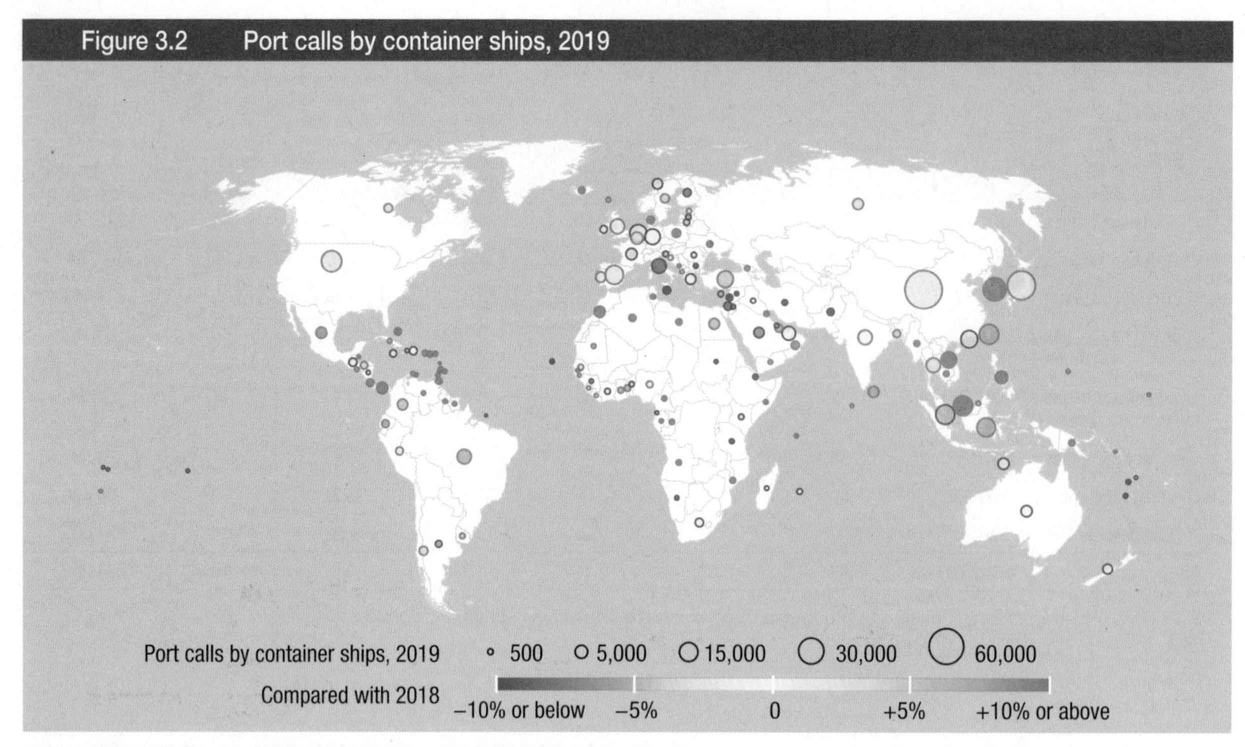

Source: UNCTAD calculations, based on data provided by MarineTraffic.
Notes: Ships of 1,000 gross tons and above. For data that include all countries, see http://stats.unctad.org/maritime.

the least time per port call in Norway (eight hours on average), and the longest in Singapore (two days).

Break bulk vessels

Norway (33,564 calls), China (30,007) and the Russian Federation (28,837) are the countries with the most port calls by break bulk general cargo vessels. Among the top 20 countries in this category, Germany and Norway have the shortest median turnaround times at 0.35 and 0.33 days respectively, while in France (1.58 days), Italy (1.98 days) and the Russian Federation (1.61 days), general cargo ships spent the longest time in their ports.[13]

3. Small island economies depend heavily on general cargo ships

Break bulk general cargo ships have a declining share in the world fleet (see also chapter 2). They remain, however, particularly important for small island economies and destinations with little port traffic, where the deployment of more specialized ships may not be justified. For small island economies or countries that are archipelagos, such as Indonesia or the Philippines, break bulk general cargo vessels account for a substantial share of the countries' total port calls.

Some small island economies are among those with the longest port turnaround times for general cargo vessels, as they may lack infrastructure or specialized port equipment. Others have very short turnaround times, owing to the lack of congestion because of low frequencies and the low cargo volumes in loading and unloading (UNCTAD, 2019a). Between 2018 and 2019, the Comoros, Maldives and New Caledonia saw significant improvements both in terms of increased port calls and shorter port turnaround times. Fiji and New Caledonia are served by the youngest and most modern fleet of general cargo ships, while French Polynesia, Maldives and Saint Kitts and Nevis receive vessels that are on average more than 30 years old (table 3.4).

4. A downturn in port calls during the COVID 19 pandemic

The COVID 19 crisis led to fewer port calls for most vessel types during the first half of 2020 (figure 3.3).

Liquefied natural gas and liquefied petroleum gas carriers and tankers (wet bulk carriers) continued to record increases in port calls during the first quarter of 2020. In the second quarter, however, all vessel types experienced a decline in the number of port calls. The hardest hit were roll-on roll-off vessels, which include ferries and other vessels that also carry passengers.

With regard to container ship port calls, the number of arrivals started to fall below 2019 levels about week 12 (mid-March 2020) and began to recover gradually about week 25 (third week of June) (figure 3.4). By mid-June, the average number of container vessels arriving weekly at ports worldwide had sunk to 8,722, an 8.5 per cent year-on-year drop. Since then, the average weekly calls started to recover, rising to 9,265 in early

[13] See http://stats.unctad.org/maritime for the complete tables concerning all vessel types.

Table 3.4 Port calls and median time spent in port, general cargo ships, 2019
(Selected small island economies)

Country or territory, break bulk cargo	Number of arrivals 2019	Number of arrivals, change 2019 over 2018 (percentage)	Median time in port, 2019 (days)	Median time in port, change 2019 over 2018, (percentage)	Average age of vessels (years)	Average size of vessels (gross tons)	Maximum size of vessels (gross tons)
American Samoa	57	-6.6	0.63	10.3	16	9 494	18 100
Antigua and Barbuda	193	12.9	0.39	3.4	22	5 797	17 644
Aruba	59	- 51.2	0.73	82.3	19	9 729	28 805
Bahamas	464	-15.3	0.41	28.9	26	4 831	91 784
Barbados	309	-5.8	0.56	4.3	22	6 813	22 698
Cabo Verde	360	36.9	0.63	-10.6	21	5 095	46 295
Cayman Islands	153	-14.0	0.56	3.8	24	7 513	27 818
Christmas Island	50	-35.1	0.43	-10.7	14	5 913	10 021
Comoros	197	32.2	1.03	-25.3	15	6 352	24 960
Curaçao	320	-31.9	0.53	1.9	18	3 285	16 137
Dominican Republic	107	-0.9	0.40	-1.2	16	6 586	14 413
Fiji	457	40.6	0.95	39.7	7	4 914	40 393
French Polynesia	555	-12.9	0.19	20.4	39	3 165	54 529
Grenada	124	-23.5	0.58	43.8	24	7 016	16 639
Guam	67	-25.6	2.11	-2.5	20	8 979	61 185
Guernsey	339	63.0	0.14	13.4	25	1 687	2 601
Haiti	384	-4.5	0.96	1.9	21	4.760	24 140
Jamaica	576	1.4	0.90	-10.6	13	9.099	29 688
Maldives	101	44.3	0.49	-89.1	31	4.041	20 965
Martinique	193	-9.0	0.40	2.5	17	8.628	27 828
Mauritius	133	-10.1	3.48	47.6	21	5 317	21 483
Mayotte	25	-66.2	2.23	-8.1	11	7 219	24 960
Micronesia	73	-24.0	0.35	-53.6	22	4 352	9 924
New Caledonia	549	52.5	1.24	-24.0	8	7 507	29 829
Reunion	53	-11.7	1.30	-13.6	12	8 323	21 483
Samoa	68	-2.9	0.54	41.9	15	9 045	18 100
Seychelles	137	-18.5	5.22	-8.7	24	5 384	16 803
Sint Maarten	179	-39.9	0.38	-25.0	18	6 374	22 698
Solomon Islands	50	-38.3	1.75	2.9	17	10 509	18 468
Saint Kitts and Nevis	207	6.2	0.27	14.3	35	3 274	14 413
Saint Lucia	287	8.7	0.41	-10.6	28	5 892	16 137
Saint Vincent and the Grenadines	116	-38.6	0.38	23.6	16	9 761	16 137
Timor-Leste	164	6.5	0.98	-2.5	16	4 339	9 719
Tonga	82	3.8	0.39	-12.3	15	8 363	18 100
Trinidad and Tobago	584	-14.4	0.91	13.6	16	7 326	30 488
Turks and Caicos Islands	197	-27.0	0.43	-3.6	19	1 749	2 191
Tuvalu	69	-4.2	11.21	-19.9	28	4 047	6 965
Vanuatu	17	-55.3	0.83	21.4	15	15 551	18 100
Word total	**446 817**	**3.8**	**1.10**	**-0.7**	**20**	**5 476**	**91 784**

Source: UNCTAD calculations, based on data provided by MarineTraffic.

Note: Ships of 1,000 gross tons and above. For data that include all countries, see http://stats.unctad.org/maritime.

Figure 3.3 Global change in the number of port calls, first and second quarters of 2020 compared with the first and second quarters of 2019, selected vessel types

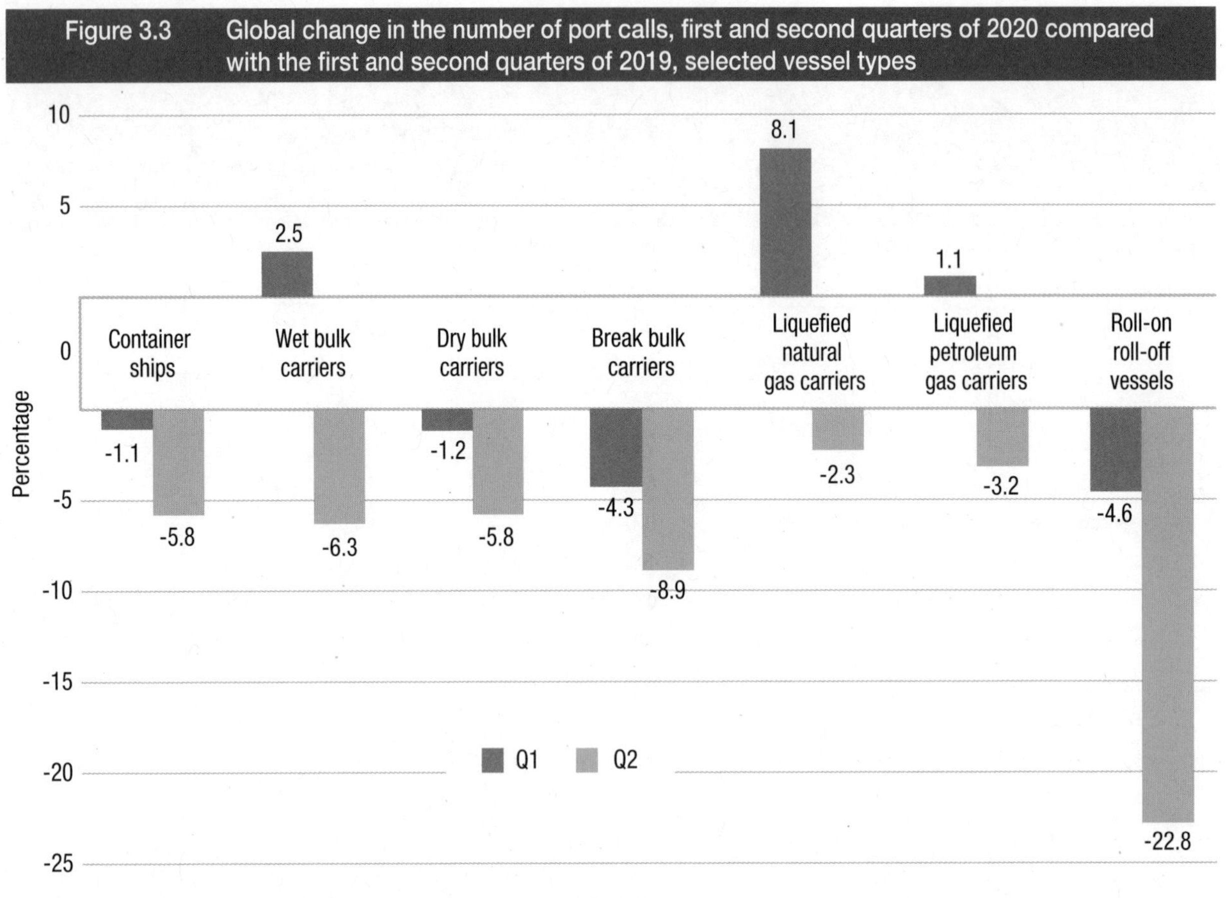

Source: UNCTAD calculations, based on data provided by MarineTraffic.
Abbreviation: Q, quarter.

Figure 3.4 Number of weekly container ship port calls worldwide, moving four-week average, 2019 and 2020

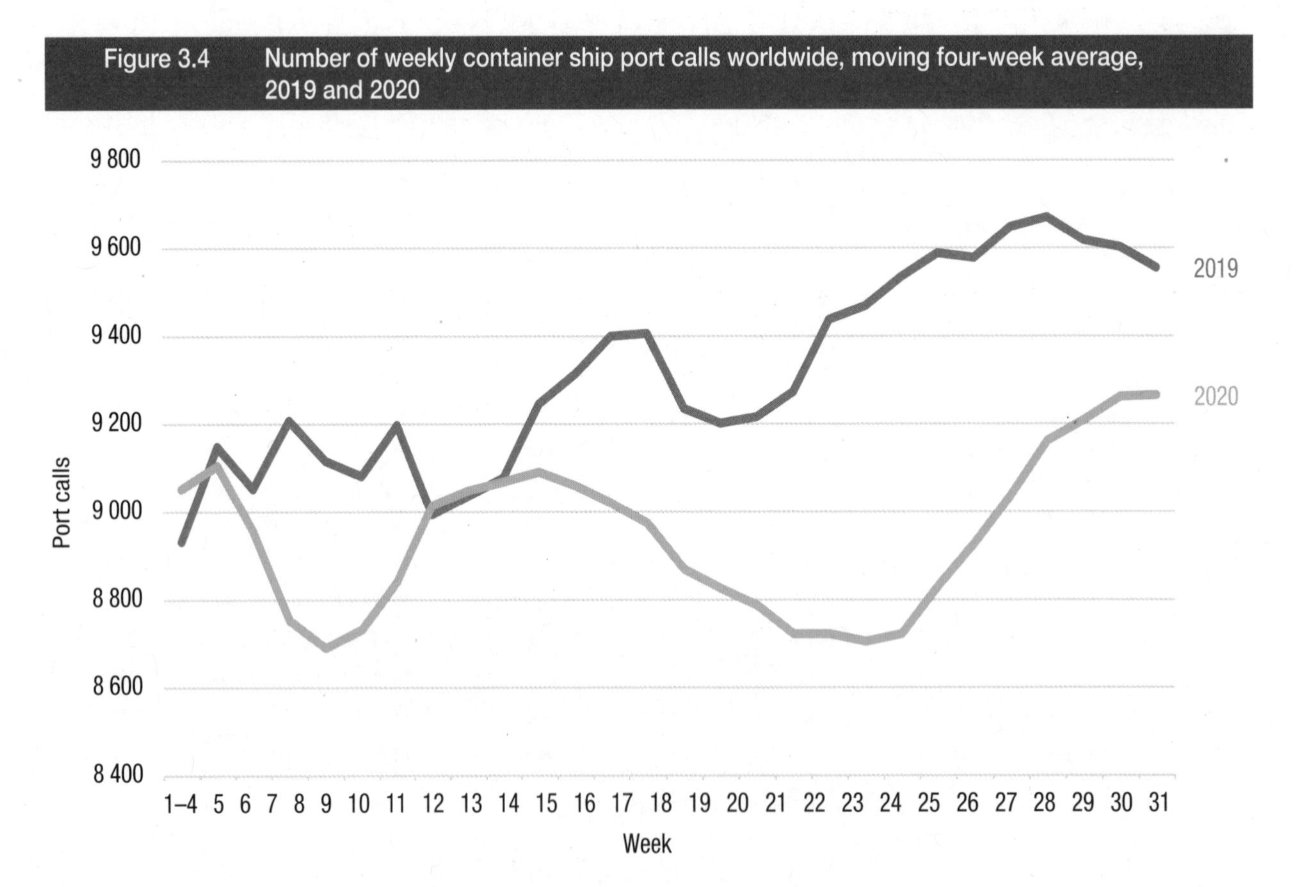

Source: UNCTAD calculations, based on data provided by MarineTraffic.

August 2020, just 3 per cent below the levels recorded 12 months earlier. For a more detailed analysis by region, see UNCTAD, 2020a (https://unctad.org/en/pages/newsdetails.aspx?OriginalVersionID=2465).

5. Future uses of automatic identification system data to assess port and shipping performance

The automatic identification system was initially developed and introduced as a tool to support navigational safety. Today, the signals transmitted through the system are used to track the movement of vessels, even if the owners of those vessels may prefer otherwise. Without publicly available data, the data and analysis presented above would not be possible. The transmission of signals from automatic identification systems is mandatory and increasingly scrutinized, and the data coverage is continuously improving. Combining automatic identification system-derived statistics with other sources of data and information can help respond to growing demands for optimization of the supply chain, monitoring of emission data and trade forecasts.

Optimizing the supply chain

Already today, initiatives such as port-call optimization benefit from automatic identification system data (UNCTAD, 2020b). Beyond the seaside of the operation, the whole supply chain can benefit from exchanging data, including automatic identification system data on ship movements, but also data on other modes of transport, ports and the goods that are being traded. In this context, digitalization, artificial intelligence, blockchain, the Internet of things and automation are of growing relevance. They help optimize existing processes, create new business opportunities and transform supply chains and the geography of trade (UNCTAD, 2019b).

Notwithstanding the potential opportunities and benefits offered by the automatic identification system, including low-cost global access, its use requires capacity-building and investments in digitalization, especially in developing countries. There is a need for policy design at the national and international levels to ensure that developing countries can benefit from the automatic identification system and the digitalization of maritime transport (UNCTAD, 2019b).

Trade statistics and forecasts

Automatic identification system data do not include information about the cargo the ships carry. However, by combining the data on vessel moves and drafts with information on vessel type, trade flows and countries of departure and destination, automatic identification system data can help obtain an increasingly exact and up-to-date picture of trade flows (Arslanalp et al., 2019; Cerdeiro et al., 2020; United Nations, 2020; World Bank, 2020). Combined with information on the speed of vessels, port departures and idle ships, this can serve to produce nowcasts and forecasts of trade and economic growth. It can also help verify trade statistics by checking published trade data against the vessel moves that would be necessary to actually transport those goods. Such efforts would benefit from further standardization of data.

Reducing emissions

Shipping will have to move away from carbon. Initiatives such as the Getting to Zero Coalition, supported by UNCTAD, aim to reduce carbon-dioxide emissions from shipping to net zero (Global Maritime Forum, 2020). A ship's emissions depend on numerous factors, including vessel size, engine type, fuel used and speed. Automatic identification system data – combined with information on the ship's engine and fuel – can help assign carbon-dioxide emissions to the country of the vessel's flag or the country's waters where the carbon dioxide is being emitted. Section E below provides an example of such use of automatic identification system data.

B. CONTAINER SHIPPING: LINER SHIPPING CONNECTIVITY

1. Countries' evolving liner shipping connectivity

In 2020, 6 of the 10 most connected economies are in Asia (China; Singapore; the Republic of Korea; Malaysia; Hong Kong, China; and Japan, 3 are in Europe (Spain, the Netherlands, and the United Kingdom), and 1 in North America (the United States) (figure 3.5). The most connected country – China – improved its liner shipping connectivity index by 56 per cent since the baseline year 2006, while the global average liner shipping connectivity index went up by 50 per cent during the same period.

Since 2020, UNCTAD, in collaboration with MDS Transmodal, reports quarterly values for the liner shipping connectivity index, both at the port and country levels.[14] The work is based on empirical

[14] UNCTAD developed the liner shipping connectivity index in 2004. The basic concepts and major trends are presented and discussed in detail in UNCTAD, 2017a and MDS Transmodal, 2020. In collaboration with MDS Transmodal, the liner shipping connectivity index was updated and improved in 2019 to offer additional country coverage, including several small island developing States, and to add a component covering the number of countries that can be reached without the need for trans-shipment. The remaining five components, notably the number of companies that provide services, the number of services, the number of ships that call per month, total annualized deployed container-carrying capacity and ship sizes, have remained unchanged. Applying the same methodology as for the country-level liner shipping

Figure 3.5 Liner shipping connectivity index of top 10 economies, first quarter 2006– second quarter 2020

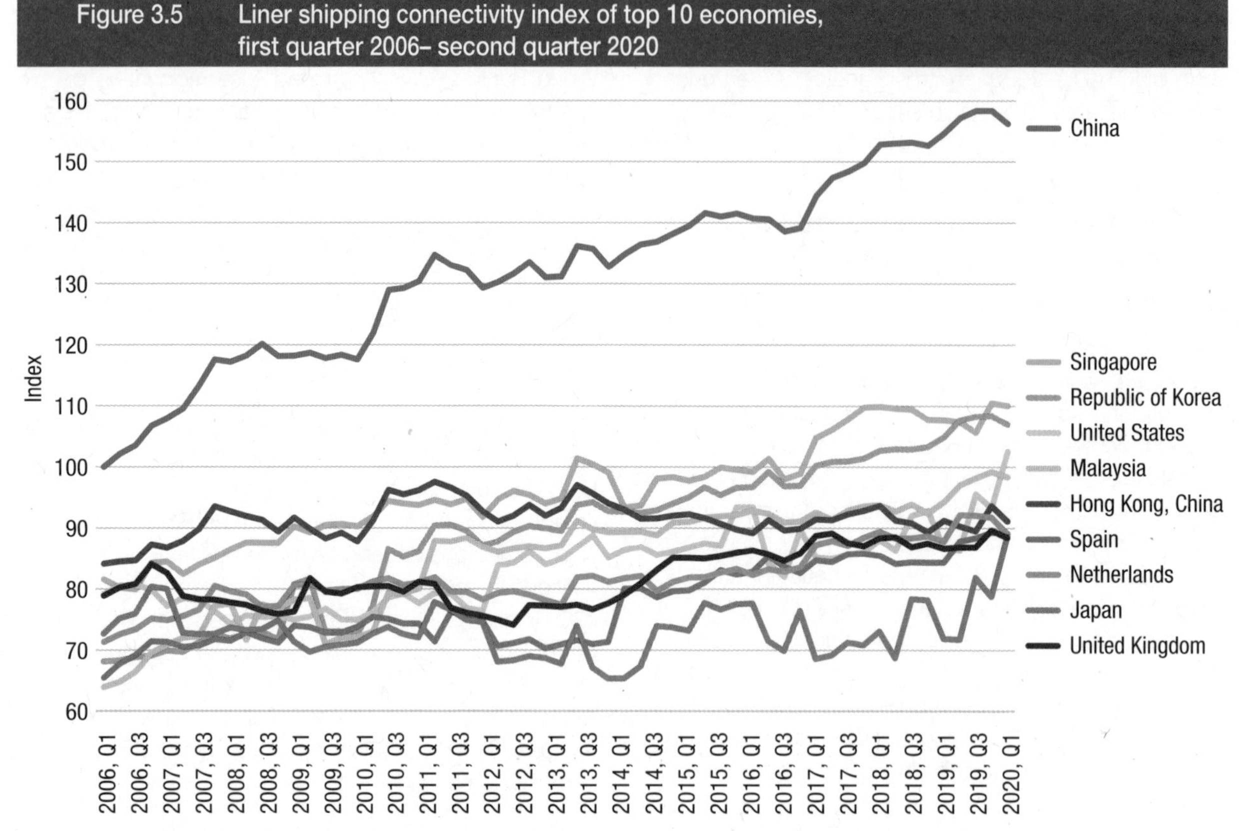

Source: UNCTAD calculations, based on data provided by MDS Transmodal. For the data set that includes all countries, see http://stats.unctad.org/LSCI.

Abbreviation: Q, quarter.

evidence that a country's competitiveness and access to overseas markets benefit from better liner shipping connectivity, which reflects access to the global container shipping network (UNCTAD, 2017a). This section first analyses trends at the country and port levels, and then goes on to discuss developments regarding the different components from which the index is generated.

connectivity index, UNCTAD has generated a new liner shipping connectivity index for ports.

Each of the six components of the port liner shipping connectivity index captures a key aspect of connectivity:

- A large number of scheduled ship calls allows for a high frequency of servicing imports and exports.
- A large deployed capacity allows shippers to trade sizable volumes of imports and exports.
- A large number of regular services to and from a port is associated with shipping options to reach different overseas markets.
- A large number of liner shipping companies that provide services is an indicator of the level of competition in the market.
- Large ship sizes are associated with economies of scale on the sea leg and possibly lower transport costs.
- A large number of destination ports that can be reached without the need for trans-shipment is an indicator of fast, reliable and direct connections to foreign markets.

Since 2020, the same methodology has been applied to country and port levels on a quarterly basis.

2. Liner shipping connectivity of many small island developing States stagnates

Many small island developing States and other small island economies have poor shipping connectivity. Yet, there is often little they can do to enhance their liner shipping connectivity, which remains limited, given their geographic position, lack of a wider hinterland and low trade volumes. Figure 3.6 depicts the liner shipping connectivity index of selected small island developing States and other small island economies where shipping schedules are reported separately.

A few small island developing States, notably the Bahamas, Jamaica and Mauritius, have been able to position their ports as trans-shipment hubs and increase their attraction as ports of call. Mauritius, for example, has more than doubled its liner shipping connectivity index since 2006. The additional fleet deployment stemming from trans-shipment can also be used for shipments of national importers and exporters. Nonetheless, most small island developing States continue to experience low levels of connectivity, with a lack of improvement over the years.

Among the leading ports in each subregion, Suva, in the Pacific, has the lowest port liner shipping connectivity index (figure 3.8). Among the 50 least connected economies, 37 are small island developing States.

Figure 3.6 Liner shipping connectivity index of selected small island developing States, first quarter 2006–second quarter 2020

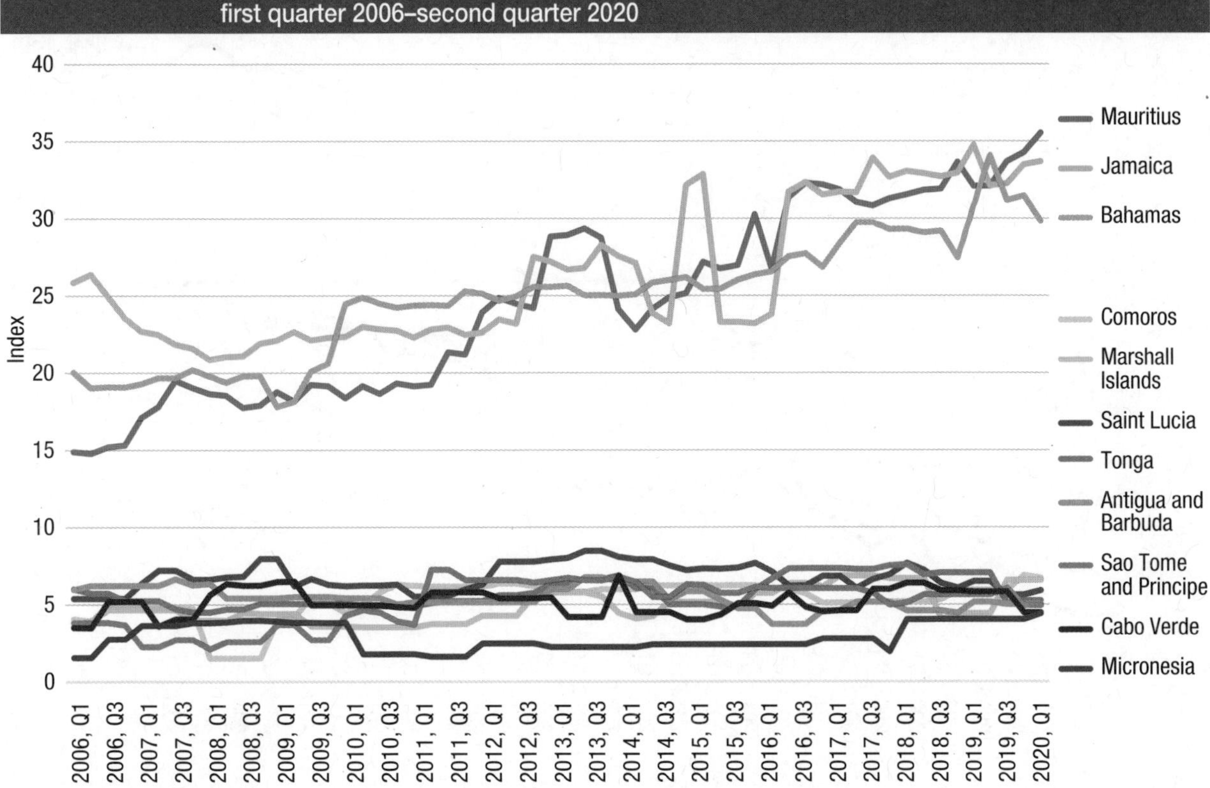

Source: UNCTAD calculations, based on data provided by MDS Transmodal. For the data set that includes all countries, see http://stats.unctad.org/LSCI.

Abbreviation: Q, quarter.

Among the 20 least connected economies, all except the Democratic People's Republic of Korea, Moldova and Paraguay are small island developing States, and the latter two are landlocked countries, whose low liner shipping connectivity index is generated from containerized river transport services.

Achieving economies of scale, while ensuring some level of competition and choice for their shippers is a difficult conundrum for many small island developing States and other small economies or remote ports. If better port infrastructure, through the use of dredging and specialized port cranes, for example, makes it possible for larger and more efficient ships to call, these same ships will then require fewer port calls to carry the same monthly volume of foreign trade. This may result in even less choice for shippers and a lower frequency of services. Put differently, it may not be possible, especially for small island developing States, to improve on all components of the liner shipping connectivity index, as illustrated in figure 3.9 (see also chapter 4, which discusses the challenge faced by small island developing States in the Pacific).

3. Developments at the port level

In 2020, five of the top 10 ports are located in China (Shanghai, Ningbo, Hong Kong, Qingdao and Xiamen), three are in other Asian countries (Malaysia, the Republic of Korea and Singapore), and two are in Europe (Belgium and the Netherlands). The liner shipping connectivity index of almost all of the top 10 ports has risen significantly since 2006, except Hong Kong, China, overtaken by four other ports (figure 3.7).

The port-level liner shipping connectivity index is generated for all container ports of the world that receive regular container shipping services.[15] In the second quarter of 2020, the database maintained by MDS Transmodal (www.mdst.co.uk) recorded regular container shipping services in 939 ports worldwide, a 12.6 per cent increase over 2006. This latest port count follows a decline of 3.6 per cent compared with the peak of the first quarter of 2019, when global liner shipping services included 974 ports in their schedules. Most of this recent decline took place during the first two quarters of 2020 and can be largely attributed to capacity management in response to the COVID-19 pandemic.

Figure 3.8 depicts the liner shipping connectivity index of the leading ports in major maritime regions. Several of the regional leaders saw a spike in the index in the second quarter of 2020, as they managed to attract additional services with larger vessels.

[15] For the complete data set providing quarterly values of the liner shipping connectivity index of more than 1,200 ports, from the first quarter of 2006 onwards, see http://stats.unctad.org/maritime.

Figure 3.7 Liner shipping connectivity index of top 10 ports, first quarter 2006–second quarter 2020

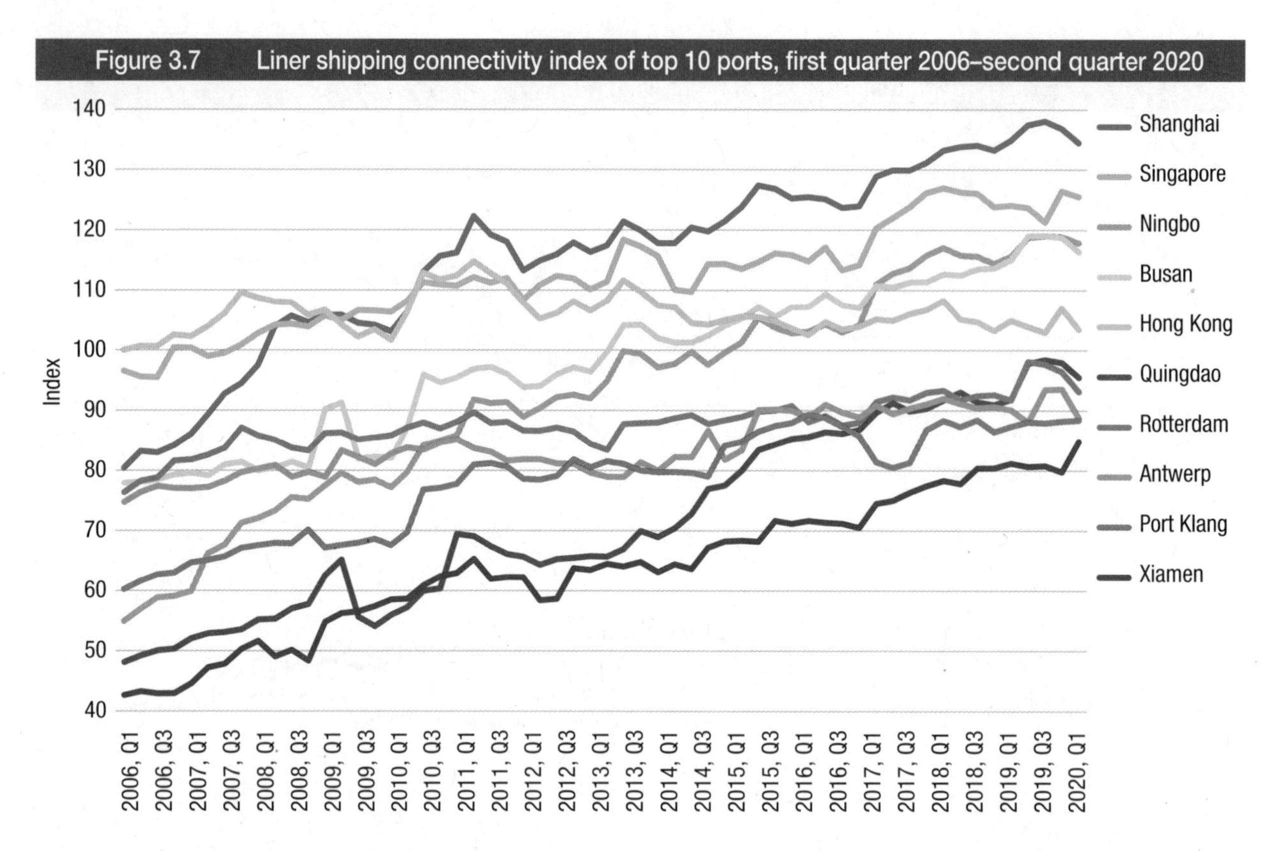

Source: UNCTAD calculations, based on data provided by MDS Transmodal. For the liner shipping connectivity index of all ports, see http://stats.unctad.org/maritime.

Abbreviation: Q, quarter.

Figure 3.8 Liner shipping connectivity index of leading regional ports, first quarter 2006–second quarter 2020

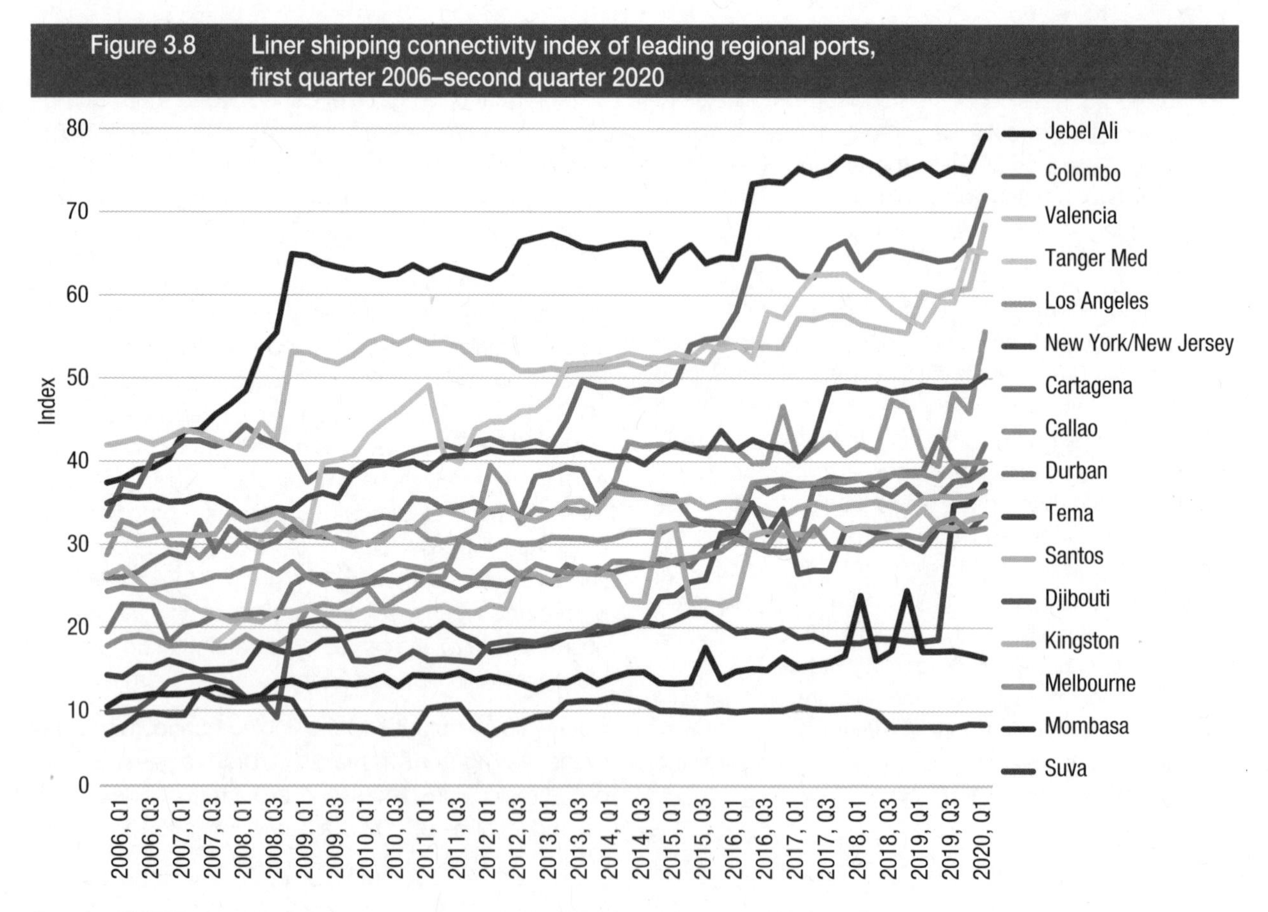

Source: UNCTAD calculations, based on data provided by MDS Transmodal. For the liner shipping connectivity index of all ports, see http://stats.unctad.org/maritime.

Abbreviation: Q, quarter.

4. Liner shipping connectivity index components: Bigger ships and fewer companies

The liner shipping connectivity index helps to analyse trends among countries and ports. A look at the six components generating the index provides insights into industry developments (figure 3.9). The average fleet deployment per country is a reflection of the long-term trend of consolidation, as vessel sizes and total capacity deployed increase sharply, while the average number of companies that provide services to and from each country continues to decrease. The number of direct connections, number of services and number of weekly calls all follow a similar, slightly downward trend.

5. Fleet deployment during the COVID-19 pandemic

During the first two quarters of 2020, carriers managed their deployed capacity by reducing the frequency of calls and number of services. The average size of the largest container ships deployed continued to grow, in line with the long-term trends analysed in chapter 2. In the first quarter of 2020, scheduled deployed capacity still stood above that of the same quarter of 2019, albeit with a larger number of blank sailings; during the second quarter of 2020, schedules were adjusted further, and total deployed capacity was reduced below 2019 levels (figures 3.9, 3.10 and 3.11).

Figure 3.9 Liner shipping connectivity index components, first quarter 2006–second quarter 2020, index of averages per country
(First quarter 2006 = 100)

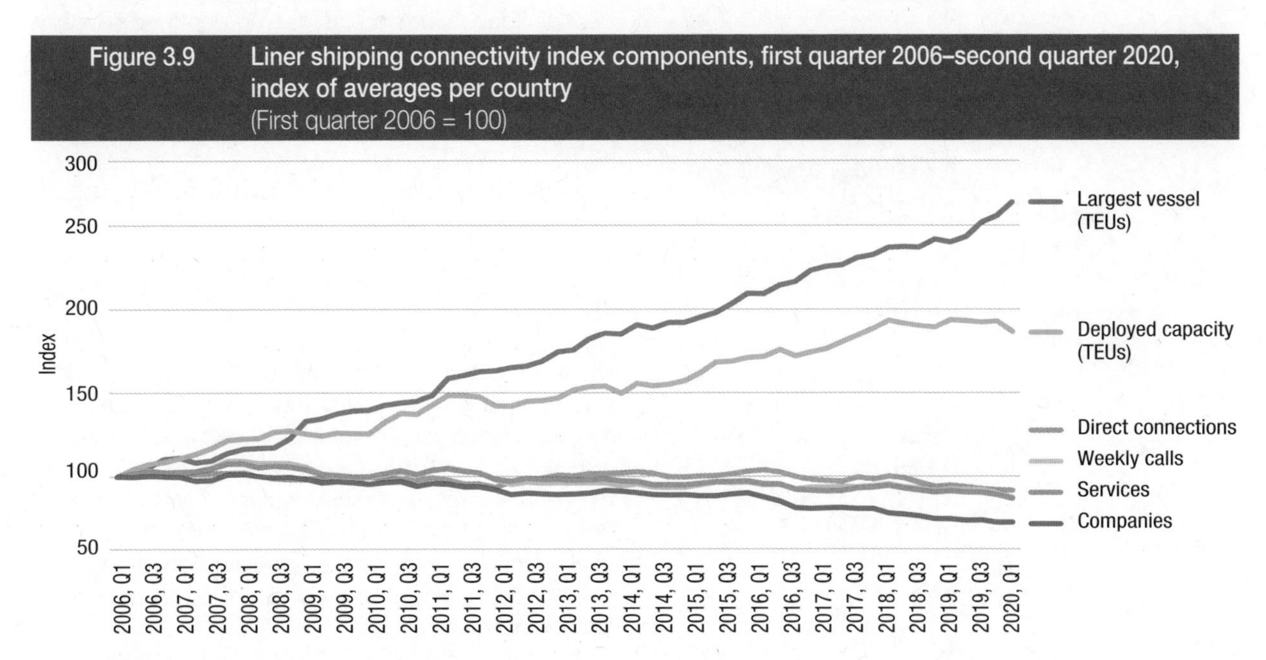

Source: UNCTAD calculations, based on data provided by MDS Transmodal.
Abbreviation: Q, quarter.

Figure 3.10 Quarterly trends in fleet deployment, first quarter 2019–second quarter 2020
(First quarter 2019 = 100)

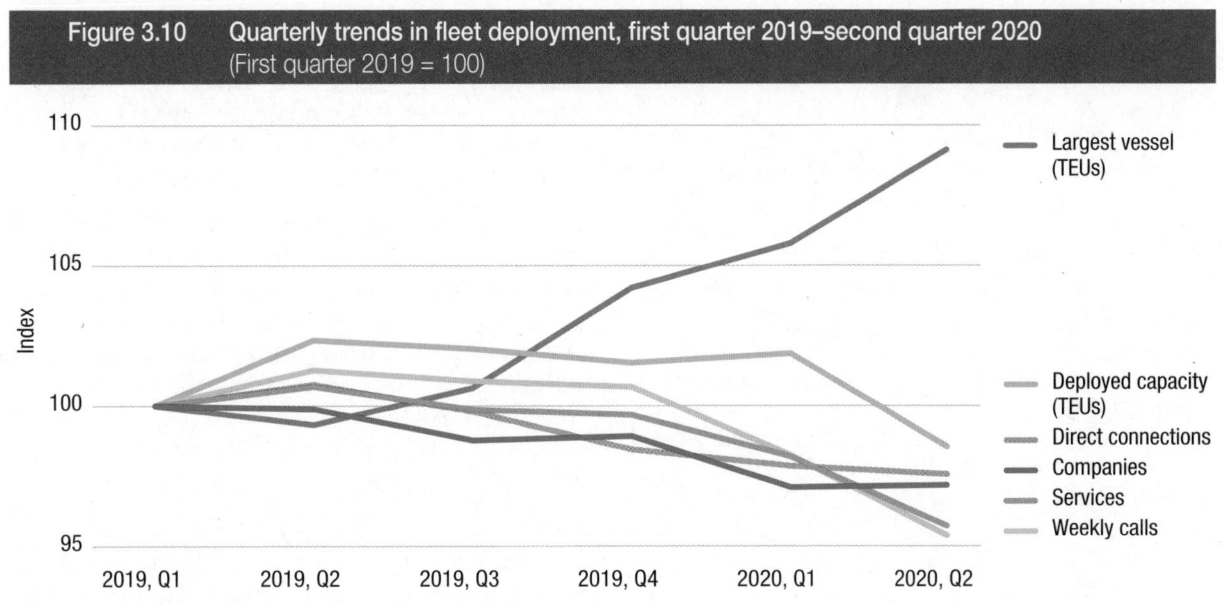

Source: UNCTAD calculations, based on data provided by MDS Transmodal.
Abbreviation: Q, quarter.

Figure 3.11 Quarterly trends in fleet deployment, selected countries, 2019–2020
(Percentage change)

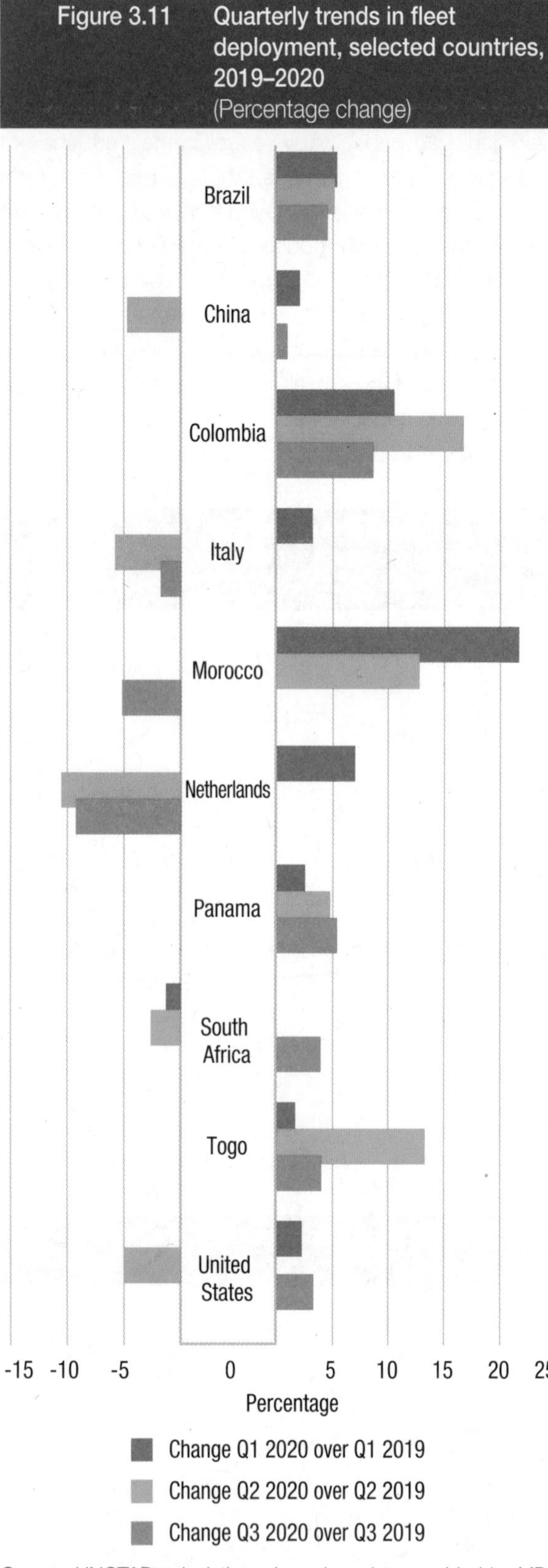

Source: UNCTAD calculations, based on data provided by MDS Transmodal.

Note: Timeline: first and second quarters of 2020 compared with first and second quarters of 2019.

Abbreviation: Q, quarter.

Container shipping schedules show that total fleet deployment during the first quarter of 2020 was still above that of the first quarter of 2019 in most economies. During the second quarter, carriers started to reduce capacity considerably. Steps taken by the shipping lines to manage capacity helped them sustain positive earnings during the first semester of 2020, in spite of less traffic (see also chapter 2).

China started 2020 with an increase of 2.1 per cent over the first quarter of 2019, recording a negative year-on-year growth of minus 4.7 in the second quarter. Growth then rebounded to more than 1 per cent in the third quarter. Most European countries underwent a steeper decline. For example, the Netherlands went from plus 7.0 per cent in the first quarter to minus 10.5 per cent in the second quarter and minus 9.3 per cent in the third quarter. Morocco experienced positive growth in in the first two quarters, but lost ground in the third quarter. Togo stands out as gaining deployed capacity, as the port of Lomé is becoming a regional hub for West African trade, especially for Nigeria, where most of the ports are draft restricted.

6. Better connectivity stimulates port traffic

The liner shipping connectivity index is an indicator of the deployment of the world's container ship fleet. It is highly correlated with a country's port traffic. If there is more demand for the shipping of containerized cargo, liner companies will deploy more and larger ships, to achieve a higher level of total fleet deployment. They are also likely to provide more services to better connect the country directly to more countries. As the demand goes up, additional companies will enter this market. These components of fleet deployment are the six components from which the liner shipping connectivity index is generated.

It is interesting to analyse the correlation between these six components, as well as the liner shipping connectivity index, and each country's port container traffic patterns. UNCTAD has been systematically gathering port traffic statistics since 2010 (http://stats.unctad.org/TEU) (see also chapter 1). Figure 3.12 depicts the correlation between the liner shipping connectivity index and the port traffic of countries in 2017, the year for which the most complete statistics are available.

Interestingly, the correlation is not linear. Each additional 1 per cent increase in the liner shipping connectivity index is associated with a 1.896 per cent increase in port traffic. In other words, as more ships and services are provided, port traffic grows exponentially. This statistical finding is in line with the data, port performance and economies of scale recorded by the shipping companies (see section C below).

Similar correlations are observed for the individual components of the index with port traffic (table 3.5). For each component, there is a high and non-linear correlation with a country's port traffic. The highest correlation and the lowest exponential growth are recorded for the total deployed container-carrying

Figure 3.12 Liner shipping connectivity index and port traffic, 2017
(20-foot equivalent units)

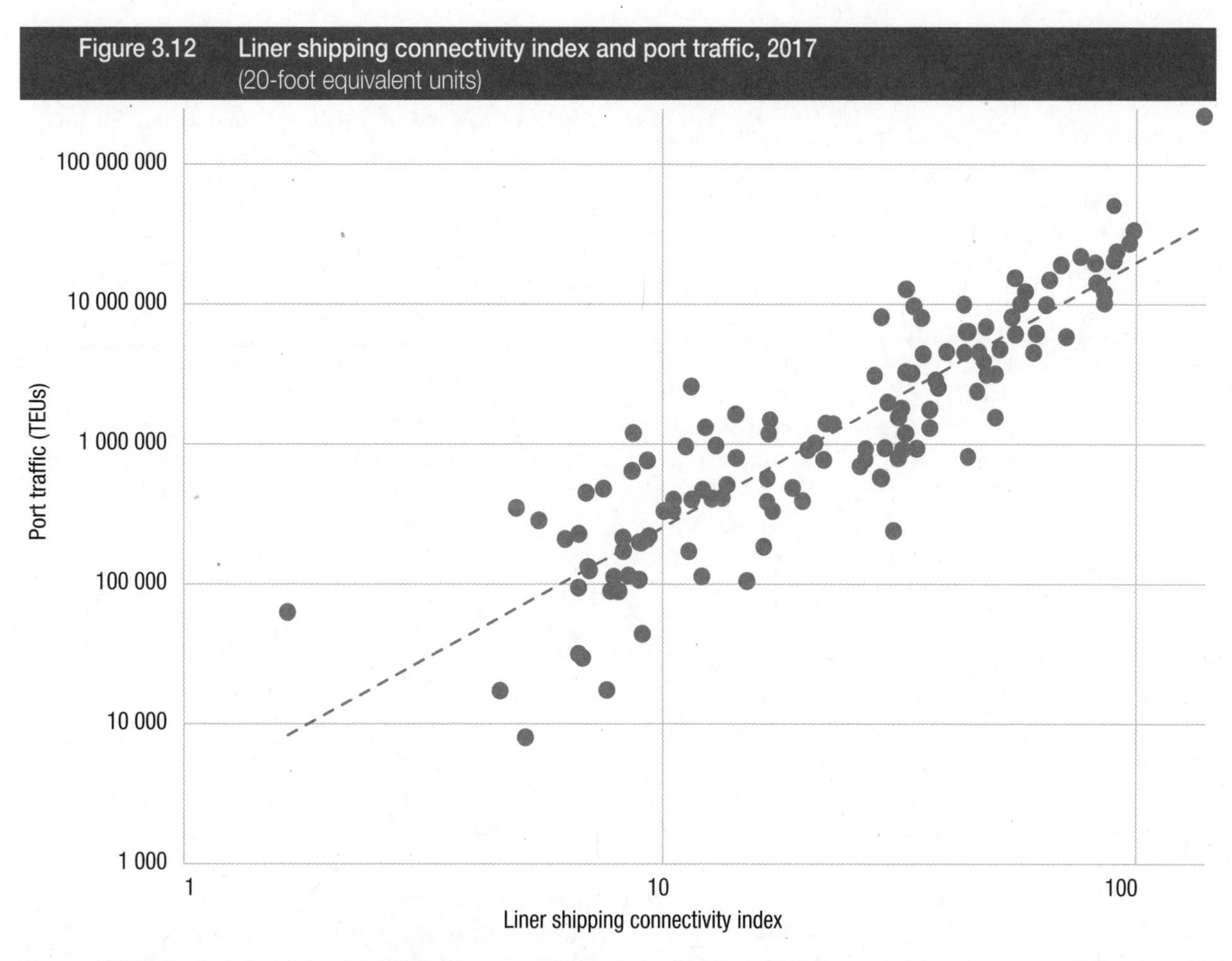

Source: UNCTAD calculations, based on UNCTAD port traffic statistics and the liner shipping connectivity index generated with data from MDS Transmodal. Values are given for the first quarter of the 2017 liner shipping connectivity index and 2017 annual port traffic volumes in TEUs.

Note: $R^2 = 0.7851$; $y = 3209.1x^{1.896}$.

Table 3.5 Correlation between components of the liner shipping connectivity index and port traffic

Liner shipping connectivity index component	Coefficient of determination (R^2)	Elasticity
Liner shipping connectivity index	0.79	1.90
Total deployed container-carrying capacity (20-foot equivalent units)	0.90	1.13
Services (number of)	0.87	1.50
Frequency of port calls (number per week)	0.86	1.43
Companies (number of)	0.82	1.90
Size of largest ships (20-foot equivalent units)	0.61	1.53
Direct connections (number of, countries)	0,56	1.96

Source: UNCTAD calculations, based on UNCTAD port traffic statistics and the liner shipping connectivity index generated with data from MDS Transmodal. Correlation and elasticity are based on a power equation (see figure 3.12). Underlying values relate to the first quarter of the 2017 liner shipping connectivity index and 2017 annual port traffic volumes in TEUs.

capacity, as the two variables should largely grow in parallel. As regards additional companies and direct connections to additional markets, exponential growth is much stronger; increasing the number of direct connections by 1 per cent is associated with an increase in the port traffic by almost 2 per cent. In other words, for a port authority that aims to boost its port traffic, it would make good sense to focus especially on attracting additional carriers that provide direct services to a large number of trading partners.

7. Connecting trading partners through the container shipping network

In the second quarter of 2020, there were 939 seaports that were connected to the global liner shipping network through regular container shipping services (figure 3.13). If all ports had direct connections with each other, there would be 440,391 port-to-port liner shipping services. In reality, only 12,748 port pairs had such direct services, that is to say, 2.9 per cent of the theoretical total. For trade between 97.1 per cent of port pairs, containers need to be trans-shipped in one or more other ports. The necessary number of trans-shipments is one or two for most port pairs. The least connected port pairs require up to six trans-shipments. For example, 7 shipping services and 14 port moves would be necessary to export a container from some Pacific island ports to some Atlantic island ports for one trade transaction.

The structure of the liner shipping network is further illustrated in figure 3.14. Through an algorithm, the illustration visualizes ports that are well connected by locating them in close proximity to each other. Ports that have more direct connections in total are represented by larger points. The more distant ports are from each other, the more trans-shipments would be required to transport a container between them. An example of low connectivity depicted in figure 3.14 would be that of connectivity between Coatzacoalcos, Mexico with Basra, Iraq or with Malacca, Malaysia or with Rarotonga, the Cook Islands. Colour schemes reflect the geographical location of the port, and as expected, ports that are geographically closer to each other tend to be better connected with each other through the container shipping network.

The port pair that is most connected through direct services is Ningbo–Shanghai, China, with 52 liner shipping companies providing 154 direct services and a total deployed annualized capacity of 50.1 million TEUs between the two ports. It is followed by Port Klang, Malaysia–Singapore, with 41 companies; Busan, the Republic of Korea–Shanghai, China, with 38 companies; and Shanghai–Qingdao, China, with 37 companies.

All the top 50 most connected port pairs are on intraregional routes, almost exclusively within Asia, except for two connections within Europe: Antwerp, Belgium-Rotterdam, the Netherlands, with 24 companies and Hamburg, Germany–Rotterdam, the Netherlands, with 23 companies.

In other regions, too, neighbouring ports are generally the most connected with each other. These intraregional connections do not necessarily carry trade between neighbouring ports, but the high connectivity is the result of being connected to the same overseas routes, in combination with feedering and trans-shipment services.

In Africa, for example, Durban and Cape Town, South Africa are connected with each other by services provided by 12 companies. In Angola, Luanda is most connected with Cape Town, South Africa with seven companies, and Mombasa, Kenya is most connected with Dar-es-Salam, the United Republic of Tanzania through direct services by 10 companies. By comparison, there are only six companies that connect Mombasa, Kenya with Ningbo, China. The connectivity

Figure 3.13 Number of seaports with regular container vessel calls, first quarter 2006–second quarter 2020

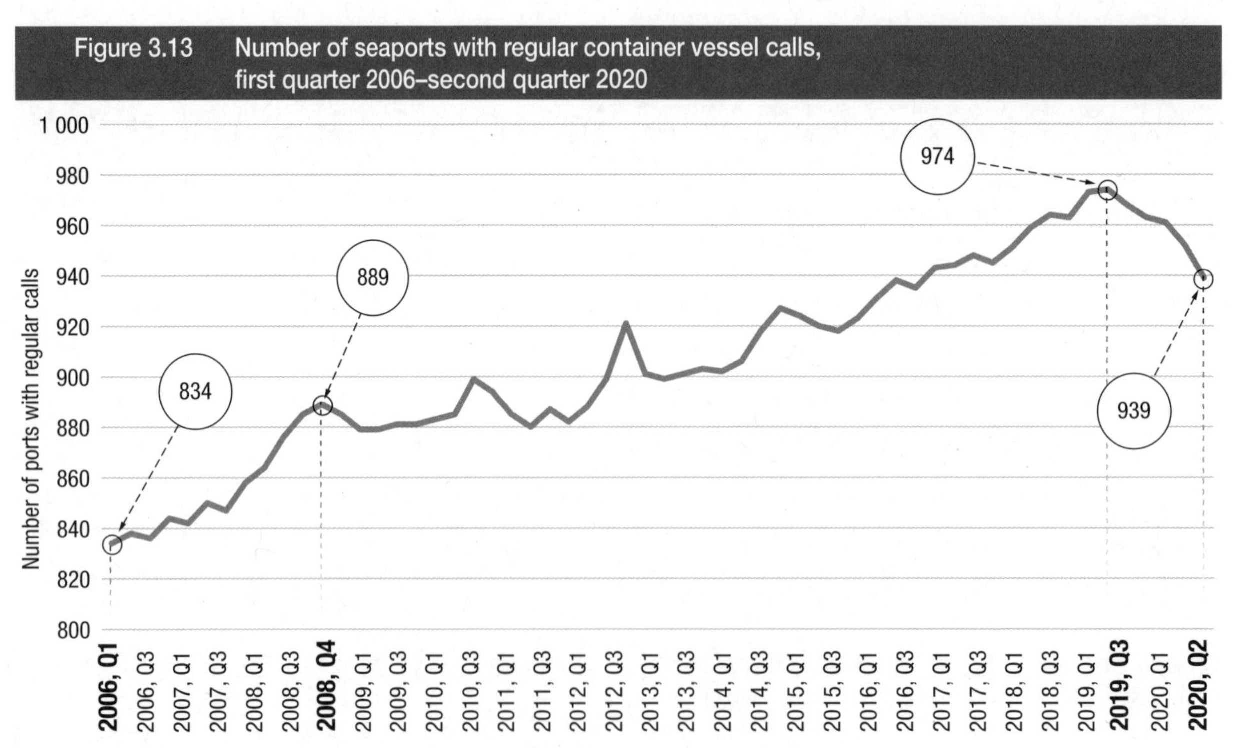

Source: UNCTAD calculations, based on data provided by MDS Transmodal.

Abbreviation: Q, quarter.

Figure 3.14 Global liner shipping network, second quarter 2020

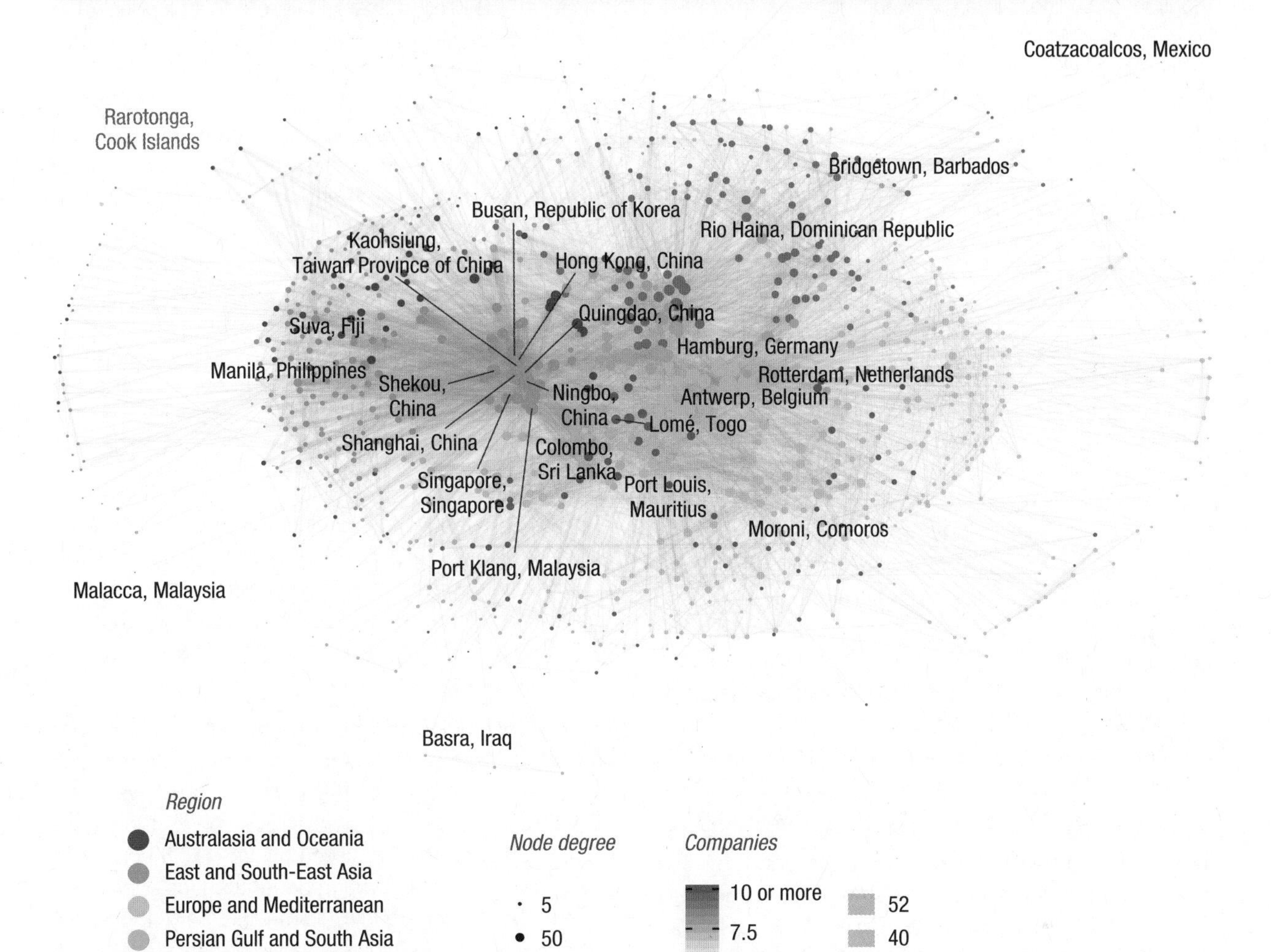

Source: UNCTAD calculations, based on data provided by MDS Transmodal; visualization by Julian Hoffmann.
Notes: Layout = stress; links = number of companies providing a direction connection.

level of Tanger Med, Morocco is highest with Algeciras and Valencia, Spain, through services provided by nine liner companies.

In South America, Buenos Aires, Argentina is most connected with Montevideo, Uruguay (13 companies) and in Brazil, 14 companies provide direct services between Paranaguá, Rio de Janeiro and Santos. There are 10 companies that connect San Antonio, Chile with Callao, Peru; 15 companies that connect Callao, Peru with Guayaquil, Ecuador and 12 companies that provide direct services between Cartagena, Colombia and Manzanillo, Panama.

In the Pacific, two ports in Fiji (Lautoka and Suva) are connected through services by seven liner companies, while Betio, Kiribati is connected with Lautoka and Suva, Fiji through services by two carriers. Also, Kosrae and Pohnpei, Micronesia have direct services with Majuro, the Marshall Islands that are provided by two companies, while only one company connects these ports with Yokohama, Japan and other ports in Asia. Honiara, Solomon Islands and Port Vila, Vanuatu are most connected with ports in Fiji (four companies) and with Yokohama, Japan and other ports in Asia (3 companies).[16]

C. CONTAINER SHIPPING: PORT PERFORMANCE

1. Container terminal performance

On average, 75–85 per cent of the port call time of container ships is taken up by container operations, that is to say, the time between the first and last container lifts, while the remaining time may be due to pilotage, mooring,

[16] Data relate to the second quarter of 2020. These are UNCTAD calculations based on data provided by MDS Transmodal. The liner shipping bilateral connectivity index for all port and country pairs is available at http://stats.unctad.org/maritime.

customs formalities and other operational or procedural requirements. The efficiency of the container operation segment is influenced by the combination of crane speed multiplied by the quantity of cranes deployed (crane intensity). Although constrained occasionally by stowage plans, a ship's overall length or available cranes, crane intensity is also largely influenced by the call size.

There are large variations in average port times, and this should be seen as an opportunity for improvement. The gaps are too large to be closed with a single giant step, so a succession of smaller but progressive steps is required in all countries located towards the bottom of table 3.6.

The lead metric for the 2019 port turnaround times is the average of total port hours per port call. For this, port hours are counted from the time a ship reaches the port limits (pilot station or anchorage) until it departs from the berth after operations are completed. It therefore incorporates waiting/idle time, steaming-in time and berth time. The time taken to steam out of the port limits is not included because first, it is very homogeneous, and second, it is not influenced by port effectiveness. Any delays in departure due to channel congestion; absence of pilots, tugs or other resources; and ship readiness are all incurred before a ship departs from the berth and the last line is released. Ships may also sit idle on departure for bunkering or repair or simply in safe waters if the next port cannot accommodate berthing on arrival.

The data used in this section are provided by IHS Markit from its extensive, proprietary Port Productivity Programme. It comprises close to 200,000 container ship port calls per year, approximately 42 per cent of the total. It combines data on the vessel calls and time in port with detailed information about the containers loaded and unloaded at each call, totalling more than 300 million TEUs, at more than 430 ports in 138 countries. The underlying data are provided by 10 of the world's largest shipping lines and are enhanced with matched port arrival times from the IHS Markit automatic identification system database.

The time ships spent in port in 2019 is reported in section A (table 3.3). It is measured in absolute numbers, without considering the number of containers loaded or unloaded during this period. For the selected ports and carriers analysed in this section, the Journal of Commerce–IHS Markit database makes it possible to adjust the port turnaround time for loading and unloading operations during this period.

For an objective overview of container ship in-port time, different factors need to be considered, including the call size and quantity of container moves per ship call. For objective benchmarking, the actual port call hours are weighted by the quantity of containers exchanged per call. The formula used to achieve this for each country is as follows:

Actual port hours/actual call size x actual call size of full benchmark group

For example, if a country takes 12 hours to handle a ship with 1,200 containers loaded and unloaded, and the average of the benchmark group is 1,500 moves per call, it is then assumed that it will take the subject port 15 hours to handle that same quantity (*12/1,200 x 1,500*). In sum, the resulting weighted port hours represent the time a ship spends in port per container loaded and unloaded, multiplied by the global average number of containers of the benchmark group.

2. Most of the countries with the best port performance are in Asia

A shorter time in port is a positive indicator of a port's efficiency and trade competitiveness. Based on the criteria explained above, container ships spent an average time of 23.2 hours (0.97 days) in port per call in 2019.

Table 3.6 lists the world's leading 25 economies in terms of total container ship port calls (as per table 3.3) and provides their average in-port time, weighted by call size. The average port-call time across these 25 economies in 2019 was 21.7 hours (0.91 days), slightly less than the global average.

Among the leading 25 countries in terms of container ship port calls, the United Arab Emirates hold the record for the shortest in-port time (14.1 hours of weighted port time), followed by China (15.5 hours), Singapore (17.4 hours) and the Republic of Korea (17.8 hours). Of the nine countries performing better than the average of the entire group, only two (Belgium and the Netherlands) are outside Asia. The lowest levels of performance are represented by France (41.8 hours), Italy (36.5 hours), Australia (34.6 hours) and Brazil (33.6 hours).

Table 3.7 lists the top and bottom 10 countries in terms of their weighted average port hours, as well as the average vessel size in terms of container-carrying capacity (TEUs). Four Middle Eastern countries were among the top 10 in 2019. Along with the Republic of Korea, Singapore and Sri Lanka, the ports of these countries handle predominantly trans-shipment containers. They generally have high crane densities on the quay walls, enabling high crane intensities. The ratio of yard to quay equipment is similar to that of most contemporary container terminals but a trans-shipment container has only one yard move per quay move, whereas that number is doubled in gateway ports.

Trans-shipment ports have some fundamental advantages, such as limited gateway cargo, with fewer outside trucks causing congestion in the yards, and potentially planned days ahead, with cargo arriving and departing in large batches. Last, but not least, most ports are operated by global terminal operators, and many are set up as cost centres or joint ventures with the ship operators.

Hub ports face other challenges, such as tight connections, fragmented discharge and roll-overs with an impact on yard integrity; in addition, the last port

Table 3.6 Weighted average port call hours in top 25 economies, 2019

Country	Number of weighted average port hours
United Arab Emirates	14.1
China	15.5
Singapore	17.4
Republic of Korea	17.8
India	18.2
Thailand	20.0
Netherlands	20.3
Malaysia	20.5
Belgium	20.7
Hong Kong, China	22.5
Germany	23.0
Viet Nam	23.0
United States	24.7
Taiwan Province of China	25.8
United Kingdom	26.5
Spain	26.8
Indonesia	27.2
Japan	28.2
Philippines	31.7
Panama	32.3
Turkey	32.5
Brazil	33.6
Australia	34.6
Italy	36.5
France	41.8
Top 25 economies	**21.7**

Source: Journal of Commerce–IHS Markit Port Productivity Programme.

Note: The top 25 countries are derived from the total number of container ship port calls shown in table 3.3.

Table 3.7 Weighted average port call hours, top and bottom 10 countries or territories

Economy	Weighted average port hours	Average vessel size
Oman	12.5	9 002
United Arab Emirates	13.8	7 619
China	15.1	8 483
Poland	16.6	6 357
Saudi Arabia	16.8	8 351
Singapore	17.0	6 183
Republic of Korea	17.4	7 425
Qatar	17.7	7 081
India	17.8	7 463
Sri Lanka	18.5	5 749
Top 10	**15.9**	**7 769**
Canary Islands	61.7	984
Mozambique	62.6	2 533
Norway	62.9	1 259
Cameroon	63.7	2 541
Bulgaria	64.1	1 162
El Salvador	64.2	2 203
Nigeria	65.0	4 379
Gabon	65.9	1 559
Namibia	71.8	3 561
Trinidad and Tobago	72.1	1 490
Bottom 10	**65.1**	**2 530**

Source: Journal of Commerce–IHS Markit Port Productivity Programme.

before a head-haul must often contend with scattered load stowage in high-profile stacks.

Five of the lowest-ranking countries in table 3.7 are in Africa, which is still catching up in terms of building sufficient infrastructure and implementing the necessary port and trade facilitation reforms to be able to handle ever-growing demand effectively. Much additional investment is required, and the performance indicators presented above suggest that this could well come from private sector operators.

3. Economies of scale in port performance

The larger container ships appear to benefit from economies of scale. As a general rule of thumb, higher move counts (call size) on the larger ships allow terminals to deploy a higher quantity of cranes (crane intensity), and therefore handle more containers per ship hour than countries with smaller average vessel calls. Larger vessels also tend to be assigned a higher priority when scarce resources within a terminal or port are being shared among multiple ships. The larger vessels tend to be deployed to modern and efficient ports where the handling efficiency is significantly more refined than ports and terminals in secondary or tertiary ports of call.

As shown in figure 3.15, the more containers loaded and unloaded per port call (call size), the longer a ship needs to stay in port (average port hours). However, thanks to economies of scale, this relationship is not linear; as the call size goes up by 1 per cent, the time spent in port increases only by 0.5 per cent. The regressions illustrated in figures 3.15 and 3.16 statistically explain 47 per cent of the variance of the time a ship spends in port ($R^2 = 0.47$), while the remainder of the differences between countries need to be explained by factors such as trans-shipment incidence, port infrastructure, management and trade facilitation, as well as other parameters often associated with economic and institutional development.

As shown in figure 3.15, the longest average port call durations are those of the Sudan and Yemen. Although both had few port calls in 2019, those port calls involved

Figure 3.15 Country averages of port time per ship and call size, 2019
(Hours in port and moves per port call)

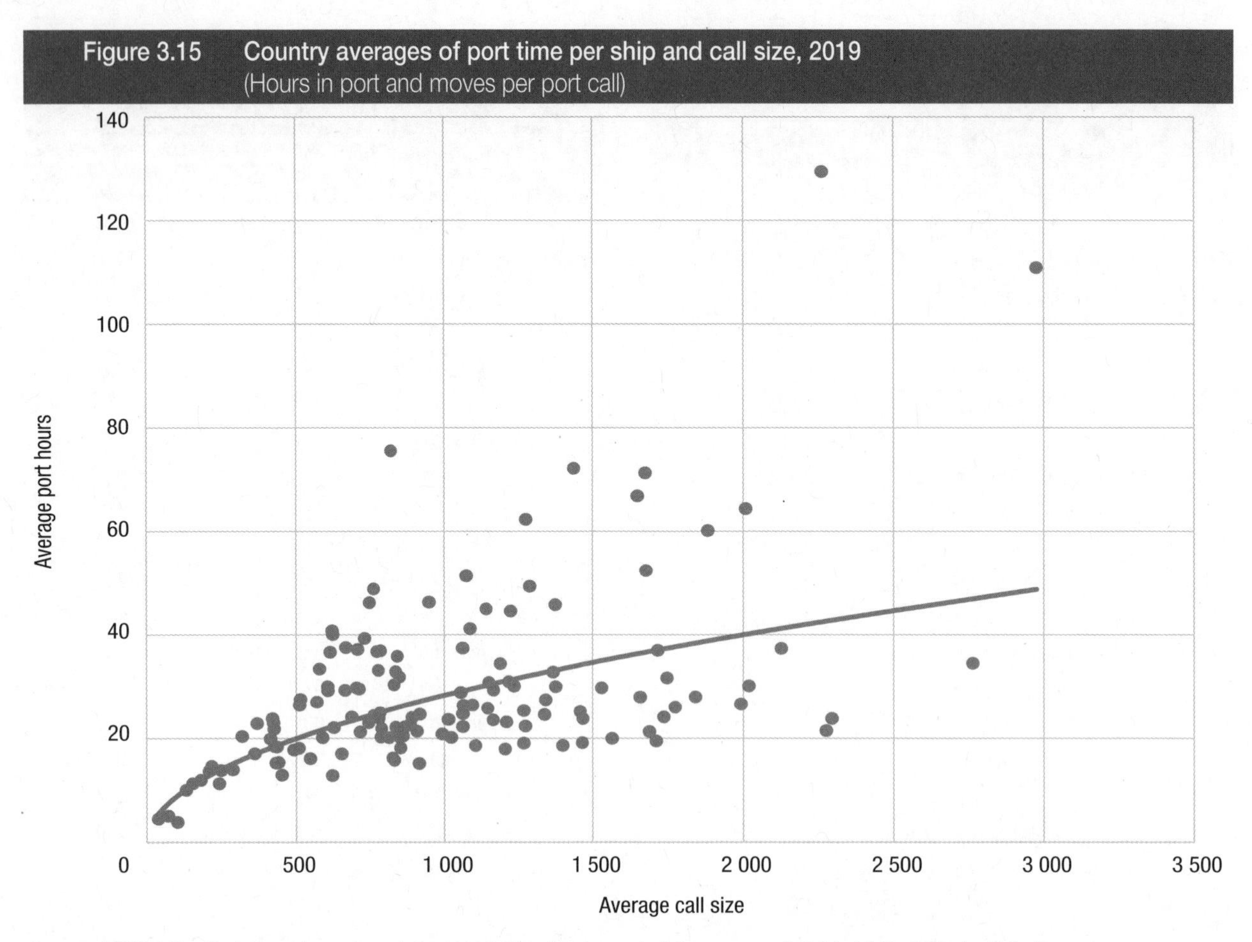

Source: UNCTAD calculations, based on data provided by the Journal of Commerce–IHS Markit Port Productivity Programme.
Note: $R^2 = 0.47$; $y = 0.90\ x^{0.50}$.

Figure 3.16 Minutes in port per container move and average call size, 2019

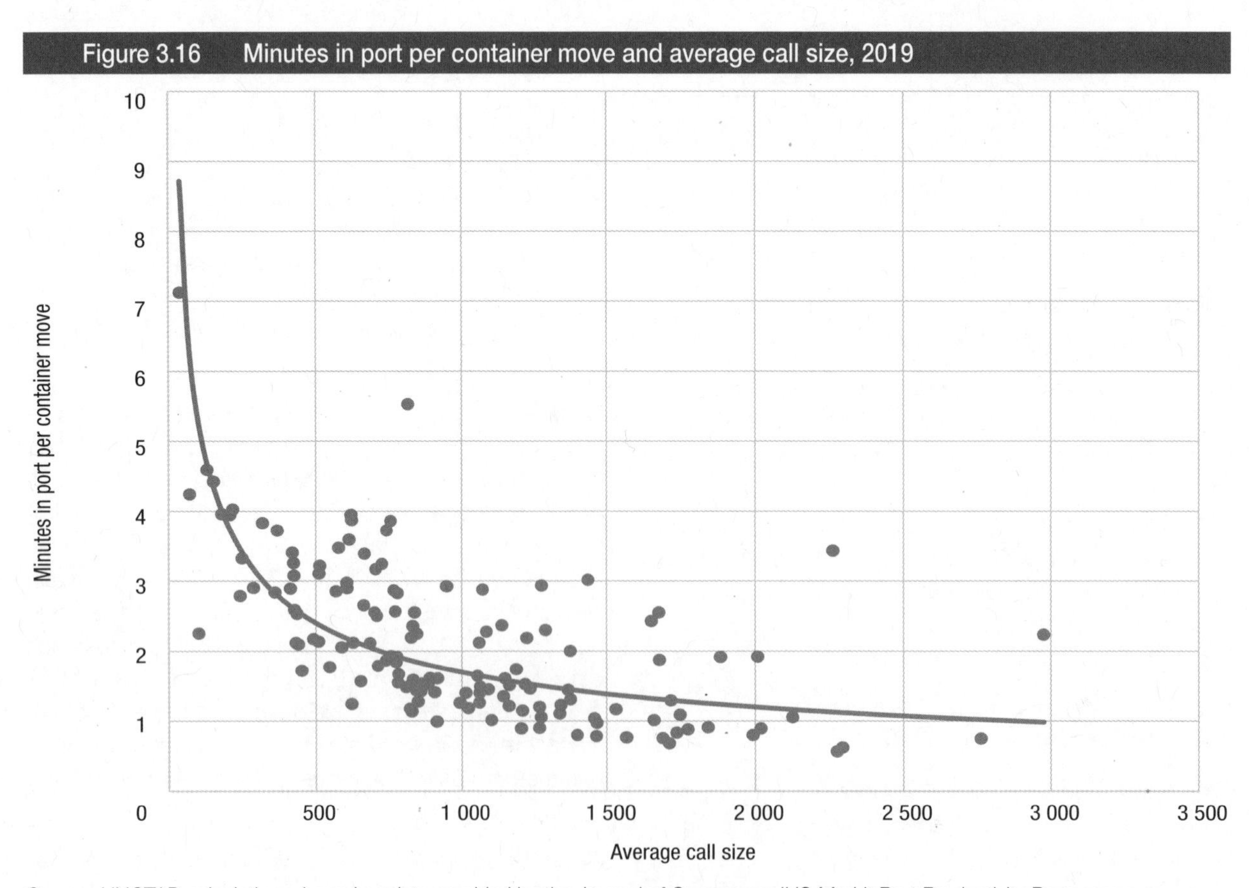

Source: UNCTAD calculations, based on data provided by the Journal of Commerce–IHS Markit Port Productivity Programme.
Note: $R^2 = 0.47$; $y = 53.83\ x^{-0.50}$.

Figure 3.17 Minutes in port per container move and number of port calls per country, 2019

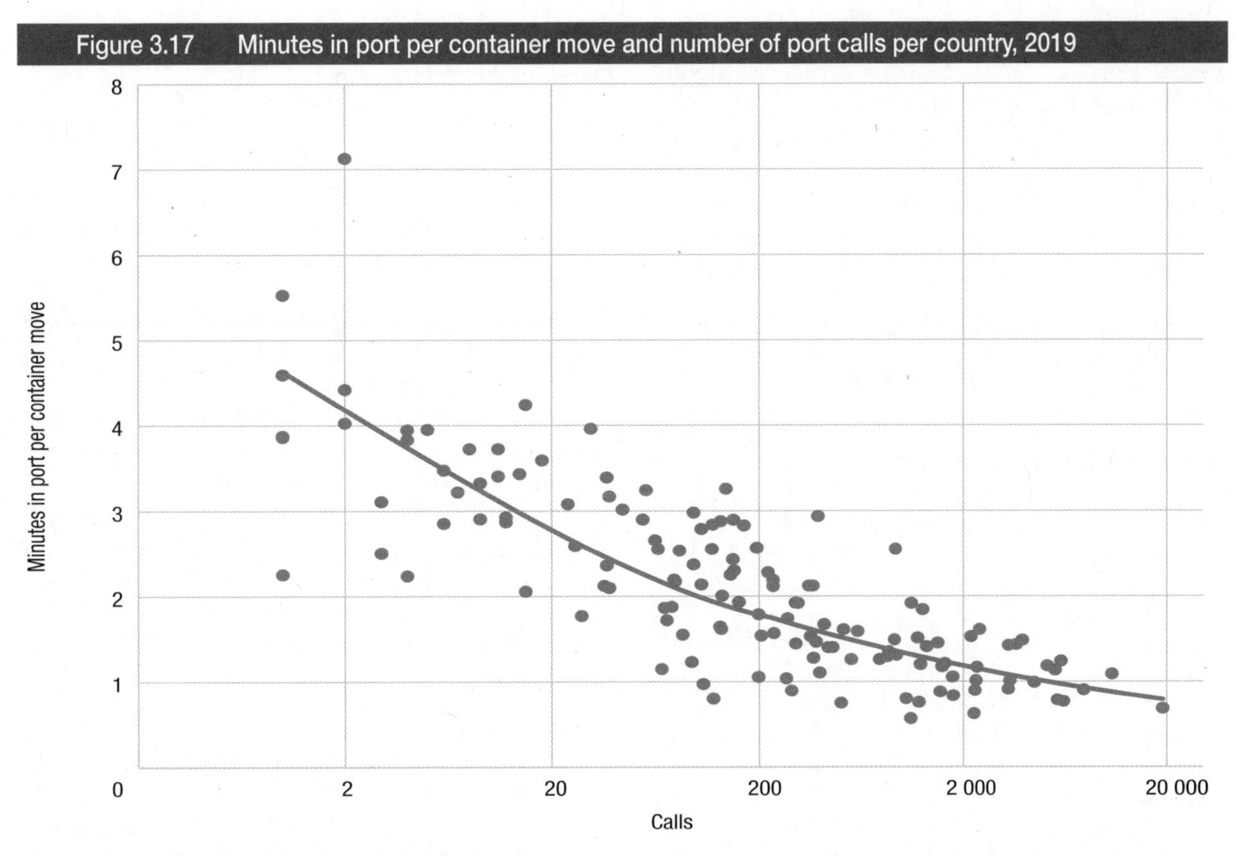

Source: UNCTAD calculations, based on data provided by Journal of Commerce–IHS Markit Port Productivity Programme.

Note: $R^2 = 0.65$; $y = 4.63 x^{-0.18}$.

Figure 3.18 Minutes in port per container move and average vessel size, 2019

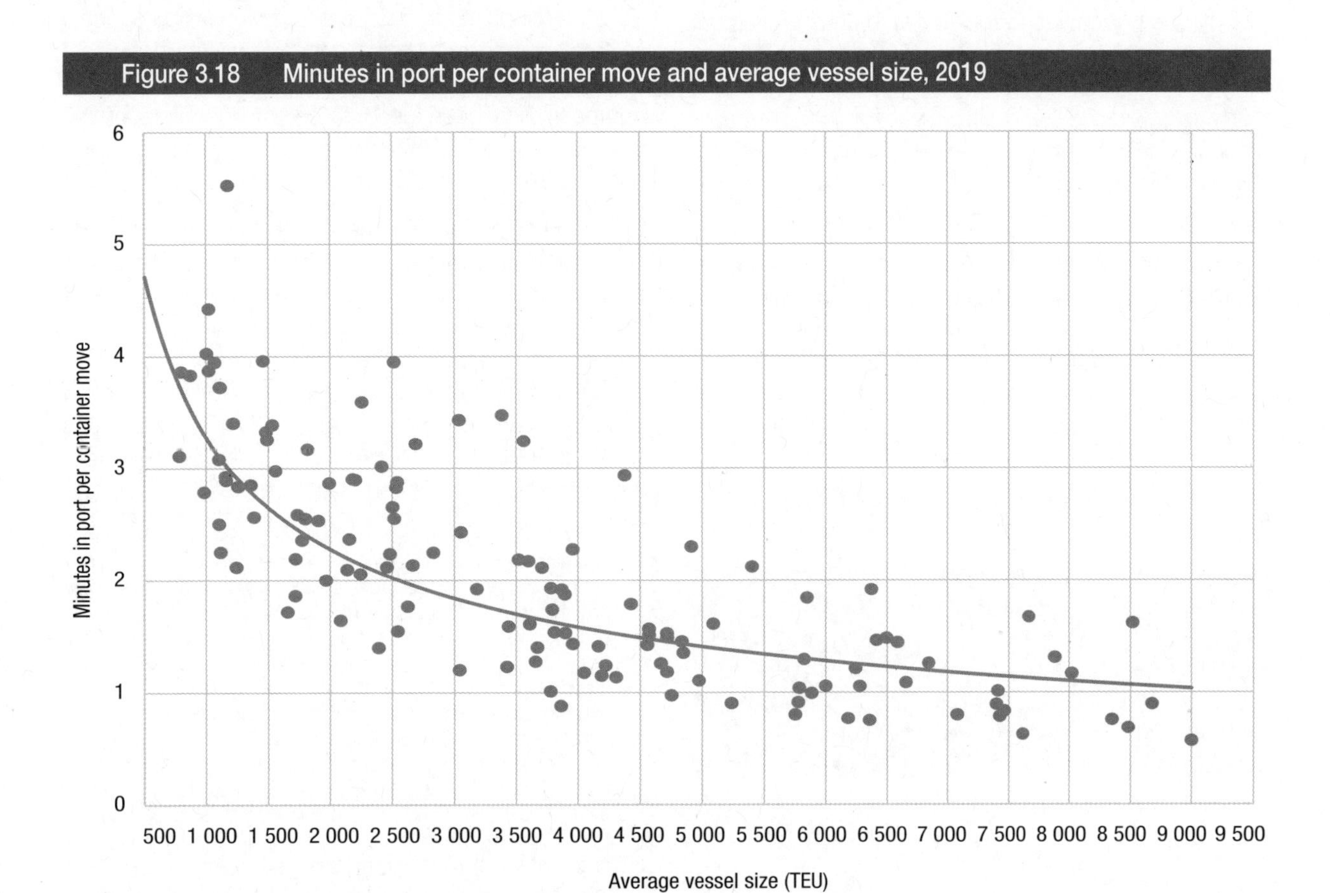

Source: UNCTAD calculations, based on data provided by Journal of Commerce–IHS Markit Port Productivity Programme.

Note: $R^2 = 0.64$; $y = 123.04 x^{-0.52}$.

Abbreviation: TEU, 20-foot equivalent unit.

a large number of loaded and unloaded import and export containers, for which the ships spent an average of more than 100 hours in port. The three countries with the highest average call size below the trend line (that is to say, they are more efficient) are Oman, Poland and the United Arab Emirates, which have a large share of trans-shipment cargo and whose main terminals are operated by international private terminal operators.

To shed further light on port performance and economies of scale, it is worth considering the time spent in port per container loaded and unloaded.

Economies of scale and efficiencies are mutually beneficial. The faster a ship can load and unload containers (the fewer minutes it needs per container in port), the more ships ports can accommodate with a given number of piers and infrastructure (figure 3.17). Increasing the number of calls by 1 per cent is associated with a decrease of the time in port per container by 0.18 per cent.

A similar picture emerges when the time in port is correlated with average ship sizes (figure 3.18). Larger ships will bring more containers and be assigned more resources (cranes, piers on arrival, yard equipment), and they will thus also spend less time in port for each container loaded and unloaded. At the same time, carriers will assign their largest and most expensive ships, preferably to those ports that can handle them in the shortest time. On average, increasing the average vessel size by one per cent is associated with an improvement in the time spent per container by 0.52 per cent. Among the five countries with the largest average vessel sizes, four are below the trend line, meaning they are more efficient. These are China, the Netherlands, Oman and Saudi Arabia. One is above the trend line: Croatia.

The economies of scale illustrated above are in line with the analysis of other data sets discussed in this chapter, in particular those relating to port traffic and fleet deployment (figure 3.12) and to the time spent in port (table 3.3). The importance of economies of scale does not bode well for small island economies (figure 3.6), which have fewer possibilities to attract more cargo, services or larger ships.

The following section will further explore the issue of port performance from the perspective of ports.

D. PORT PERFORMANCE: LESSONS LEARNED FROM THE TRAINFORTRADE PORT MANAGEMENT PROGRAMME OF UNCTAD

1. TrainForTrade port performance scorecard

Within the framework of the port network of the TrainForTrade Port Management Programme, over 3,600 port managers have been trained in the last two decades in 60 countries in Africa, Asia, Europe, Latin America and the Caribbean.[17]

This section reports on the latest developments regarding the port performance component of the TrainForTrade Port Management Programme. The initiative started in 2012 with a series of international conferences held in cities belonging to the TrainForTrade network (Belfast, Northern Ireland; Ciawi, Indonesia; Geneva, Switzerland; Manila, the Philippines; and Valencia, Spain). Thereafter, the port performance scorecard has gone through enhancements and upgrades to respond to four main technical requests from port members. The new pps.unctad.org website now features a more user-friendly interface, incorporated data-consistency checks, an automated past-entry function and advanced analysis tools by regions and categories with automated graphics and filters. The process captures data through annual surveys (starting with the year 2010) sent to focal points in each port entity around April, to report for the previous calendar year.

In 2020, 24 port entities (out of the 50 ports which reported data since the inception of the port performance scorecard) completed the 2019 survey, reporting a total of 2,509 data points with an average of 72 data points for the five-year rolling back average of the global results. The data were collected through a series of questions (82) from which the port performance scorecard derives 26 agreed indicators under the following six categories: finance, human resources, gender, vessel operations, cargo operations, and environment (table 3.8). This approach has been used since the inception of the port performance scorecard to ensure consistency and comparability of measures over time.

With the newest development of the port performance scorecard platform and the digital strengthening of the backbone information technology architecture, UNCTAD expects to increase the participation of port entities beyond the scope of the TrainForTrade network to provide more and more accurate and relevant data and analysis over time. Simultaneously, UNCTAD pursues efforts to include more port entities and countries from the TrainForTrade network that are not yet reporting

[17] See also TrainForTrade Port Management Series (volumes 1 to 7) featuring best case studies and actionable recommendations in line with the Sustainable Development Goals (https://tft.unctad.org/tft_documents/publications/port-management-series). The impact of the programme is measured regularly using two indicators from the TrainForTrade methodology: the performance rate (75 per cent global average) and the satisfaction rate (88 per cent global average) collected over time and for each activity conducted in the TrainForTrade network. Given the long-standing success of the Port Management Programme, which capitalizes on training and capacity-building for port managers and strengthening port institutions equally through the implementation of good governance mechanisms and best practices, it is now time for a deeper analysis of its long-term impact. Based on this assumption and with the support of member ports in the TrainForTrade network, Irish Aid and port partners (France, Ireland, Portugal, Spain and the United Kingdom), steps were taken at the operational level in 2012 to identify the necessary metrics for such an analysis.

Table 3.8 Port performance scorecard indicators, 2015–2019

Category	Indicator number	Description	Mean	Number of values
Finance	1	EBITDA/revenue (operating margin)	38.8%	85
	2	Labour/revenue	22.3%	89
	3	Vessel dues/revenue	15.7%	90
	4	Cargo dues/revenue	34.9%	90
	5	Concession fees/revenue	14.7%	83
	6	Rents/revenue	6.4%	84
Human resources	7	Tons per employee	62 649	94
	8	Revenue per employee	$202 476	88
	9	EBITDA per employee	$104 812	80
	10	Labour cost per employee	$35 760	82
	11	Training cost/wages	1.6%	82
Gender	12	Female participation rate (global)	17.6%	96
	12.1	Female participation rate (management)	38.0%	95
	12.2	Female participation rate (operations)	13.2%	84
	12.3	Female participation rate (cargo handling)	5.5%	60
	12.4	Female participation rate (other employees)	29.4%	27
Vessel operations	13	Average waiting time (hours)	13	83
	14	Average gross tonnage per vessel	18 185	94
	15.1	Average oil tanker arrivals	10.4%	80
	15.2	Average bulk carrier arrivals	10.9%	81
	15.3	Average container ship arrivals	31.8%	79
	15.4	Average cruise ship arrivals	1.4%	78
	15.5	Average general cargo ship arrivals	23.6%	82
	15.6	Average other ship arrivals	24.2%	80
Cargo operations	16	Average tonnage per arrival (all)	7 865	103
	17	Tons per working hour, dry or solid bulk	416	60
	18	Tons per hour, liquid bulk	428	40
	19	Boxes per ship hour at berth	27	44
	20	20-foot equivalent unit dwell time (days)	7	54
	21	Tons per hectare (all)	140 408	91
	22	Tons per berth metre (all)	10 091	102
	23	Total passengers on ferries	1 458 596	57
	24	Total passengers on cruise ships	126 976	61
Environment	25	Investment in environmental projects/total CAPEX	7.2%	35
	26	Environmental expenditures/revenue	2.3%	50

Source: UNCTAD calculations, based on data provided by selected member ports of the TrainForTrade network.

Abbreviations: CAPEX, capital expenditure; EBITDA, earnings before interest, taxes, depreciation and amortization.

in the port performance scorecard component. Major advances in the port performance scorecard tools, enhanced in terms of how the data are validated, as well as comparisons with external data, essentially on gross tonnage and total time in port, add considerable value.

The number of participating ports across the regions has varied over the 10 years of reporting now held in the data set.[18] There are 23–26 ports that report comprehensively every year. This provides a basis for comparative financial and operational benchmarks. These reports can be applied by member ports in a range of planning and performance-based analyses. Table 3.9 provides a summary for the five-year period from 2015 to 2019 of the average port by region and size in each category using the traditional throughput performance measure.

The key elements of the data set are as follows:

- In 2019, port sizes ranged from 1.5 million tons to 80.7 million tons.
- The average port has handled 19.2 million tons per annum since 2015.
- The median value for the same period is 8 million tons.
- Twenty-five per cent of ports averaged less than 3.3 million tons over the 2015–2019 period.

[18] A partnership with MarineTraffic has been established to share data concerning the port entities participating in the port performance scorecard to ensure consistency of data provided by ports.

Table 3.9 Average annual throughput volume, 2015–2019
(Million tons)

Region	Category				Average
	Small <5m	Medium <10m	Large <20m	Very large <20m	
Africa	4.4	8.7	14.2	22.7	11.9
Asia	3.3	7.2		61.5	11.1
Europe	1.5			47.1	41.4
Latin America and the Caribbean	2.2	8.7	14.4	31.9	14.3
Average	**3.0**	**8.5**	**14.3**	**43.4**	**19.2**

Source: UNCTAD calculations, based on data provided by selected member ports of the TrainForTrade network.

2. Financial sustainability

The financial analysis presented on the port performance scorecard platform shows the range of values for ports between 2015 and 2019. Over that period, the average of the annual total revenues of all participating ports was $1.97 billion or 417 million tons. The average revenue per ton varies widely, depending on a port's financial profile, including port dues, port estate, concessions and other services or investment income. Figure 3.19 shows the income categories of interest used in the data (indicators 3–6). The analysis of port revenue by region shows the expected dominance of cargo-related income for port entities, especially when compared with vessel-related income. Thus, ports generate a higher return on working quays for cargo and relatively less on marine assets such as dredged berths and channels.

The ports that show higher values in the concessions category tend to be larger ports with container terminals. Europe has the largest proportion of revenue for this income category.

Figure 3.20 represents the mean values for earnings before interest, taxes, depreciation and amortization as a proportion of revenue (indicator 1), while figure 3.21 shows labour costs as a proportion of revenue (indicator 2). Profit levels, represented here by indicator 1, were reported each year in a consistent range of 36–40 per cent as a global average; it appears reasonable to suggest that this average is a baseline required for a sustainable modern port.

Between 2015 and 2019, the average revenue per port was $88.9 million; 50 per cent of ports brought in less than $49 million in revenue. The ports in quartile 1 (25 per cent of sample) averaged $13.3 million, whereas the large ports in quartile 3 (25 per cent of sample) averaged above $80 million per annum. It is not possible to share the results per individual port, but UNCTAD analysis finds evidence of average rates being closely aligned when similar ports in the same regional group are compared. For example, publicly available data for Irish ports shows this when gross revenue per ton is compared across Ireland. The financial indicators are useful benchmarks by region and by size when forecasting revenue for development projects.

Figure 3.19 Revenue mix of ports by region, 2015–2019

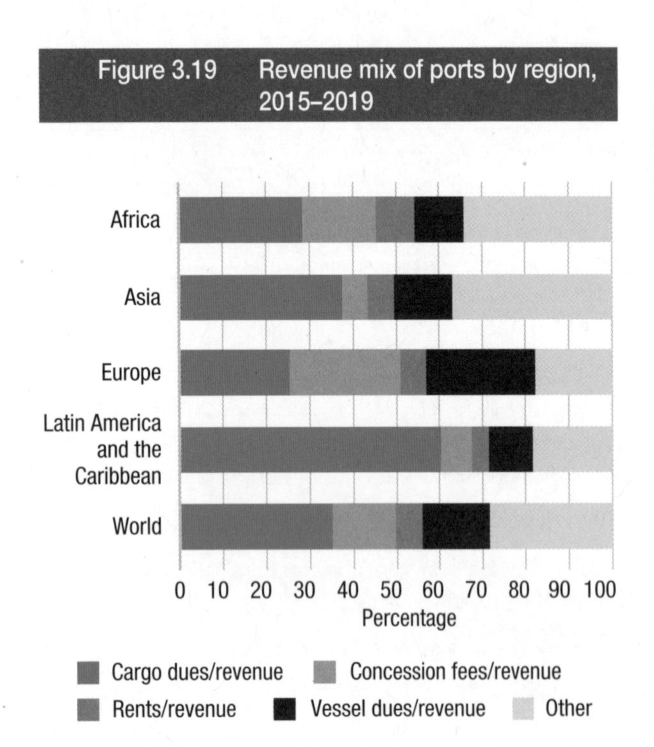

Source: UNCTAD calculations, based on data provided by selected member ports of the TrainForTrade network.

Figure 3.20 Earnings before interest, taxes, depreciation and amortization as a proportion of revenue, 2015–2019
(Percentage)

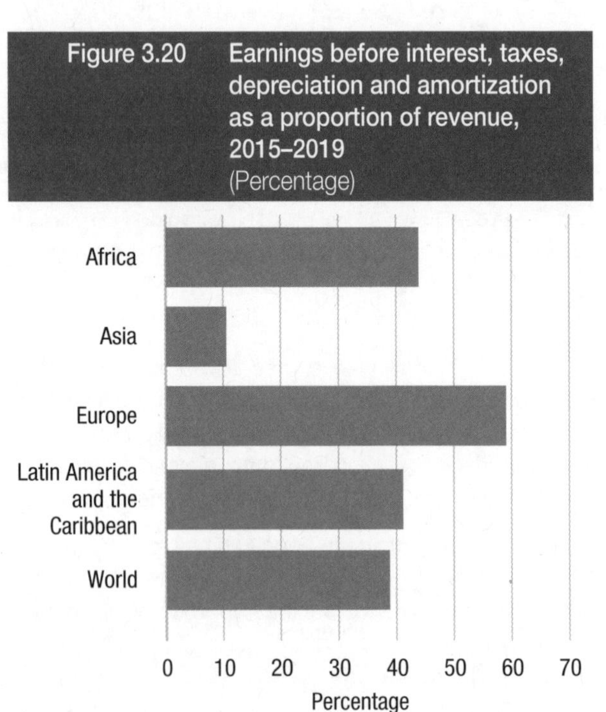

Source: UNCTAD calculations, based on data provided by selected member ports of the TrainForTrade network

Abbreviation: Earnings before interest, taxes, depreciation and amortization.

Figure 3.21 Labour costs as a proportion of revenue, 2015–2019

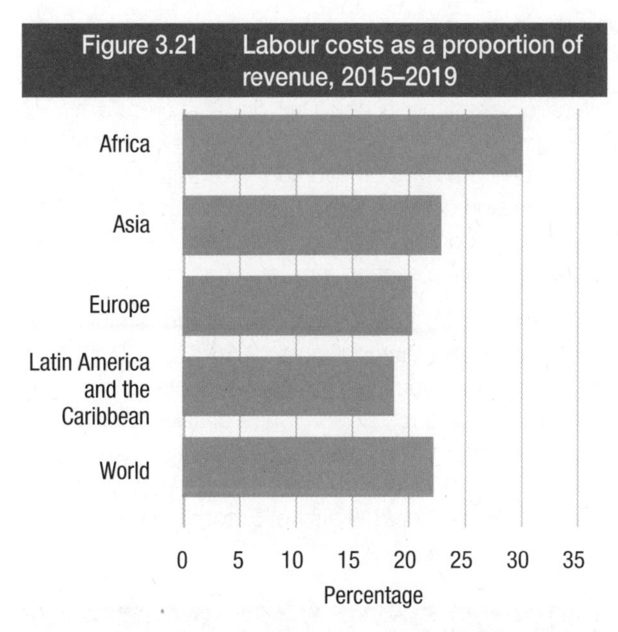

Source: UNCTAD calculations, based on data provided by selected member ports of the TrainForTrade network.

Labour costs have recorded a stable average over the 10 years covered by the port performance scorecard. Values have settled at around 20 to 22 per cent as a proportion of gross revenue (indicator 2). When analysed by region (figure 3.21) and as a proportion of the number of employees, there is a significant range across mean values. For Africa, the value is relatively high and for Latin America and the Caribbean, it is low. It is not clear at this level of data abstraction if this is attributable to rates of pay or employee numbers, which in turn may reflect levels of private supply to port entities as contractors. In the case of Latin America and the Caribbean, the average rate is lower than the global mean, suggesting that ports have relatively high staffing levels (figure 3.22, indicator 10). However, the analysis is less clear with regard to Africa, where labour rates are at the higher end of the spectrum. Europe shows the highest rate per employee – $67,705 per annum.

The average proportion of total capital expenditure on investment in environmental projects (indicator 25) is 7.2 per cent, with 2.3 per cent of operating expenditures reported being devoted to environmental requirements (indicator 26). This is a difficult number to isolate, and therefore the reported benchmarks come with a note of caution. However, throughout the data-collection period, the recorded numbers have been consistent. This suggests a relatively low proportion of total spending, and it will be useful to note any upward trend, should new regulatory requirements be implemented as the effects of climate change increase.

3. Gender participation

The gender profile remains low in terms of female participation in the port workforce (figure 3.23, indicators 12–12.4). The category that is not very far from a gender-balanced distribution is management and administration. However, much remains to be done across the participating ports to achieve greater female participation.

Figure 3.22 Average wages per employee, 2015–2019 (Dollars)

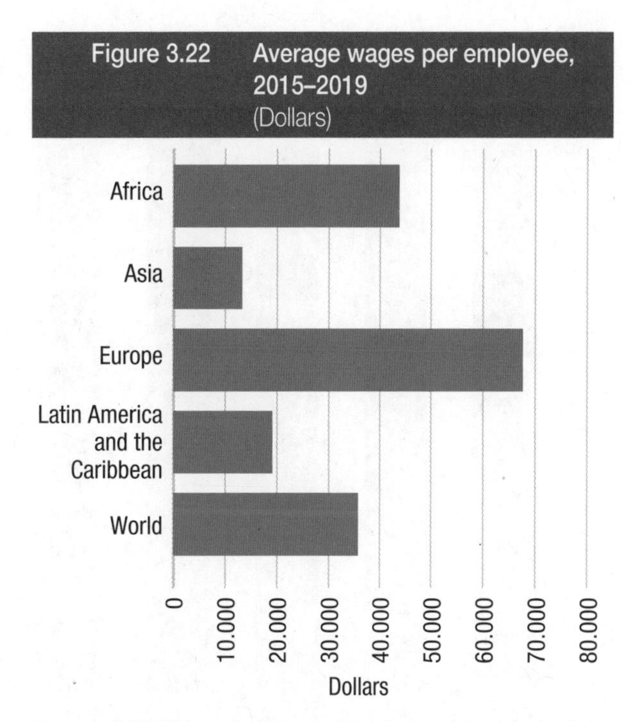

Source: UNCTAD calculations, based on data provided by selected member ports of the TrainForTrade network.

Figure 3.23 Female participation rate in the port workforce, 2015–2019

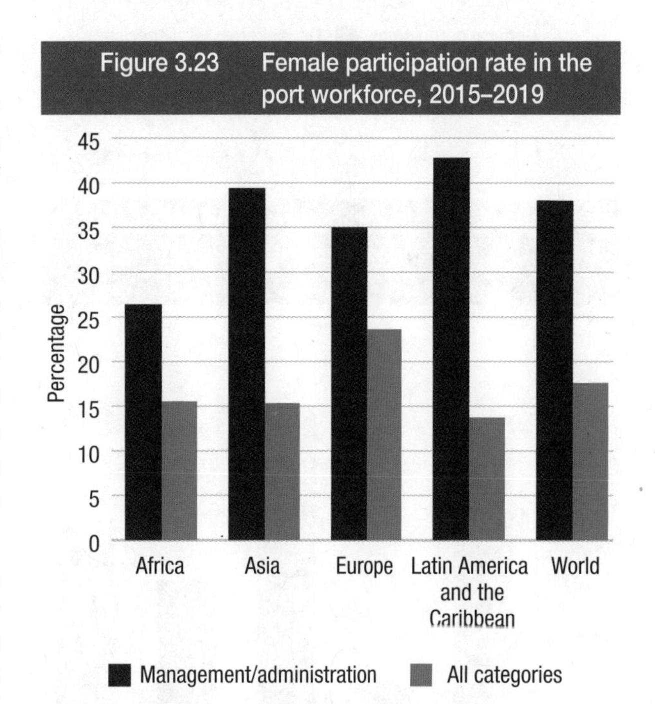

Source: UNCTAD calculations, based on data provided by selected member ports of the TrainForTrade network.

4. Vessel and cargo operations

Figures 3.24 and 3.25 illustrate the profile of participating ports in terms of vessel type (indicators 15.1–15.6) and cargo volumes handled (indicator 16). The graphics show once again that there are no two ports with the same vessel and cargo mix. Both Africa and Europe have

Figure 3.24 Share of vessel arrivals, 2015–2019

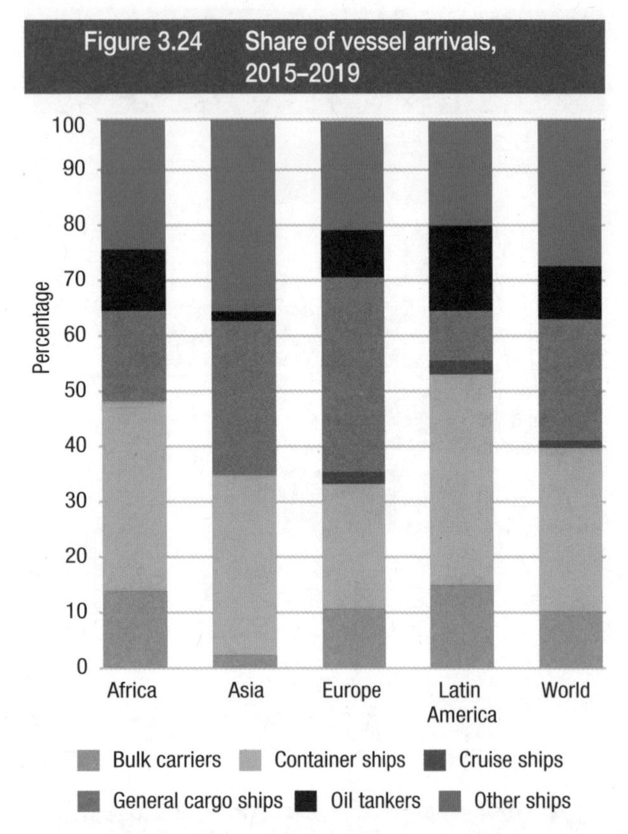

Source: UNCTAD calculations, based on data provided by selected member ports of the TrainForTrade network.

Figure 3.25 Average cargo per arrival or departure, 2015–2019
(Tons)

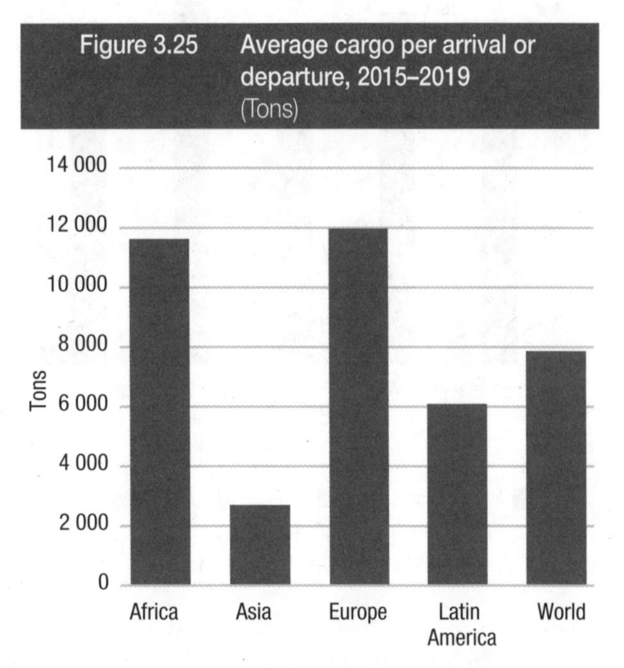

Source: UNCTAD calculations, based on data provided by selected member ports of the TrainForTrade network.

the largest average cargo tons per arrival or departure but arguably for different reasons, given their different vessel mix.

Relating the average time in port to the varied cargo size per vessel can be a useful comparison. There is a tight range of 1.5–2 days in port, on average. Therefore, the larger cargo lots are handled by higher labour and equipment output. With container vessels taking, on average, less time in port (1.2 days), there are higher averages in dry and wet bulk carriers. Dry bulk carriers stay in port 3.5 days on average. Overall, data from the TrainForTrade network show values similar to the global statistics recorded through automatic identification system data (see section A of this chapter).

The online port performance scorecard shows little change in waiting times. Figures 3.26 and 3.27 provide some insights into the efficiency of container-handling operations. There are a wide range of values across the standard performance metrics of dwell time and crane lifting rates, and the overall results are in line with the data presented in section 3.C above. Europe has particularly

Figure 3.26 Maximum 20-foot equivalent unit dwell time, 2015–2019
(Days)

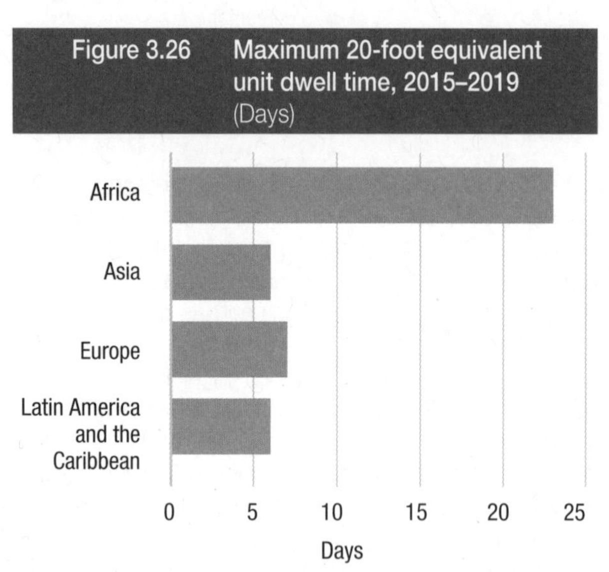

Source: UNCTAD calculations, based on data provided by selected member ports of the TrainForTrade network.

Figure 3.27 Average box-handling rate, 2015–2019
(Boxes per ship-hour)

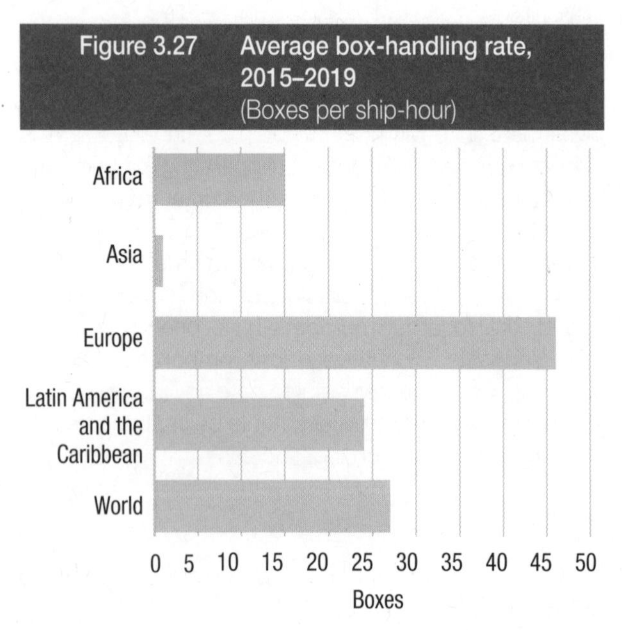

Source: UNCTAD calculations, based on data provided by selected member ports of the TrainForTrade network.

higher lifting rates that perhaps reflect equipment capacity rather than labour efficiency (figure 3.27; indicator 19). Figure 3.26 shows the highest dwell time in days for each region (indicator 20). This topic requires sophisticated analysis to identify the reasons for slow processing, for example, customs procedures, storage agreements, port-container stripping, multiple-user facilities and congestion in road network at or near the port.

E. SHIPPING: EMISSIONS OF THE WORLD FLEET

1. Initiatives to reduce carbon emissions from shipping

Member States of IMO agreed in 2018 "to reduce the total annual greenhouse gas emissions by at least 50 per cent by 2050 compared with 2008" as part of the Initial IMO Strategy on reduction of greenhouse gas emissions from ships (IMO, 2018; UNCTAD, 2020c; UNCTAD, 2020d) (see also chapter 5.B. for additional background information).

To help achieve this objective, the International Chamber of Shipping and other maritime industry associations propose the establishment of a research and development fund to help cut emissions (BIMCO et al., 2019). For heavy fuel oil, this would correspond to a carbon price of $0.63 per ton of carbon dioxide. The project would raise about $5 billion over 10 years. This fund is to be financed by a contribution of $2 per ton of marine fuel oil purchased for consumption. The private sector-led Getting to Zero Coalition suggests that "[S]hipping's decarbonization can be the engine that drives green development across the world" (Global Maritime Forum, 2020).

The falling costs of net zero-carbon energy technologies make the production of sustainable alternative fuels increasingly competitive. Determined collective action in shipping can increase confidence among suppliers of future fuels that the sector is moving in this direction. UNCTAD supports the Getting to Zero Coalition and promotes efforts to achieve sustainability, helping developing countries adapt and build resilience in the light of the climate emergency.

According to Parry et al., 2018, "[T]he environmental case for a maritime carbon tax is increasingly recognized". According to the Environmental Defence Fund (2020), "meeting the IMO's 2050 target represents $50 billion to $70 billion per year for 20 years' spending, but this is also a revenue opportunity". Englert and Losos, 2020 (from the World Bank), also a supporter of the Getting to Zero Coalition, state that a large share of this investment opportunity could lie in developing countries. A sizable part of these investments will have to be made ashore, including in energy infrastructure and in seaports. Shipowners will have to invest in the renewal of the fleet and new technologies (UNCTAD, 2020e).

Engine power limit is a short-term measure proposed by Japan that would enable shipowners to meet requirements relating to the energy efficiency index for existing ships and to reach the IMO target in 2030. Engine power limit decreases vessel speed with minimal changes in ship performance, thus reducing fuel use and emissions based on the cube law (relationship between engine load and vessel speed). In a recent study, the systematic assessment of vehicle emissions model of the International Council on Clean Transportation is used to evaluate different scenarios of engine power limit focusing on container ships, bulk carriers and oil tankers, with 2018 automatic identification system data being utilized as a baseline. The study argues that carbon-dioxide "reductions will not be proportional to engine power limit because ship engines are already operating far below their maximum power" (Rutherford et al., 2020). This model shows the negligible effect of engine power limit of less than 20 per cent on a ship's carbon-dioxide emissions. As for an engine power limit ranging between 30 and 40 per cent, emissions reduction is between 2 and 6 per cent. However, the study shows a significant reduction of carbon-dioxide emissions (by 8–19 per cent) for a larger engine power limit of 50 per cent or more.

2. Emissions by vessel type and other determinants

A wide range of parameters influences the amount of carbon dioxide a ship emits per ton-mile. These include vessel type, speed, size, hull design, ballast, technologies and types of fuel used. A larger ship will naturally emit more carbon dioxide per mile, but thanks to economies of scale, it will emit less carbon dioxide per ton-mile; the smallest container ships of up to 999 TEUs emit about twice as much carbon dioxide per container carried as the largest container ships. Container ships tend to transit at higher speeds than dry bulk carriers, thus – all other things being equal – emitting more carbon dioxide per ton-mile than the latter. Liquefied natural gas and cruise ships are on average far larger than offshore or service vessels, such as tugs, and will thus emit more carbon dioxide per ship than the smaller vessels (see figure 3.28).

The shift toward larger tankers, bulk carriers and container vessels over the past decade, combined with multiple efficiency gains and the scrapping of less efficient vessels, has meant that carbon-dioxide emissions growth has trailed behind the increase in fleet dead weight. This has been most noticeable for container ships, where modest speed reductions have materially lowered fuel consumption and associated emissions. Whereas container fleet capacity rose by 45 per cent between 2011 and 2019, carbon-dioxide emissions are only 2 per cent higher. Over the same period, carbon-dioxide emissions from tankers and bulk carriers increased by 19 per cent and 17 per cent, respectively, well below the 38 per cent and 51 per cent growth in respective fleet capacity (see figures 3.29 and 3.30).

Figure 3.28 Annual carbon-dioxide emissions per vessel by vessel type, 2019

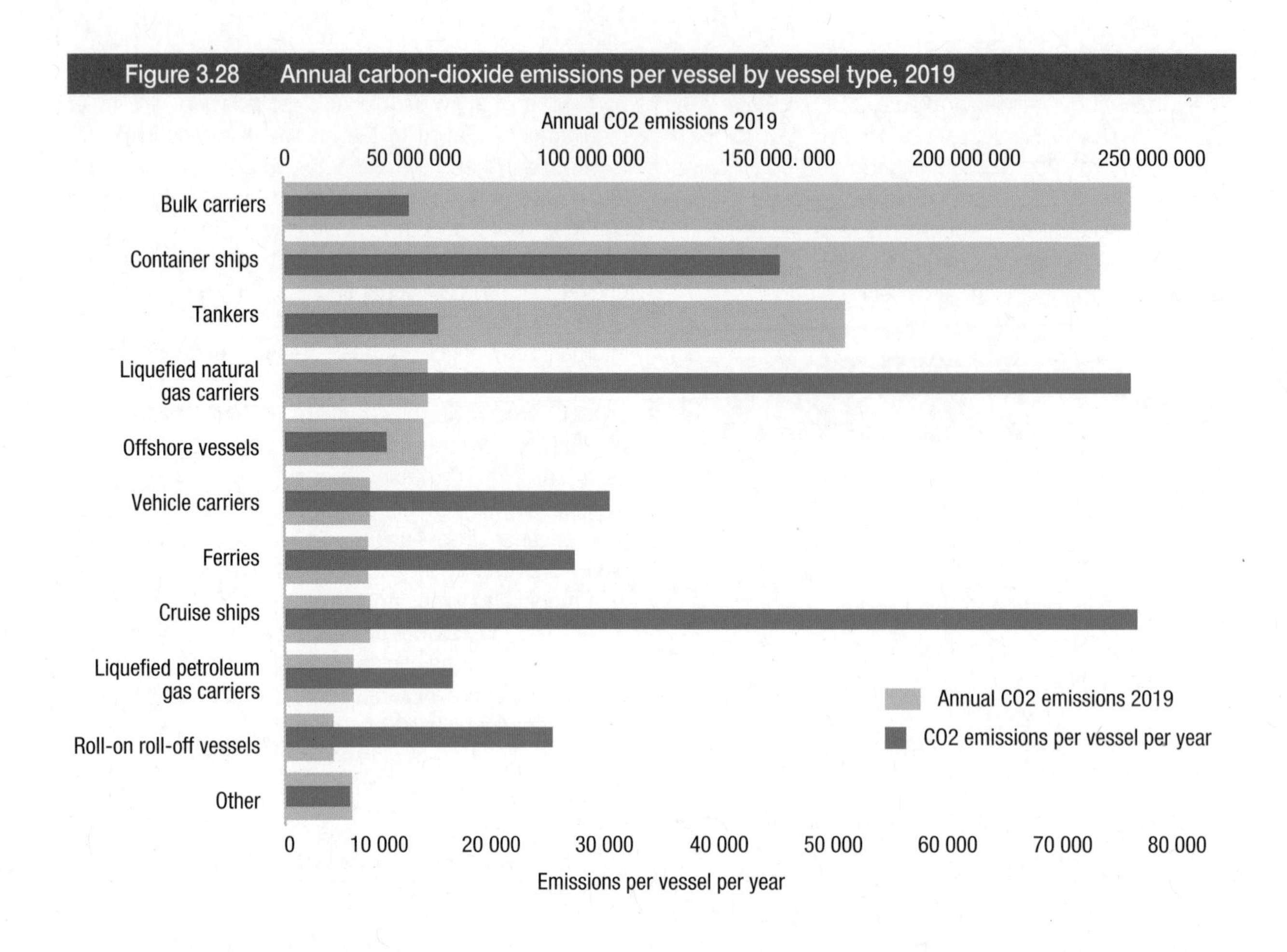

Source: UNCTAD calculations, based on data provided by Marine Benchmark.

Figure 3.29 Comparison of dead-weight tonnage of respective fleet and carbon-dioxide emissions from bulk carriers, container ships and tankers, 2011–2019
(2011 = 100)

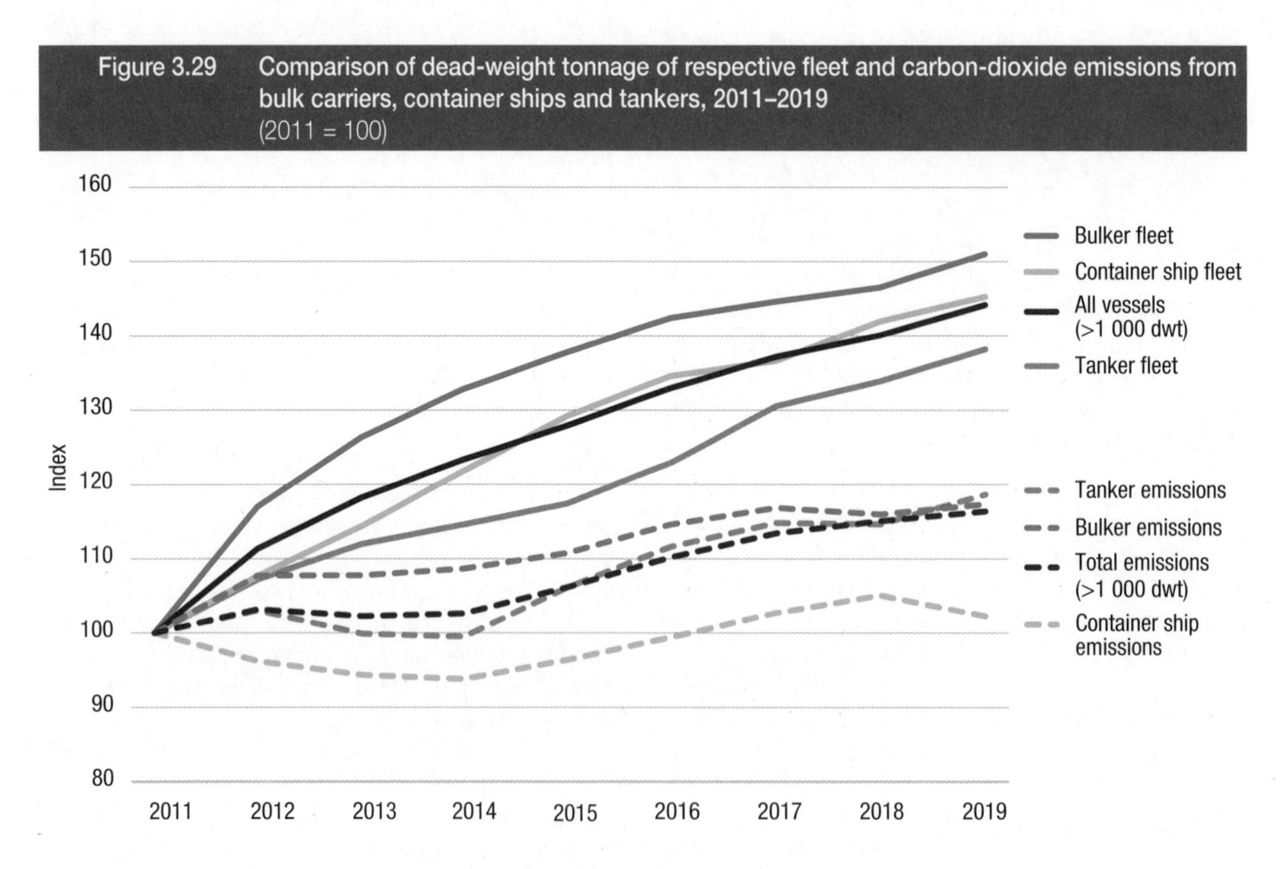

Source: UNCTAD calculations, based on data provided by Marine Benchmark.

Figure 3.30 Annual carbon-dioxide emissions per vessel by vessel type, 2011–2019

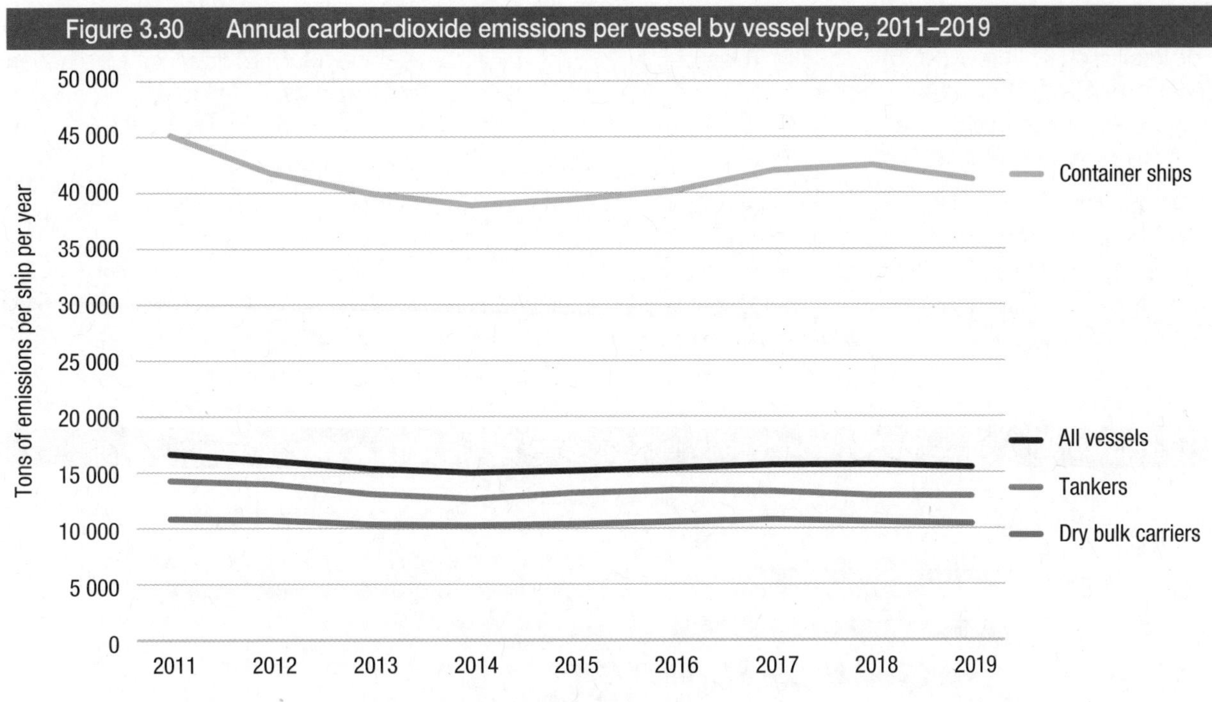

Source: UNCTAD calculations, based on data provided by Marine Benchmark.

Despite larger average vessel sizes, carbon-dioxide emissions per vessel have declined slightly over the past decade. While further gains can reasonably be expected over the next decade, as modern eco-designs continue to replace older, less efficient designs, and with some further increases in average vessel size likely, these will not be enough to meaningfully reduce overall carbon-dioxide emissions in line with the 2050 targets of IMO. Achieving these targets will require radical engine and fuel technology changes.

According to Shell International (2020), more than 90 per cent of interviewees of a survey on the industrial perspectives of shipping decarbonization stated that such a policy was a main priority of their organization. They also considered the economic disruption induced by the COVID-19 pandemic as an opportunity to accelerate the decarbonization progress. Eighty per cent of the persons interviewed stated that the lack of technology alignment (especially alternative fuels) was a major barrier to decarbonization. Hydrogen and ammonia were considered the most promising long-term fuel alternative, despite its present unviability, due to its significantly lower energy density as compared with heavy fuel oil, challenges relating to its storage and the immaturity of fuel cell technology.

Some shipowners are turning towards liquefied natural gas as an alternative to meet IMO targets for 2030, as liquefied natural gas is 20–25 per cent less carbon-intensive than heavy fuel oil. However, other interviewees are more reserved about the long-term perspectives of liquefied natural gas. Owing to methane slip and other challenges arising during extraction and transport, there is no life-cycle greenhouse gas emission benefit to be derived from liquefied natural gas for any engine technology (Pavlenko et al., 2020).

3. Emissions by flag of registration

Flag States have an important role to play in enforcing IMO rules. They exercise regulatory control over the world fleet, applying the law and imposing penalties in case of non-compliance, on diverse issues. These range from ensuring safety of life at sea to protection of the marine environment and the provision of decent working and living conditions for seafarers.

With regard to the implementation of the initial strategy on reduction of greenhouse gas emissions of IMO, flag States will have to ensure that ships are compliant with applicable IMO rules. In addition, they could also provide incentives for the ships registered under their flag to reduce carbon-dioxide emissions and help ensure the collection of future fees or contributions associated with such emissions. For example, the International Chamber of Shipping proposal mentioned above suggests that contributions to the proposed fund will be made commensurate with the ship's annual fuel oil purchased for consumption, as verified by the flag State.

Flag States could also consider such involvement a business opportunity, where more transparent and reliable flag States provide better services than others. In addition, many major flag States are affected by the impacts of climate change. For example, the Panama Canal is confronted with a shortage of fresh water; Liberia has developed a national adaptation plan to mainstream climate change adaptation into planning and budgets; and the Marshall Islands are among the low-lying small island developing States most at risk from sea-level rise (UNCTAD, 2020f). Therefore, it should be in these countries' interest to support the reduction of global greenhouse gas emissions, including from shipping (UNCTAD, 2017b).

Data generated from the automatic identification system tracking system for ships, including the above-mentioned information on vessel characteristics, speed, type of fuel and ballast situation, makes it possible to calculate estimates for carbon-dioxide emissions from each ship and aggregate those estimates. On this basis, ships registered in the Marshall Islands, Liberia and Panama accounted for almost one third (32.96 per cent) of carbon-dioxide emissions from shipping in 2019 (figure 3.31).

Using the same metrics, in 2019, ships (commercial vessels of 1,000 dwt and above) registered in the top 10 economies accounted for 67.15 per cent of total maritime carbon-dioxide emissions. As of 1 January 2020, these 10 flags represented 48.52 per cent of the

Figure 3.31 Annual carbon-dioxide emissions per vessel by flag of registration, 2019

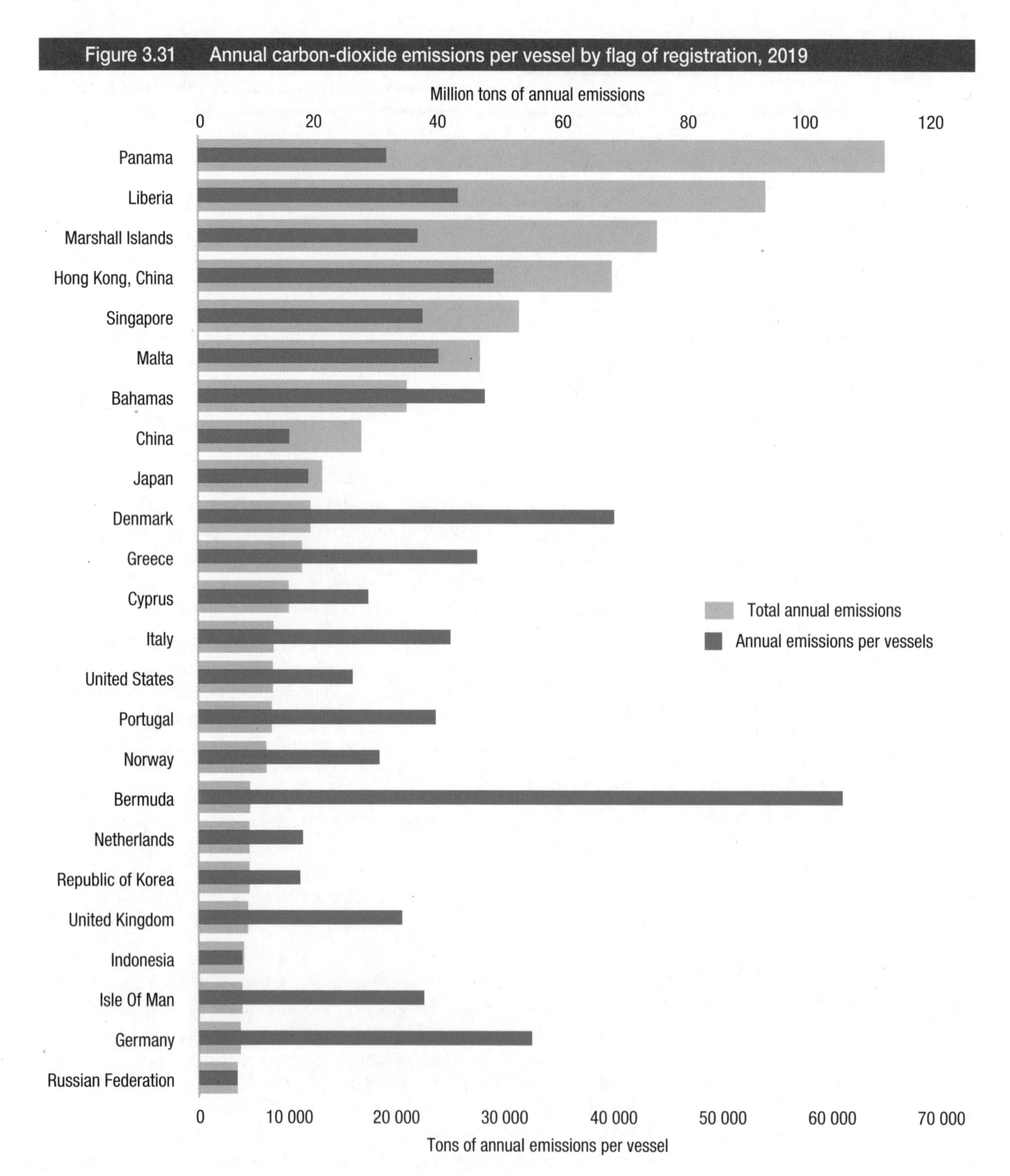

Source: UNCTAD calculations, based on data provided by Marine Benchmark.

world fleet and 65.73 per cent of world gross tonnage. World maritime carbon-dioxideemissions rose by 8 per cent between 2014 and 2019, based on the latest analysis by Marine Benchmark.[19]

F. SUMMARY AND POLICY CONSIDERATIONS

The growing availability of port and shipping data helps the maritime industry to monitor and improve its performance. It also allows analysts to compare and report on differences among ports, countries and fleets, which in turn helps Governments and port and maritime authorities to make adjustments to their activities and policies, if necessary. Based on the performance indicators discussed above, the five points set out below would merit consideration by analysts and policymakers:

- **First, economies of scale are important, but they do not benefit all stakeholders.**

  The different data sets covering port and shipping performance all show that larger ports, with more ship calls and bigger vessels, also report better performance and connectivity indicators. Clearly, economies of scale are still relevant to maritime transport and port performance.

  At the same time, for those ports that aim to attract ever larger ships and call sizes, a note of caution is warranted. The economies of scale presented above reflect averages: they do not cover the total costs of door-to-door logistics. While a shipowner will be satisfied if a ships spends less time in port (sections A to D) and is more fuel efficient (section E), the shippers, ports and intermodal transport providers may well be confronted with diseconomies of scale.

  If the average call size goes up without any corresponding increase in the total cargo throughput, the higher call size will lead to more peak demand for trucks, yard space and intermodal connections, with additional investment needed for dredging and bigger cranes. Those costs will have to be borne by shippers, ports and inland transport providers, while shipowners will reduce the number of ship calls to deliver the same volume of trade. The concentration of traffic in fewer major ports may also imply that shippers could suffer from the choice of fewer ports and costs of trucking extra distances.

  All things being equal, the concentration of cargo in bigger ships and fewer ports with a given cargo volume often implies that there is business for fewer companies in the market. The resulting reduction in competition levels may lead to a situation where not all cost savings made on the seaside will be passed on to the clients in terms of lower freight rates, especially in markets with only few service providers to start with, such as in the case of many small island developing States.

- **Second, small island developing States continue to face challenges in maritime trade.**

  Some small island economies are among those with the longest port ship turnaround times and lowest service frequencies, as they may lack infrastructure or specialized port equipment, and they will not attract more ship calls if there is not much cargo to carry. These States are thus confronted with diseconomies of scale and – at the same time – low levels of competition and limited options in choosing their importers and exporters.

  Often there is little small island developing States can do to improve their liner shipping connectivity, owing to their geographic position, lack of a wider hinterland and low trade volumes. At times, it is possible to attract trans-shipment services, and the resulting additional fleet deployment can then be used for shipments of national importers and exporters. A small number of island economies become hub ports for third countries' trade, and the resulting higher connectivity also benefits those countries' own importers and exporters.

- **Third, emissions reductions will require radical technological changes.**

  Larger vessel sizes, combined with multiple efficiency gains and the scrapping of less efficient vessels, has led to lower growth of carbon-dioxide emissions compared with global fleet tonnage. Container ship fleet capacity, for example, increased by 45 per cent between 2011 and 2019, while carbon-dioxide emissions from container ships went up by only 2 per cent during the same period. Despite the trend towards larger container ships, annual emissions per ship have effectively declined.

  Some further gains can reasonably be expected over the next decade, as modern ecological designs continue to replace older, less efficient designs. However, these marginal improvements will not suffice to meaningfully reduce overall carbon-dioxide emissions in line with IMO targets for 2050. Achieving these targets will require radical engine and fuel technology changes.

  As shown in the Review, thanks to new technologies that help track vessels and identify fuels, combined with reporting requirements of vessel operators, it is possible today to assign carbon-dioxide emissions to vessels and flags of registration. The resulting statistics and insights

[19] Data provided electronically on 2 August 2020 by Marine Benchmark (www.marinebenchmark.com/).

may contribute to discussions on market-based measures to reduce carbon-dioxide emissions.

- **Fourth, nowcasts, forecasts and monitoring pandemics have a growing role to play in the maritime industry.**

  Ship movements, schedules and port traffic data are often available at short notice, before official statistics on economic growth or trade are published. There is an opportunity to make use of maritime data to obtain an early picture of physical trade in goods.

  The trends reported above show that during the first quarter of 2020, the total fleet deployment in most economies was still above that of the first quarter of 2019. For the second quarter, carriers started to significantly reduce capacity. China, for example, started with positive growth in the first quarter of 2020, compared with the first quarter of 2019, but then recorded a negative year-on-year growth in the second quarter. Most European and North American countries saw a steep decline between the first and second quarter.

  Such data is being used and analysed by international organizations and professional forecasters aiming to predict the economic and trade growth of upcoming weeks. Ports and shipping companies will at least to some extent plan their fleet deployment for the same upcoming period, based on such predictions. It is important not to fall into circular reasoning, where pessimistic forecasts may lead to a further withdrawal of shipping capacity, which in turn may lead to further worsening predictions of growth.

- **Fifth, there is a need to standardize maritime data.**

  For ports and shipping companies to benefit from benchmarking, data should be comparable. Ship types, key performance indicators, definitions and parameters need to be standardized. In the long run, the UNCTAD port performance scorecard has the potential to become an industry standard and thus, a globally accepted benchmark, helping the port sector to continuously improve its efficiency. For example, a port entity member of the TrainForTrade Port Management network stated that when it prepares or updates a strategic submission to the Government, port performance scorecard values are useful in drawing up baseline metrics for a proof-of-concept appraisal, in particular when forecasting profit levels, wage profiles, employment numbers and revenue profiles.

  UNCTAD is pursuing efforts to include more port entities and countries from the TrainForTrade network that are not yet reporting in the port performance scorecard component and to collaborate with international partners, such as the International Association of Port Authorities, to further contribute to the standardization of data and tracking of port performance.

REFERENCES

Arslanalp S, Marini M and Tumbarello P (2019). Big data on vessel traffic: Nowcasting trade flows in real time. Working Paper. No. 19/275. International Monetary Fund. Washington, D.C.

BIMCO, Cruise Lines International Association, International Association of Dry Cargo Shipowners, International Chamber of Shipping, Interferry, International Parcel Tankers Association, Intertanko, World Shipping Council (2019). Reduction of greenhouse gas emissions from ships: Proposal to establish an international maritime research and development board (IMRB). MEPC 75/7/4. 18 December.

Cerdeiro DA, Komaromi A, Liu Y and Saeed M (2020). World seaborne trade in real time: A proof of concept for building AIS [automatic identification system] -based nowcasts from scratch. Working Paper. WP/20/57. International Monetary Fund. Washington, D.C.

Englert D and Losos A (2020). Zero-emission shipping: What's in it for developing countries? World Bank Blogs. 24 February.

Environmental Defence Fund (2020). Shipping's green $1trn is a profitable investment, not a cost. 30 January.

Global Maritime Forum (2020). Getting to Zero Coalition. Available at www.globalmaritimeforum.org/getting-to-zero-coalition.

IMO (2018). Initial IMO Strategy on reduction of GHG [greenhouse gas] emissions from ships. MEPC 72/17/Add.1. Annex 11. April.

MDS Transmodal (2020). Port liner shipping connectivity index. Available at www.portlsci.com/index.php.

Parry I, Heine D, Kizzier K and Smith T (2018). Carbon taxation for international maritime fuels: Assessing the options. Working Paper. WP/18/203. International Monetary Fund. Washington, D.C.

Pavlenko N, Comer B, Zhou Y, Clark N and Rutherford D (2020). The climate implications of using LNG [liquefied natural gas] as a marine fuel. Working Paper 2020-02. International Council on Clean Transportation.

Rutherford D, Mao X, Osipova L and Comer B (2020). Limiting engine power to reduce [carbon dioxide] CO_2 emissions from existing ships. Working Paper 2020-10. International Council on Clean Transportation.

Sánchez RJ, Hoffmann J, Micco A, Pizzolitto GV, Sgut M and Wilmsmeier G (2003). Port efficiency and international trade: Port efficiency as a determinant of maritime transport costs. *Maritime Economics and Logistics*. 5(2):199–218.

Shell International (2020). *Decarbonizing Shipping: All Hands on Deck – Industry Perspectives*.

United Nations (2020). International Trade Statistics Database (Comtrade). AIS [automatic identification system]: Trade volume. Trade, transport and travel dashboards. Available at https://public.tableau.com/views/CerdeiroKomaromiLiuandSaeed2020AISdatacollectedbyMarineTraffic/AISTradeDashboard; https://comtrade.un.org/.

UNCTAD (2017a). *Review of Maritime Transport 2017*. (United Nations publication. Sales No. E.17.II.D.10. New York and Geneva).

UNCTAD (2017b). *UNCTAD Framework for Sustainable Freight Transport* (United Nations publication. New York and Geneva).

UNCTAD (2019a). Container ports: The fastest, the busiest, and the best connected. 7 August.

UNCTAD (2019b). Digitalization in maritime transport: Ensuring opportunities for development. Policy Brief No. 75.

UNCTAD (2020a). Navigating through the coronavirus crisis and uncertainty: How maritime transport data can help. Transport and Trade Facilitation Newsletter No. 87.

UNCTAD (2020b). *Digitalizing the Port Call Process*. Transport and Trade Facilitation Series. No. 13 (United Nations publication, Geneva).

UNCTAD (2020c). Decarbonizing shipping: What role for flag States? Transport and Trade Facilitation Newsletter No. 86.

UNCTAD (2020d). Towards the decarbonization of international maritime transport: Findings from a methodology developed by ECLAC [Economic Commission for Latin America and the Caribbean] on shipping [carbon-dioxide] CO2 emissions in Latin America. UNCTAD Transport and Trade Facilitation Newsletter No. 86.

UNCTAD (2020e). Decarbonizing maritime transport: Estimating fleet renewal trends based on ship-scrapping patterns. Transport and Trade Facilitation Newsletter No. 85.

UNCTAD (2020f). *Climate Change Impacts and Adaptation for Coastal Transport Infrastructure: A Compilation of Policies and Practices*. Transport and Trade Facilitation Series No. 12 (United Nations publication. Sales No. E.20.II.D.10. Geneva).

World Bank (2020). COVID-19 trade watch #3 – Signs of recovery? 29 June.

4

THE CORONAVIRUS DISEASE 2019 PANDEMIC: LESSONS LEARNED FROM FIRST-HAND EXPERIENCES

This chapter highlights selected maritime stakeholder experiences with regard to the COVID-19 pandemic, including challenges faced, related response measures and potential lessons learned. Five stakeholders, from across regions and representing a mix of public authorities and maritime transport industry actors, directly involved in operating and managing maritime transport and logistics, were invited to share their respective experiences. While not exhaustive and only intended as illustrative examples, the reflections received generate additional insights into the implications of the pandemic for maritime transport and trade. Key findings are consistent with the data and analysis detailed in the preceding chapters on the impact of the pandemic on maritime trade; the supply of maritime transport infrastructure and services; and the performance of the sector.[20]

[20] The experiences presented in this chapter are based on the inputs received by UNCTAD from five entities. They are illustrative in nature and may not reflect the experiences of a broader set of stakeholders.

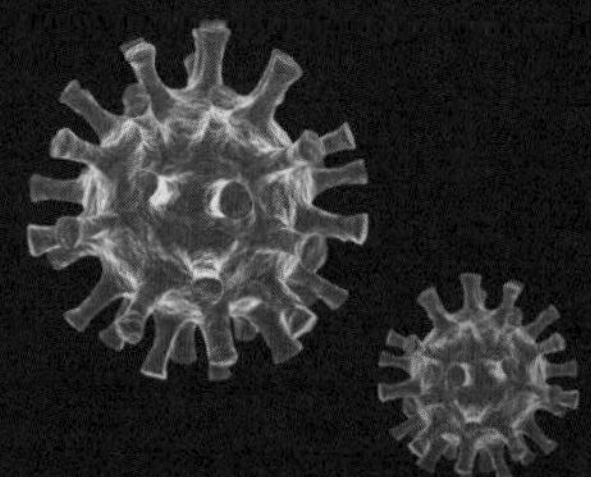

A. INVITED REFLECTIONS ON THE CORONAVIRUS DISEASE 2019 PANDEMIC IN MARITIME TRANSPORT AND HINTERLAND CONNECTIONS

The COVID-19 pandemic is an unprecedented global challenge with significant consequences for all economies, sectors and industries, including maritime transport and logistics. Data and analysis presented in the preceding chapters have underscored the magnitude of the disruption caused by the pandemic. By tracking changes in maritime trade, port traffic, port calls, liner shipping connectivity levels and deployed vessel capacities, the various data sets, including automatic identification system data, have shown the magnitude of the impact of the pandemic on maritime transport and trade. The disruption triggered a sudden slowdown in seaborne trade and increases in blank sailings, delays at ports and closures of ports, as well as reductions in working hours, shortages of equipment, shortages of labour and capacity constraints in truck and other inland transport systems.

An important takeaway from the research and analysis detailed in the preceding chapters is related to the strategic role of maritime transport and logistics in ensuring the continuity and reliability of global supply chains and cross-border trade. Beyond ensuring the smooth delivery of the essential goods and services required to manage crises, the sector is crucial in keeping trade flows moving. Another conclusion from the analysis concerns the need to ensure the integrity, connectivity and smooth functioning of maritime transport for all economies, both developed and developing, in particular small island developing States and the least developed countries. The latter already have disproportionately high transport costs and low levels of shipping connectivity, which makes their trade uncompetitive, volatile, unpredictable and costly. Finally, risk assessment and management and emergency and disaster response planning have emerged as key for business continuity and robustness.

To complement these findings, UNCTAD sought contributions from the field concerning the experiences of some of the main actors involved in maritime supply chains. Building on synergies arising from ongoing collaboration with UNCTAD, selected stakeholders representing a mix of public authorities and maritime transport industry actors were invited to share their experiences with regard to the impact of the pandemic, the measures applied to date and potential lessons learned and good practices. Stakeholders were also invited to share their perspectives regarding the impact on the maritime supply chain and challenges faced and to elaborate on ways in which they have acted to mitigate risks and address challenges generated by the pandemic.[21]

The contributions received provide further clarity on the immediate impact of the disruptions caused by the pandemic on various stakeholders, while taking into account differences in the functions and roles of each stakeholder in the maritime supply chain. They also specify additional efforts that may be required to build the resilience of the maritime transport system and supply chain in the future. While not exhaustive and not meant to be representative of all public authorities and industry actors, the views and experiences shared by the stakeholders provide useful insights into specific occurrences, the related responses and the lessons to be learned, all with a view to any future disruption. The stakeholders shared the perspectives of small island developing States; landlocked, transit and coastal countries, through the lens of a transit and transport corridor; international maritime passage authorities; port authorities; and global shipping companies. These views cover different regions, namely, Africa, Europe, Latin America and the Pacific. The shared experiences are presented in detail in the subsequent sections in this chapter.

The following overview of the various experiences highlights some key aspects that have been crucial to the stakeholders in navigating the crisis. One trend, identified in the analysis in the preceding chapters and reiterated in the experiences shared, is the importance of keeping trade moving during and beyond crisis conditions through well-functioning and resilient maritime supply chains. Despite the difficulties and challenges faced during the pandemic, most shipping companies, ports and other relevant stakeholders have remained operational and put in place immediate measures to facilitate trade and the movement of goods, in particular vital commodities and products. This was highlighted in the experiences shared. Stakeholders reported that

[21] This chapter is informed by contributions from the following stakeholders: Mediterranean Shipping Company; Micronesian Centre for Sustainable Transport; Northern Corridor Transit and Transport Coordination Authority; Panama Canal Authority; Port Authority of Valencia; and Sailing for Sustainability, Fiji.

while experiences varied depending on pre-existing conditions and levels of preparedness, overall, maritime transport and logistics helped to keep essential goods and trade moving. The digitalization of interactions and information-sharing were also emphasized as critical to the continuity of maritime transport operations during the pandemic, and the stakeholders noted that digitalization would be a key component of resilience-building efforts. Finally, the stakeholders stated that awareness was required of the potential changes in trade patterns resulting from the disruptions caused by the pandemic, along with the need to prepare and adapt infrastructure and operations accordingly and to promote the sustainability and resilience of the sector.

Key findings from the experiences shared include the following:

- **The pandemic directly impacted the maritime supply chain and hinterland connections.** Returning to normal will take time and this normality will likely differ from that expected before the pandemic.
- **Responses and adjustments to pandemic-related disruptions spanned various areas**, including operations; financial and economic areas; sanitary and safety protocols and processes; and working practices and organizational aspects.
- **Some of the responses entailed a substantial reorganization of operations**, including prioritization of essential services; reorganization of operations and working conditions due to sanitary and safety protocols; and advancement of digitalization and communications strategies.
- **Sanitary and safety protocols and related measures had to be urgently implemented in a short time.** The capacity to coordinate with local and/or national authorities and communicate with other actors in the maritime supply chain were critical to responses and coping strategies.
- **Work-related and operational adjustment measures that helped the sector adapt were transformational for maritime supply chain stakeholders.** The digitalization of processes and the use of technology by much of the workforce triggered the need to revisit operations and upgrade knowledge and skills.
- **Challenges related to crew changes highlighted the need to orchestrate an integrated approach by all relevant stakeholders.** This was one of the major issues faced in the maritime supply chain. Stakeholders included ministries of health and third parties, for example with regard to public policies that implemented restrictions on travel.
- **Ports managed to avoid significant disruptions to cargo operations.** This was facilitated by the reduced number of port calls by vessels and maritime trade flows.
- **The revision of capacity management plans and the adaption of services were key.** These were significant features of the adjustment measures introduced by shipping lines.
- **Maintaining landside operations was difficult for transit and transport corridors.** Long queues at borders highlighted the importance of reliable chains during a crisis such as the pandemic. Such difficulties affected not only coastal countries but also landlocked and transit countries, which needed to maintain access to seaports. The pandemic exposed potential limitations in trade facilitation measures applied in the context of cross-border transport by land.
- **Business continuity plans emerged as key to acting swiftly.** Such plans are important and likely to be further developed and revised, to integrate lessons learned and help better prepare for any future disruption from events such as pandemics or those due to climate change-related factors.
- **Responding to pandemic-related challenges required collaboration and coordination, as well exchanges of information among all stakeholders.** Wherever they had been established, collective actions were more effective in combating risks and improving decision-making and resilience. Adjustments to the governance and communications strategies of the parties involved, as well as exchanges of information and the sharing of experiences, were important.
- **Furthering systemic, coordinated responses and building the capacities of staff were important.** In future, for example, there is a need for transboundary disaster management

strategies that are well coordinated including, for transit and transport corridors, a harmonized disaster response mechanism. Coordination and collaboration could also focus on sharing intelligence from early-warning systems, conducting capacity-building for personnel involved in the transport logistics chain and embedding disaster responses into national and regional policies that affect trade, transport and other related infrastructure.

- **The pandemic may have had a less obvious impact on small island developing States in the Pacific.** However, the impact may be longer lasting and more critical, in particular as multiple crises or shocks could occur at the same time. The decision to divert a single ship from some countries, the absence of vessels calling at certain ports or even the availability a single operator, due to reductions in the cargo available at a destination at a key export market, has tested the ability of maritime transport to deliver essential goods. There has also been an increase in shipping costs for small island developing States. Such States need to develop risk mitigation capabilities and resilience-building, including through green shipping solutions, at the national, regional and international levels.
- **Small island developing States remain a vulnerable country grouping.** They often experience a combination of disruptive factors and shocks. For example, in April 2020, small island developing States in the Pacific region also experienced the impact of a tropical cyclone. Losses and damages were significant and the pandemic made the delivery of emergency support and relief more challenging. In this context, climate change mitigation and adaptation remain important priorities and efforts to address the challenge, including under the auspices of IMO, should be further enhanced.

B. EXPERIENCE OF SMALL ISLAND DEVELOPING STATES: SMALL ISLAND DEVELOPING STATES IN THE PACIFIC

The coronavirus disease and Cyclone Harold: Lockdown in the Pacific

In 2019, Samoa experienced a measles epidemic and when news of COVID-19 first emerged, small island developing States in the Pacific were therefore cautious and some restricted travel as of January 2020, following which a period of lockdown was instituted. As at June 2020, of the 15 small island developing States in the Pacific, only Fiji and Papua New Guinea had recorded cases of COVID-19 (see https://www.ncbi.nlm.nih.gov/pmc/articles/PMC7348597/). The remoteness of the small island developing States in the Pacific has been beneficial in this instance, as increased case numbers would have put a strain on the limited health-care systems and possibly been further increased by poor sanitation levels and the often overcrowded urban areas.

In April 2020, severe tropical Cyclone Harold struck Fiji, Solomon Islands, Tonga and Vanuatu, causing significant loss of life and damage to crops and buildings. In one instance, 27 lives were lost from an overcrowded interisland vessel leaving Honiara due to the pandemic (see https://apnews.com/article/f15a56f7b85f79c9f22fc28d055c78ec). Cyclone Harold caused damage worth millions of dollars to port infrastructure and jetties, and pandemic-related restrictions put additional pressure on responses, with relief goods and teams from abroad having to comply with quarantine requirements; multiple states of emergency impacting international responses; and pandemic-related restrictions on interisland shipping limiting the reshipment of emergency relief to remoter islands and communities.

The coronavirus disease: Impact on shipping, food, fishing and tourism

In the period January–April 2020, the impact on shipping was mixed. Most small island developing States in the Pacific did not have processes or policies in place to deal with a global pandemic. Some countries instituted a total ban on the arrival of ships or certain types of ships, in particular cruise liners. Other countries imposed varying periods of quarantine and still others allowed access to ships only if they had not come from specific countries or ports and had been at sea for varying periods, of 5 to 28 days. This resulted in blank sailings, reductions in cargo throughput, ships being diverted from some countries and trans-shipment mainly through Fiji. The World Food Programme activated a COVID-19 pandemic response team to collect data on the impact on shipping and to share information with stakeholders among small island developing States in the Pacific, in line with the recommendations of the International Chamber of Shipping, ILO, IMO and others. The Global Logistics Cluster, for which the World Food Programme is the lead agency, provides a weekly update on the international shipping situation in the Pacific, identifying national quarantine requirements, ship schedules and sources of information for advice.[22] Government plans and systems for dealing with the pandemic were put in place and teams of government officials were trained and briefed. Shipping issues were no longer predominantly due to quarantine restrictions, but due to the significant reductions in demand as the international tourism industry slowed down and to the lack of goods in key resupply hubs such as Hawaii.

When the first case of COVID-19 was recorded in Lautoka, Fiji, the international port was closed, as of March 2020, and all ships were diverted to Suva or elsewhere. There has been a significant drop in throughput at international ports in small island developing States in the Pacific and there have been a few instances of food shortages. For example, in June, certain islands in Kiribati began to experience shortages of foodstuffs, as ships had not called there since March 2020 (see https://logcluster.org/document/pacific-shipping-operations-update-20-may-2020). As demand for cargo declined, the shipping industry applied measures such as, among others, blank sailings, reduced frequencies of services and alterations to scheduled routes. For example, sailings of the Pacific Direct Line New Zealand feeder service to Fiji were reduced from four to three per month, Mariana Express Lines removed Bairiki Tarawa from its Majuro South Pacific Service schedule until end-2020 and there are blank sailings across the region (see https://logcluster.org/document/pacific-shipping-operations-update-25-june-2020). Domestic interisland shipping was initially confined to ports, with interisland travel not allowed; later, commercial interisland vessels began to be allowed to operate, in stages, beginning with cargo only, then with limited numbers of passengers. The Government of the Cook Islands subsidized interisland shipping to the northern islands to ensure that essential cargo was delivered. Other States, such as the Marshall Islands, did not experience an impact on interisland shipping. As at September 2020, ships from Samoa to Tokelau – which does not have an airport and can only be reached by ship – still do not permit passengers.

The economic impact of the pandemic has also included high levels of unemployment in tourism-dependent economies such as Fiji, Vanuatu and the Cook Islands. Governments initiated plans for dealing with the pandemic and many citizens left urban centres and returned to villages to farm. Import and export volumes dropped, but community resilience was seen in self-sufficiency with regard to food and in the increased use of barter systems that helped to reduce the demand for imported goods. Some countries have experienced shortages in fresh food, while others have surpluses due to a drop in demand from the tourism industry and increased use of local gardens.[23] There are thus opportunities for regional trade between States that are free of COVID-19, which are not fully being explored, in part due to the lack of appropriate shipping services.

There has also been a major impact on seafarers, with crew from small island developing States in the Pacific serving on international ships, in particular cruise liners, being stranded abroad. The fishing fleet has been much less affected, with foreign flagged vessels continuing to fish in the region and calling at ports for trans-shipment and resupply, although restrictions are beginning to affect the sector. For example, Samoa restricts the docking of fishing vessels to two per day and crew are not allowed to disembark, while the date since the last port and crew change must not be less than 28 days previously, and compliance with other requirements related to quarantine and notification are also in place. Various surcharges and increases to shipping costs have been put in place by carriers, which have increased the costs of international shipping to the customer, despite significantly lower fuel prices (see table 4.1).

As at September 2020, quarantine restrictions were beginning to be relaxed. For example, in the Marshall Islands, crew with no record of disembarkation and vessels that regularly serviced small island developing States in the Pacific were exempt from the 14-day quarantine period. However, crew changes were still not permitted. Schedules were being altered to reduce quarantine periods in ports, in particular for shorter voyages, for example between Papua New Guinea and Solomon Islands.

[22] See https://logcluster.org/search?f%5B0%5D=field_raw_op_id%3A33587&f%5B1%5D=field_document_type%3A156&f%5B2%5D=field_logistical_category%3A16.

[23] See https://www.fijitimes.com/lautoka-market-sales-plummet/ and https://pacificfarmers.com/resource/pacific-farmers-have-their-say-survey-report/.

Table 4.1 Examples of surcharges and shipping costs

Shipping line	Additional charges	Application
Neptune Pacific Line	$349/TEU, $25/revenue (break bulk)	Temporary quarantine surcharge for Pacific ports
	$100/TEU	Freight cost increase on shipments from Australia and New Zealand to Fiji, from 3 and 5 July 2020, respectively
Pacific Direct Line	$100/TEU	Rate restoration charge on shipments from Asia to Pacific island ports, from 15 July 2020
China Navigation Company	$150 (20-foot full container load), $300 (40-foot full container load) and $8.50/ revenue ton (break bulk)	Rate restoration charge on shipments to Fiji
	$163–285 (20-foot full container load), $326–570 (40-foot full container load) and $10–16.75/m3 (break bulk)	Quarantine surcharge applied to vessels calling at Honiara

Source: Global Logistics Cluster data.

Conclusion and way forward

The pandemic may have had a less obvious impact on small island developing States in the Pacific. However, the impact may be longer lasting and more critical. The pandemic was a new setback for small island developing States in the Pacific already experiencing climate change-related and extreme weather events, such as severe tropical Cyclone Harold. Building the resilience of small island developing States, including with regard to maritime transport chains, in preparing for, responding to and recovering from significant multi-hazard threats such as pandemics and climate change-related events is therefore critical.

As small island developing States in the Pacific are among the most vulnerable with regard to the impact of climate change, achieving reductions in emissions from international shipping, in line with efforts to limit the global temperature increase to 1.5°C above pre-industrial levels, is essential to the survival of these States in the next few decades and they cannot afford any delay. In 2018, IMO adopted an initial strategy on the reduction of total annual greenhouse gas emissions by at least 50 per cent by 2050 compared with 2008 while, at the same time, pursuing efforts towards phasing them out entirely (see http://www.imo.org/en/MediaCentre/PressBriefings/Pages/06GHGinitialstrategy.aspx). The current delay in the adoption of the short-term reduction measures in the strategy will in turn defer debates on the medium-term measures, such as market-based measures and, in particular, a carbon tax, which are key if international shipping is to deliver on the vision of decarbonization as soon as possible. There may also be a delay in the review of the initial targets, agreed prior to the release of the special report by the United Nations Intergovernmental Panel on Climate Change in 2018. More recent data demonstrates that significantly greater emissions reduction levels are required in all sectors if limiting the global temperature increase to 1.5°C is to remain a viable option (see Bullock et al., 2020). The fourth IMO greenhouse gas study was submitted to the Marine Environment Protection Committee in July 2020 and shows that emissions from shipping increased by 9.6 per cent in 2012–2018, with methane emissions increasing by 151 per cent (IMO, 2020). Shipping is not yet on the pathway needed to achieve limiting the global temperature increase to 1.5°C; in fact, the trend is in the opposite direction, with a projected 50 per cent increase in emissions by 2050 (see https://www.cedelft.eu/en/publications/2488/fourth-imo-greenhouse-gas-study). As economic recovery and stimulus packages are being put in place worldwide, the inclusion measures related to the decarbonization of shipping is essential if shipping is to meet the emissions reductions targets in the initial strategy.

The global investment opportunity and initiatives in greener shipping, both nationally and internationally, are available now, and the small island developing States in the Pacific cannot afford to be left behind. The pandemic has demonstrated their resilience, but also their dependence on shipping. In this regard, for example, the Pacific Blue Shipping Partnership is a country-driven initiative for large-scale blended finance investments, to catalyse a multi-country transition to sustainable, resilient and low-carbon shipping, including in appropriate low-carbon domestic and interregional shipping driven by small island developing States in the Pacific (see: www.mcst-rmiusp.org/index.php/projects/current-projects/pacific-blue-shipping-partnership).

C. EXPERIENCE OF AN AUTHORITY COORDINATING A TRANSIT AND TRANSPORT CORRIDOR: NORTHERN CORRIDOR TRANSIT AND TRANSPORT COORDINATION AUTHORITY, EAST AFRICA

Importance of maritime transport for regional and international trade

The importance of maritime transport and the role of international shipping cannot be underestimated in current global economic and market conditions, with transport by sea becoming ever more prominent. Maritime shipping connects suppliers and producers, buyers and sellers. It is therefore one of the most important transport activities for the northern corridor and the continent of Africa as a whole. Current issues related to the status of regional maritime shipping should therefore be discussed in relation to the rest of the world, and factors crucial in sustaining the industry should be analysed.

Port of Mombasa: Gateway to regional trade

Ports serve as important transportation hubs that facilitate the movement of goods to regional markets, businesses and, in particular, landlocked countries. The port of Mombasa, for example, connects goods to consumers through the northern corridor, which includes road networks, railways, inland waterways and pipelines. The port is a gateway to East Africa and Central Africa and is one of the busiest and largest ports in East Africa. It provides direct connectivity to over 80 ports worldwide and is linked to Burundi, the Democratic Republic of the Congo, Ethiopia, Rwanda, Somalia, South Sudan, Uganda and the United Republic of Tanzania. The port comprises Kilindini Harbour, Port Reitz, Old Port, Port Tudor and the whole of the tidal waters encircling Mombasa Island and has a capacity of 2.65 million TEUs (Kenya Port Authority Strategic Plan 2018-2022). Kilindini Harbour is a natural deep-water inlet with a depth of 45–55 meters at its deepest; the controlling depth is the outer channel, with a dredged depth of 17.5 meters.

The coronavirus disease: Impact on port and northern corridor performance

The COVID-19 pandemic has had profound effects on transport and the entire logistics sector. The pandemic, a situation that was sudden and unanticipated, exposed the vulnerability of trade facilitation in the northern corridor region. Key challenges in facilitating cross-border trade included a lack of preparation and a lack of transboundary disaster management strategies. The abrupt nature of the pandemic coupled with the absence of tailored strategies affected, and to some extent continue to affect, the performance of the port of Mombasa and the northern corridor.

Declines were noted during the pandemic with regard to performance indicators for the northern corridor, with border crossing time affected the most. By May 2020, queues of trucks awaiting clearance at common border crossing points were reported to have stretched to over 50 kilometres (see http://www.ttcanc.org/documents.php). Congestion was also experienced at various crossing points due to some of the measures put in place for testing drivers for the virus. For example, transit time between two crossing points at a distance of 948 kilometres increased from an average of 3 days to 8 days. Such disruptions led to delays, in particular in the return of empty containers to the port of Mombasa, and the delays often led to retention charges set by shipping lines, posing a burden on the cost of doing business.

A number of measures have been put in place at the port of Mombasa to help curb the spread of the virus, including, among others, fumigation of key equipment, operational areas, offices and workshops; temperature checks of all individuals accessing the port; and sanitization and hand washing at gates and entrances to all buildings. The port health authority ensures that all necessary protocols are observed by ships scheduled to call at the Port. Such measures cannot be implemented without affecting normal port operations. The new interventions, coupled with the blank sailings and vessel cancellations, explain in part the changes with regard to performance indicators for both the port and the northern corridor.

The directives executed by various Governments to allow people to remain at home or telecommute also affected performances at both the port and along the northern corridor by disrupting working systems. Many adjustments had to be made and, as coping mechanisms were instituted to mitigate negative impacts, improvements began to be made in delivering services at both the port and along the northern corridor. The impact of the pandemic on transport and trade patterns along the northern corridor was apparent, as fewer cargo trucks were in operation. In addition, there were shortages of staff to operate equipment at the Port, which caused delays in the transfer of cargo. This may help explain variations in ship turnaround time and other performance indicators, as noted by the Northern Corridor Transport Observatory (see http://www.kandalakaskazini.or.ke). Some positive trends during the pandemic have been noted with regard to indicators such as vessel waiting time before berth, ship turnaround time and port dwell time. This may be attributed to decreased volumes and the reduced number of vessels calling at the Port, compared with in 2019. Time taken to pick up cargo after release from customs has increased in 2020, compared with in 2019, mainly due to the length of time taken by trucks to return from their destinations due to pandemic-related measures. The increase in transit time in January–May 2020 with regard to various destinations may also be attributed to such measures, implemented by various member States of the Northern Corridor Transit and Transport Coordination Authority (see figure 4.1).

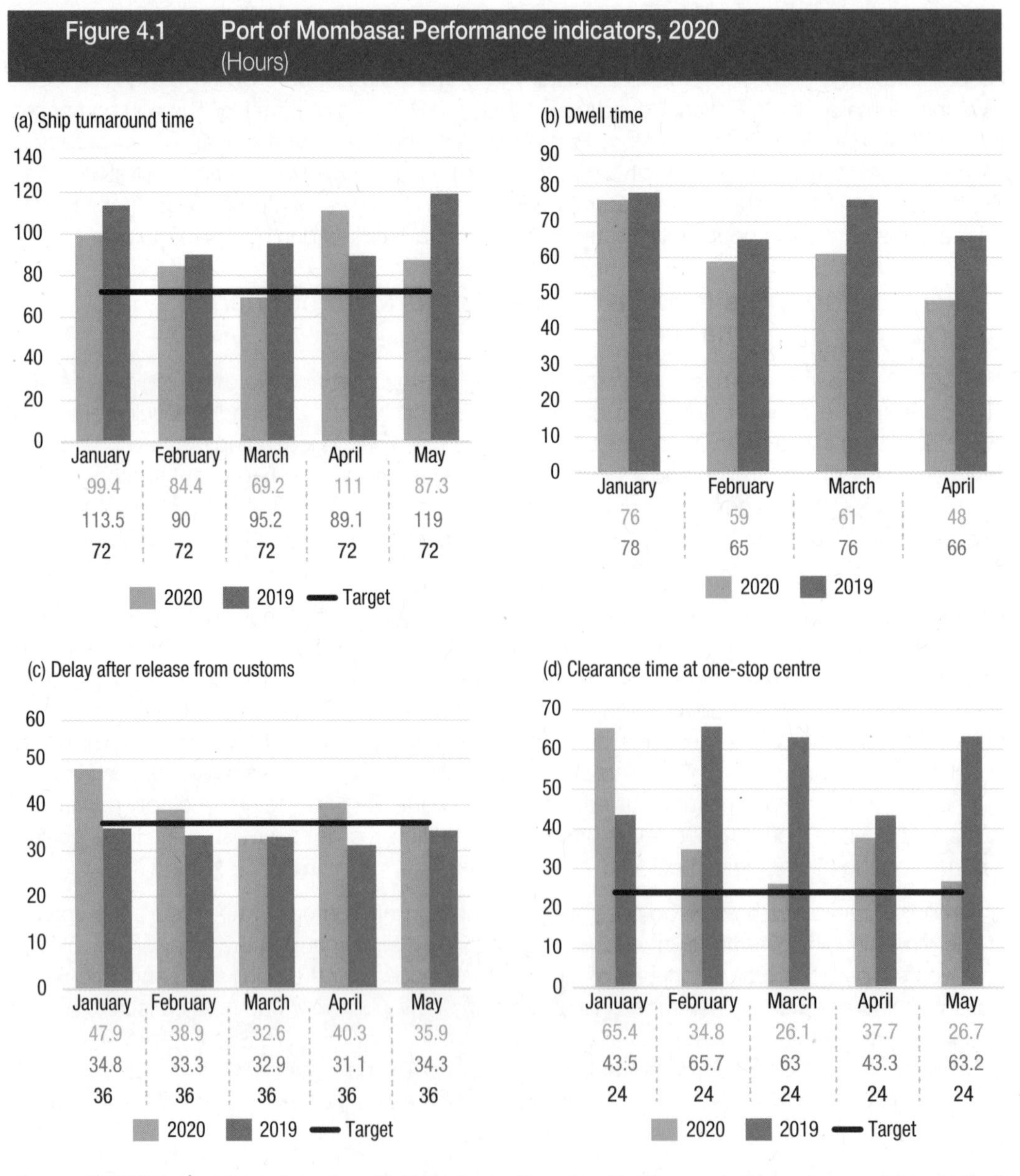

Figure 4.1 Port of Mombasa: Performance indicators, 2020
(Hours)

Source: UNCTAD calculations, based on data from the Northern Corridor Transport Observatory, available at http://top.ttcanc.org/downloads.php.

Northern Corridor Transit and Transport Coordination Authority and East African Community: Current interventions

In an attempt to address the numerous challenges affecting transport and trade logistics due to the pandemic, the secretariat of the Northern Corridor Transit and Transport Coordination Authority initiated an online platform for key stakeholders to meet and discuss issues related to the corridor and trade facilitation. Meetings bring together stakeholders from all member States of the Authority, with the aim of sharing experiences, challenges and opportunities. The platform also provides real-time updates on events in each member State, in particular at transit or transport nodes along the corridor, including ports, weighbridges, border crossing points, inland container depots and truck transit parking yards.

The East African Community is putting in place a surveillance tracker, to contribute towards dealing with the pandemic (see http://www.ttcanc.org/news.php?newsid=117). The initiative, currently in pilot testing, will provide a platform for the exchange of information in real time about tests taken by drivers and crew and about the transit movements of drivers and trucks. It will also support the tracking and tracing of drivers and their contacts.

Advocacy is being made for mutual recognition of COVID-19 testing certificates between member States of the East African Community and efforts are being made to establish testing centres at all points of origin of cargo and in other locations along the northern corridor.

Conclusion and way forward

Member States of the Northern Corridor Transit and Transport Coordination Authority have ratified various protocols and strategic responses, at both the national and international levels, aimed at enhancing safe trade in the region. However, there is a need for a detailed assessment of regional vulnerability, so that national and transboundary disaster mitigation measures may be put in place. Member States therefore need to adopt a harmonized disaster response mechanism to safeguard the transport corridor; share intelligence from early-warning systems; conduct capacity-building for personnel involved in the transport logistics chain; and embed disaster responses into national and regional policies that affect trade, transport and other related infrastructure.

D. EXPERIENCE OF AN AUTHORITY MANAGING AN INTERNATIONAL MARITIME PASSAGE: PANAMA CANAL AUTHORITY

First stage of the pandemic

On 25 March 2020, the Government of Panama declared a full quarantine and lockdown in the country. At that time, the Panama Canal Authority identified 3,700 employees as a critical minimum to maintain safe and continuous operations along the waterway and efficient services for clients. Physical distancing was enforced to protect the well-being of employees, and technology played a key role in enabling critical administrative personnel to telecommute. Systems were adapted for remote access, in a secure and stable manner. One positive aspect for the Panama Canal was that, as operations relied heavily on a culture of safety first, protocols were already in place to handle infectious diseases, such as the Regulation on Sanitation and Prevention of Communicable Diseases, last updated in 2016, which noted several diseases that required a period of quarantine and procedures to handle crews, passengers and vessels under such conditions. This regulation was the basis of the initial approach of the Panama Canal Authority to dealing with the pandemic, since it established the procedures to follow prior to the arrival of a vessel and general requirements upon its arrival, as well as protocols for inspections and health measures that included procedures designed for infectious diseases.

First challenge to transit operations

With regard to the COVID-19 pandemic, the Regulation on Sanitation and Prevention of Communicable Diseases was first applied following the notice that the cruise ship *Zaandam*, owned by Holland America Line, was on its way to Panama, carrying a number of passengers and crew that had fallen ill. The vessel had been denied entry at other ports and needed to transit the Panama Canal in order to proceed to Port Everglades, United States. The transit operation was led by the Ministry of Health of Panama, which issues recommendations on whether to allow vessels to transit based on health conditions. One key aspect in the operation was the constant communications among all parties concerned, namely, the Ministry of Health, the captain of the vessel, the local agent and head office of Holland America Line and the different offices of the Panama Canal Authority, including port captain, marine traffic control, market analysis and customer relations. Holland America Line sent a second vessel, the *Rotterdam*, and, in coordination with the Ministry of Health, COVID-19 tests were administered to the crew and passengers of the *Zaandam*; those who tested negative were transferred to the *Rotterdam*, from which oxygen tanks and medical supplies were moved to the *Zaandam*. The transfer was executed following the protocols agreed upon between the Ministry of Health, the Panama Canal Authority and Holland America Line. The transit of both vessels was approved for humanitarian reasons. Panama Canal Authority personnel only boarded the vessels after the thorough sanitation and disinfection of all of the areas to which they would have access. They wore full personal protective equipment, under the supervision of the Ministry of Health. The transit of both vessels was successfully completed on 29 March 2020. This experience raised the standards for handling similar situations in the future.

Internal protection

With the increasing number of cases in Panama, the main concern of the administration was the well-being of the workforce of the Panama Canal Authority. During the lockdown period, procedures were put in place to reduce the exposure of essential personnel. Working periods were changed to 12-hour shifts, over seven consecutive days, to help reduce contagion and secure physical distancing, and the Authority reserved hotel rooms for personnel who lived at a certain distance from working stations, to ensure their safety and availability. Private transportation was also provided, in order to maintain a group of Panama Canal Authority personnel with close contacts. At the same time, vessel arrivals were still high and putting in place such measures was key in ensuring the safety and availability of the workforce, while maintaining efficient and seamless operations for clients.

When normal operations were resumed in May 2020, all offices were cleaned and disinfected, following the recommendations of the World Health Organization and the Ministry of Health of Panama. The administration established a centre for crisis management as the official point of contact for all consultations with regard to the pandemic; the section for health, well-being and occupational safety handles questions related to health and the safety of equipment and installations. The administration issued a protocol on industrial hygiene and occupational health, a protocol on cleaning and ensuring the safety of equipment and installations and a protocol on administering COVID-19 tests. As part of the plan for a safe return to work, all employees were tested before they could return to their working stations. Employees who could telecommute were allowed to do so and, as at September 2020, there remained a number of Panama Canal Authority teams that were telecommuting.

Impact on traffic

The first impact of the COVID-19 pandemic on the Panama Canal was experienced in March 2020, when the cruise ship season was cut short. The impact with regard to commercial cargo was experienced later. In the period April–June 2020, 2,707 transits through the Panama Canal were registered, compared with 3,013 transits in 2019, a difference of 10.2 per cent. Passenger vessels, vehicle carriers, refrigerated containers, tankers and liquefied natural gas carriers were affected the most (see table 4.2).

The Panama Canal Authority, in its annual traffic projections for container ships, takes into account the blank sailings associated with the low season that generally takes place in February. In January–June 2020, the canal registered 51 blank sailings linked to the pandemic not included in the canal forecasts. With regard to container traffic, this represented a decline of around 3 per cent in April–June 2020, compared with the same period in 2019. Vehicle carriers were significantly impacted by the pandemic as car manufactures in Asia were shut down and demand in the United States soared. Similar patterns were observed in refrigerated products and the demand for oil and oil products was significantly reduced because of lockdown measures and consequent declines in the need for electricity generation. The traffic of liquefied natural gas carriers had already been affected because of the oversupplied market, but the pandemic exacerbated the situation. Of note, in the period April–June 2020, traffic through the locks for Neo-Panamax ships continued

Table 4.2 Number of oceangoing vessel transits through the Panama Canal			
Vessel type	**April–June 2019**	**April–June 2020**	**Percentage change**
Container	629	611	-2.9
Dry bulk	643	630	-2.0
Roll-on, roll-off vehicle carrier	217	111	-48.8
Tanker and/or chemical tanker	699	549	-21.5
Liquefied petroleum gas carrier	281	341	21.4
Liquefied natural gas carrier	95	89	-6.3
Refrigerated containers	163	126	-22.7
Passenger	35	6	-82.9
General cargo	163	160	-1.8
Other	88	84	-4.5
Total	**3 013**	**2 707**	**-10.2**

Source: Panama Canal Authority.

to increase slightly, while traffic through the locks for Panamax ships decreased compared with traffic in the period April–June 2019. An overall reduction of 10 per cent in transits did not have a significant impact on the operations of the canal and crew continued to work as usual, with adjustments to schedules and provisions for private transportation and shelter due to the extended lengths of shifts. The reduction in transits also helped the canal to recover the water levels necessary for operations. A drought at the start of 2020, with rain levels below historical averages, had led the Panama Canal Authority to institute water conservation measures. As traffic slowed down, water usage declined, and fewer transits were favourable for water availability purposes in the short term, while the Authority worked on implementing solutions for the long term. The Panama Canal fiscal year runs from 1 October through 30 September. It is expected that the last quarter, July–September, will behave similarly to the last quarter of 2019. However, as traffic in the first half of the fiscal year was strong, at the time of writing, performance for the full fiscal year 2020 is expected to be positive.

Lessons learned

The Panama Canal Authority maintained regular communications with customers to keep them up to date with the situation in Panama and used diverse channels to send information to employees with regard to both operational and administrative matters, to share methods and tips on preventing community transmission and to provide psychological support. Innovation also contributed to the maintenance of operations and the upkeep of morale. This involved, for example, the sharing of physical exercise routines and virtual concerts via social media, as well as the development of a number of applications, including a travel application that helped the Authority to keep track of employees using the internal transport system.

The pandemic has been anything but predictable. All procedures and measures have had to be constantly reviewed for improvement and strong and humane leadership has been necessary in order to make difficult and timely decisions with limited information. Collaboration and solidarity within as well as outside the Authority have proven helpful in decision-making processes, bringing in support and different experiences and benchmarks. Shared information and experiences have also been key for port authorities and shipping companies and communications and technology have played a key role.

The Panama Canal Authority is as resilient as its personnel, and they adapted to the new normal quickly, including the new safety protocols, the challenges related to telecommuting and, in particular, the uncertainty. One important lesson learned to date is that everything is subject to constant and ongoing improvement. A fluid situation requires frequent adjustments.

E. EXPERIENCE OF A PORT AUTHORITY: PORT AUTHORITY OF VALENCIA

The operation of ports is of vital importance in dealing with the COVID-19 crisis, as it helps to ensure that essential goods such as food, medical supplies and fuel, as well as raw materials and manufactured goods, continue to reach their intended destinations. This section provides details on the experience of a port authority in handling the crisis and the early measures applied.

The Port Authority of Valencia is a public body responsible for the management of three State-owned ports in eastern Spain, namely, Valencia, Sagunto and Gandía. To help minimize the impact of the COVID-19 pandemic, the Port Authority of Valencia applied a set of measures with regard to internal activity at the ports as well as in connection with the activity of the entire logistics chain. These measures comprised four fundamental aspects, namely, operational, sanitary, economic and social.

Operations

The Port Authority of Valencia distinguished between internal and external operations. To ensure the continuous internal operations of the ports managed by the Authority, a contingency plan was developed involving three progressive levels of emergency. Essential and non-essential jobs were clarified according to their roles in the continuity of port operations. Non-essential workers were progressively transferred to telecommuting, with over 200 employees telecommuting during the national state of alarm declared in Spain. The information technology department prepared a protocol to ensure broader, secure access to the digital resources of the Authority. Electronic data interchanges through the port community system were enhanced to ensure information management in all operating procedures (see https://www.valenciaportpcs.com/en/). Essential workers were expected to comply with strict measures concerning the use of personal protective equipment and other protocols when interacting with other employees and third parties while conducting their duties at ports. An appropriate frequency of disinfection of working areas was also maintained.

With regard to terminal operations, similar recommendations were made with regard to port services, including using personal protective equipment; maintaining physical distancing in the working environment, including on board vessels; disinfecting working spaces; and ensuring that more vulnerable employees could remain at home. Pilots followed protocols with regard to access to vessels and requirements when on board to help ensure protection from infection. Stevedores were encouraged to form stable groups with the same members to help limit community transmission. Port services were considered essential services; companies were therefore permitted to continue operations under national regulations in accordance with the national state of alarm. Port personnel were considered essential workers and therefore permitted to participate in daily operations. During the period of the state of alarm, ports managed by the Port Authority of Valencia remained fully operational. The adapted measures caused a reduction in productivity in the first few weeks, until the procedures and protocols had been adjusted to. Port services recovered ground, with productivity reaching the same maximum levels recorded before the pandemic. All measures were coordinated through the Ports of the State, the State-owned company responsible for the management of State-owned ports in Spain.

Sanitation

The Port Authority of Valencia applied the rules and recommendations established by the Ministry of Health when defining the protocols for both internal and external operations. A key recommendation related to the use of personal protective equipment, the maintenance of physical distancing and the disinfection of all installations, as well as the establishment of protocols for interactions between personnel.

Economy

With regard to the economic impact of the pandemic, the Port Authority of Valencia provided support to ports by facilitating around €10 million ($11.24 million) as an urgent compensatory measure to mitigate the impact. Since such support was implemented, in March 2020, the Authority has streamlined the payment of €7.33 million (around $8.24 million) to provide liquidity to 250 suppliers and service providers working for the ports managed by the Authority. The Authority anticipates that total advance payments to suppliers in 2020 will amount to €51 million (around

$57.3 million). The objective was to provide weekly payments until the end of the national state of alarm period, to minimize treasury-related difficulties that suppliers might be facing. This measure required the Authority to establish internal mechanisms to process invoices as quickly as possible. The Authority also provided to port clients an advance of €2.64 million (around $2.97 million) for rebates (that is, discounts on port taxes) pending from 2019, in order to reduce the impact of port taxes on both customers and port operators.

Social

The Port Authority of Valencia set up a solidarity campaign titled Al pie del cañon, an initiative launched after the declaration of the national state of alarm, which sought to shed light on the important work carried out by port personnel to guarantee the supply of goods and the smooth functioning of supply chains during the pandemic. The campaign resulted in the sharing of over 100 videos by people from all along the transport logistics chain, in Spain and worldwide, who wished to explain their work and send messages of encouragement and solidarity.[24]

Conclusion

It is too early to determine the full impact of the pandemic on trade and the economy; returning to normal will take time and this normality will likely differ from that expected before the pandemic. The Port Authority of Valencia witnessed declines traffic as lockdown measures were instituted worldwide; in January–May 2020, total accumulated traffic by volume for the ports managed by the Authority had dropped by 7.92 per cent, compared with in the same period in 2019. With regard to operational matters, the pandemic has had an impact on the way port operations are carried out, in particular with regard to passenger ships and cruise liners. Sanitary measures continue to be applied, along with new border control procedures. These processes will shape port infrastructure and operations in the coming years. Resilience will become an even more relevant concept with regard to supply chain management and the development of business continuity plans will be critical, to help better prepare for any future disruption from events such as pandemics or those due to climate change-related factors. Digitalization has been a driving force in the sustainability of business during the pandemic. The integration of port community systems along supply chains may be a development to pursue in the future, to foster resilience and innovation based on new technologies, which is a key element of competitiveness in an environment of traffic scarcity. Ports should also be aware of new trade patterns that will emerge and prepare infrastructure and operations accordingly. In this regard, the Authority has launched a new strategy that considers such changes in order to be better prepared for a new normal centred on a more digital, more innovative, more responsible, more resilient and carbon neutral port world. With regard to contributions to achieving the Sustainable Development Goals, this crisis could provide an opportunity to achieve more sustainable and inclusive development.

F. EXPERIENCE OF A GLOBAL SHIPPING COMPANY: MEDITERRANEAN SHIPPING COMPANY

The spread of COVID-19 is an unprecedented global health issue, that has triggered unexpected shocks for societies and economies. The Mediterranean Shipping Company has continued to implement health protection measures to mitigate the risk to its crew and its employees worldwide and to help curb the spread of the virus. The Company has enacted established business continuity plans and switched to telecommuting for office-based employees in most countries, all of which has helped to limit disruptions to global supply chains.

Speed of reaction

One of the biggest lessons from the first half of 2020 was the importance of acting quickly and with conviction. As soon as reports of the outbreak were received in January 2020, the Mediterranean

[24] The Chair of the Authority stated as follows: "These weeks that we have been experiencing a major crisis have brought to mind various elements for reflection… There is a change in the scale of values of the professions and also an update of values. Solidarity has come to play a fundamental role in these days. The crisis has unfortunately brought about job losses and dramatic situations for many families. Solidarity is vital." The President of the Authority highlighted that "the logistics sector has lived up to what was expected of it, has responded by contributing what it knows, bringing goods, arranging it for citizens" and conveyed a message of optimism by stating that "it is worth thinking that these lived experiences can help us plan for a better future".

Shipping Company immediately implemented robust health protection measures across its ships, infrastructure and offices, in line with guidance from the World Health Organization and in compliance with the recommendations of national authorities. The Company was also swift to implement a global ban on business travel and to cancel visits to headquarters from colleagues, customers and suppliers from end-January 2020.

Telecommuting

According to a new instruction from headquarters, international meetings would be held via videoconferencing and this instruction has remained in place since then. Since the start of the pandemic, the Mediterranean Shipping Company has seen a record number of staff working in an agile way using technology and, in many instances, telecommuting. This began in January in offices in China, then extended to headquarters in Geneva and many locations worldwide. Shifting to telecommuting is part of the established business continuity plans, and this experience demonstrated, to some extent, that these processes worked. However, this form of staff deployment resulted in new experiences in implementing company plans. There has been some new understanding of the value of videoconferencing. For many, the crisis triggered an advancement in skills and knowledge with regard to videoconferencing and the efficient use of online workspaces. Guidance on taking care of one's health, while keeping up productivity levels, was regularly shared across all company agencies. In addition, the global intranet was used to disseminate information and news about the pandemic. The Mediterranean Shipping Company aims to emerge from the pandemic with a heightened internal awareness of the benefits of the use of digital tools and, as a result, greater resilience given any business continuity shocks in future.

Operational flexibility

Implementing existing business continuity plans ensured that operations and customer service could continue, while company staff avoided travel and practiced confinement or physical distancing. In China, for example, the Mediterranean Shipping Company maintained operations by shifting certain functions to other offices and relying on the support of shared services centres in other regions, as part of a plan determined before the pandemic. Preserving close contacts and relationships with customers was essential. The challenge of maintaining contact with customers without face-to-face meetings was easily overcome, as most customers were in the same situation in terms of telecommuting. In addition, the Company worked continuously to adapt contingency plans and regularly advise customers of the online booking platform myMSC on how to manage changes, relying on its internal information sharing system to collect data from 155 countries. Digitalization has been slow to be adopted in container shipping. Only recently have significant changes begun to take place in documentation and booking processes, the incorporation of electronic business tools and the online connectivity of equipment. The case for investing in digital platforms and processes has become clearer and more compelling, even if the availability of funds for such investments may be affected in the short term by the impact of the pandemic on trade.

Essential workers

In addition to maintaining services to support cargo flows, supporting employees that could not easily telecommute was a challenge. Seafarers were among the groups of workers most significantly affected by the pandemic, due to border closures and other restrictions on movement, which led to long shifts at sea. Among the necessary measures introduced at the height of the crisis in certain countries, ships in the Mediterranean Shipping Company fleet of 550 vessels were equipped with personal protective equipment. In addition, new company policies restricted crew from going ashore at ports. The most significant impact on seafarers were the restrictions by Governments that limited crew changes on ships in many ports worldwide. In this regard, the Company extended contracts for container shipping crew and provided social and financial support in relevant cases to help mitigate the challenges for crew at sea and to facilitate crew changes in support of seafarers and their families. Governments that took steps to designate seafarers as key workers, in line with a request from IMO, made a positive difference to the situation (see http://www.imo.org/en/MediaCentre/HotTopics/Pages/Coronavirus.aspx). As a company founded by a ship captain, the Company places a high value on the contribution of seafarers to its business and aims to ensure that the key role of seafarers in the economy and their contribution to well-functioning societies may be better understood. A similar label of importance and expression of gratitude should be directed, by policymakers and the general public, to employees at port terminal

depots and warehouses, as well as the drivers of trucks, trains and barges carrying containers, who have continued to work during the pandemic as and when permitted under national rules.

Adapting services

To help ensure the minimum level of disruption to customers, the Mediterranean Shipping Company adapted its shipping services networks to help companies ship goods more easily. The sudden slowdown in trade resulted in necessary reductions in the capacities of container shipping networks in order to match the lower level of demand for cargo shipments. However, subsequent rebounds in trade flows following the easing of lockdown measures underscored the importance of flexible network management. In the first half of 2020, the Company helped shippers use its short-sea shipping networks, in Europe in particular, as a reliable alternative to road transport. This helped mitigate later delays at border crossing points on land that were due to restrictions on movement. The Company also introduced a suspension of transit programme for container shipping at dedicated trans-shipment hubs, as follows: Bremerhaven, Germany; PSA Panama International Terminal; Port of Busan, Republic of Korea; King Abdullah Port, Saudi Arabia; Port of Lomé, Togo; Asyaport, Tekirdağ, Turkey. This programme provided for flexibility and substantial cost savings as it enabled shippers to better control storage costs at the point of booking, while allowing them to adapt the delivery date to their needs. It also helped minimize congestion at ports of discharge and improve efficiency, as products were placed closer to distribution networks. One of the lessons learned from the crisis is to innovate not only through the provision of new services and storage solutions, but also by employing solutions from past incidents, such as reintroducing a discontinued service to help enable the partial recovery of cargo volumes on a particular route.

Keeping the world moving

Despite the difficult operating conditions during the pandemic, the Mediterranean Shipping Company, as a major shipping and logistics services provider, has contributed to ensuring the high priority transport of essential goods such as food, agricultural products, raw materials and medical equipment. Container shipping lines and their customers have a crucial role in the global economy and in enabling well-functioning societies. In future, the Company aims to strengthen business continuity planning and the technology and processes related to telecommuting and digitalization, as well as to raise awareness of the essential role of all personnel in container supply chains, in particular at sea, to keep the blood flowing in these arteries of the global economy.

REFERENCES

Bullock S, Mason J, Broderick J and Larkin A (2020). Shipping and the Paris climate agreement: A focus on committed emissions. *BMC Energy.* 2:5.

IMO (2020). Reduction of greenhouse gas emissions from ships: Fourth IMO greenhouse gas study 2020 – Final report. MEPC 75/7/15. London. 29 July.

This chapter provides a summary of important international legal and regulatory issues, as well as some related technological developments during the period under review, and presents some policy considerations.

Among the issues worth highlighting is the need to implement IMO resolution MSC.428(98) of 16 June 2017 on maritime cyberrisk management in safety management systems, which encourages Administrations to ensure that cyberrisks for shipping are appropriately addressed in safety management systems, effective 1 January 2021. Thus, in preparation for its implementation – ahead of the first inspection by the international safety management auditors after 1 January 2021 and particularly during 2020 – shipping companies need to assess their risk exposure and develop information technology policies to include in their safety management systems, in order to mitigate increasing cyberthreats. Owners who fail to do so may risk having their ships detained by port State control authorities. Strengthening cybersecurity is likely to increase in importance, given that cyberrisks have grown, with greater reliance on virtual interaction as a result of the ongoing COVID-19 crisis.

In addition, work is progressing with respect to the development, testing and operation of maritime autonomous surface ships, and their market value is growing. Industry collaboration on the use of autonomous drones is also continuing, including with regard to inspections and commercial drone delivery to vessels anchored in port. The use of electronic trade documentation has increased in importance, particularly in the context of the COVID-19 pandemic, and international organizations and industry bodies have issued calls for Governments to remove restrictions on the use and processing of electronic trade documents, and where possible, ease requirements for any documentation to be presented in hard copy.

Other important regulatory developments relate to the reduction of greenhouse gas emissions from international shipping and other ship-source pollution control and environmental protection measures. Issues covered include shipping and climate change mitigation and adaptation; air pollution, in particular sulphur emissions; ballast water management; biofouling; pollution from plastics and microplastics; safety considerations of new fuel blends and alternative marine fuels; and the conservation and sustainable use of marine biodiversity of areas beyond national jurisdiction. In addition, an important development covered in this chapter includes a decision by the European Commission to extend the liner shipping Consortia Block Exemption Regulation[25] until 25 April 2024.

[25] Commission Regulation (EC) No 906/2009 of 28 September 2009 on the application of article 81(3) of the Treaty to certain categories of agreements, decisions and concerted practices between liner shipping companies (consortia).

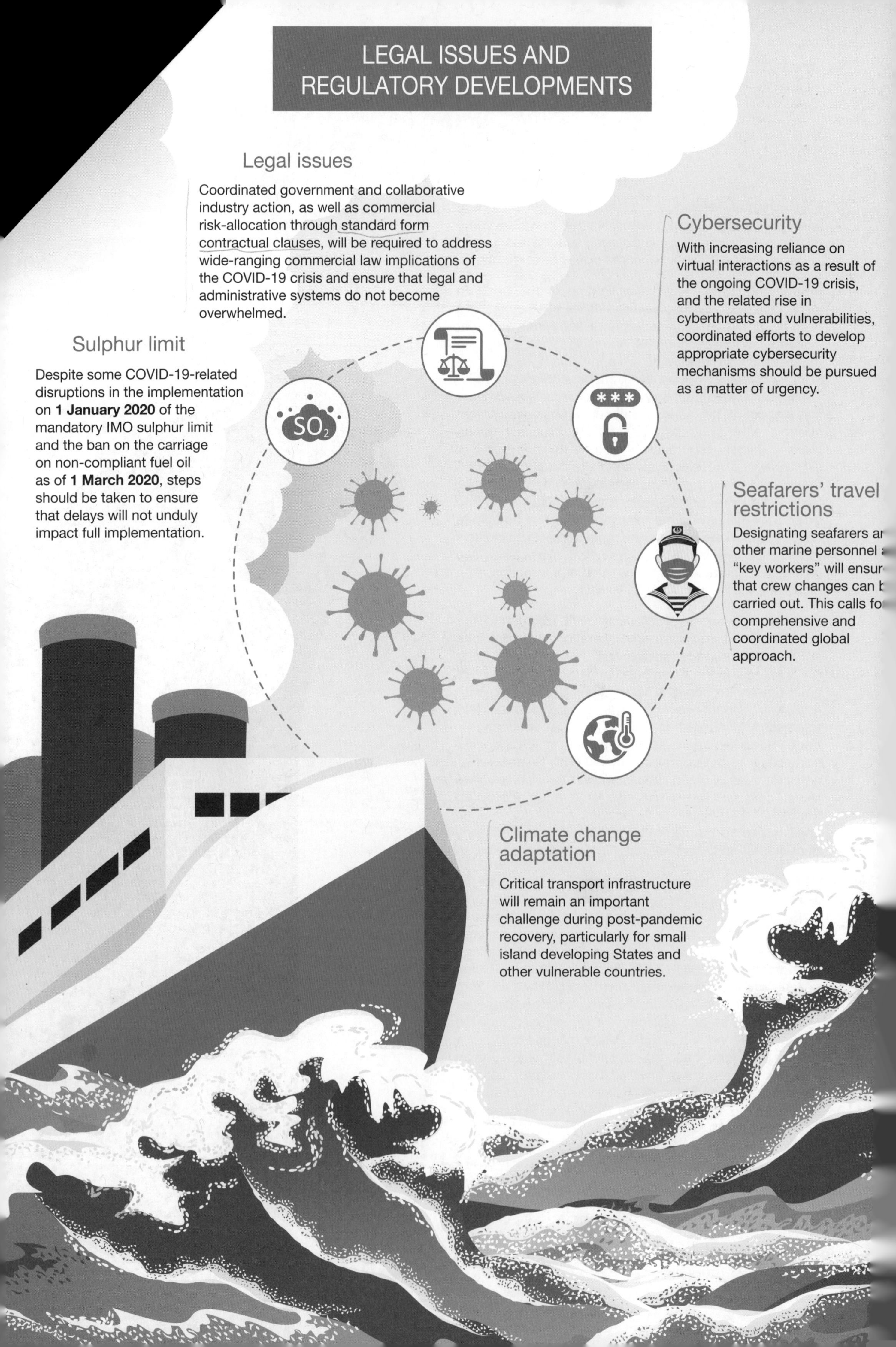
LEGAL ISSUES AND REGULATORY DEVELOPMENTS
Legal issues
Coordinated government and collaborative industry action, as well as commercial risk-allocation through standard form contractual clauses, will be required to address wide-ranging commercial law implications of the COVID-19 crisis and ensure that legal and administrative systems do not become overwhelmed.
Cybersecurity
With increasing reliance on virtual interactions as a result of the ongoing COVID-19 crisis, and the related rise in cyberthreats and vulnerabilities, coordinated efforts to develop appropriate cybersecurity mechanisms should be pursued as a matter of urgency.
Sulphur limit
Despite some COVID-19-related disruptions in the implementation on 1 January 2020 of the mandatory IMO sulphur limit and the ban on the carriage on non-compliant fuel oil as of 1 March 2020, steps should be taken to ensure that delays will not unduly impact full implementation.
SO2
***
Seafarers' travel restrictions
Designating seafarers ar
other marine personnel
"key workers" will ensur
that crew changes can b
carried out. This calls fo
comprehensive and
coordinated global
approach.
Climate change adaptation
Critical transport infrastructure will remain an important challenge during post-pandemic recovery, particularly for small island developing States and other vulnerable countries.

A. TECHNOLOGICAL DEVELOPMENTS AND EMERGING ISSUES IN THE MARITIME INDUSTRY

1. Ensuring maritime cybersecurity

Ship cybersecurity

Ships have become better integrated into information technology networks. Moreover, communication and operational processes have been further digitalized, and smart navigation and advanced analytics are being used to optimize ship operations and reduce fuel consumption and greenhouse gas emissions. In line with these recent trends, implementing and strengthening cybersecurity measures has become a priority for shipowners and managers. In 2019, cyberincidents were rated second among the top five risks for the maritime and shipping sector, according to a major industry survey (Allianz, 2019). While cyberrisks had already become a major concern, the COVID-19 crisis has compounded existing problems and provided a new impetus for action. The importance of cybersecurity is expected to grow further, given the increasing reliance on virtual interactions as a result of the pandemic, and the related rise in cyberthreats and vulnerabilities.

The Digital Container Shipping Association – a consortium of nine container lines[26] – recently published a cybersecurity implementation guide to ensure vessel preparedness for relevant IMO regulations, outlining best practices that would provide all shipping companies with a common language and a manageable, task-based approach for meeting the IMO implementation deadline of January 2021 (Digital Container Shipping Association, 2020a). The guide is in line with BIMCO and National Institute of Standards and Technology cyberrisk management framework guidelines, enabling shipowners to effectively incorporate cyberrisk management into their existing safety management systems. The guide aims to provide a management framework that can be used to reduce the risk of cyberincidents that could affect the safety or security of vessels, crews or cargo. It breaks down the BIMCO framework into themes and maps them to the controls that underpin the functional elements of the Institute: identify, protect, detect, respond, recover (Digital Container Shipping Association, 2020b). In January 2020, the first cybersecurity management system – that of the Nippon Yusen Kabushiki Kaisha Group – had already been certified by industry classification society Nippon Kaiji Kyokai, commonly known as ClassNK, as being compliant with the latest IMO guidelines (Nippon Yusen Kabushiki Kaisha Line, 2019).

Among the relevant IMO instruments, the above-mentioned IMO resolution on maritime cyberrisk management in safety management systems affirms that an approved safety management system should take into account cyberrisk management in accordance with the objectives and functional requirements of the International Safety Management Code[27] and encourages Administrations to ensure that cyberrisks are appropriately addressed in safety management systems no later than the first annual verification of the company's document of compliance after 1 January 2021 (IMO, 2017a).

The International Safety Management Code, in force since 1 July 1998, is now more important than ever to ensure that vessels become cyberresilient and report any identified cyberrisk, given that the underreporting of cybersecurity incidents is considered a problem in the maritime industry (Safety4Sea, 2019a). Many issues may be identified on board ships that make them more vulnerable to cyberattacks, including unsecure networks and software, lack of seafarer training and insufficient protection of data. Shipping companies will need to consider these issues and include cyberrisk into their safety management systems, so they know how to deal with and approach a cyberincident. As this will require some time, all work should be completed ahead of the first inspection by International Safety Management auditors after 1 January 2021. Owners who fail to comply may risk having their ships detained by port-State control authorities that will aim to enforce the requirement in a uniform and equitable manner. At the same time, implementing cybersecurity is important to protect shipping assets and technology from mounting cyberthreats, in particular given that cyberrisks are expected to grow, with greater reliance on virtual interaction as a result of the ongoing COVID-19 crisis.

Cybersecurity is covered under the International Ship and Port Facility Security Code, in force since 1 July 2004 (see BIMCO et al., 2018 for related guidance). Thus, as set out in part A, section 8.4 of the Code, ship security assessment shall include, inter alia, "2. the identification and evaluation of key ship board operations that it is important to protect; 3. identification of possible threats to the key ship board operations and the likelihood of their occurrence, in order to establish and prioritize security measures; and 4. the identification of any weakness, including human factors in the infrastructure, policies and procedures".

[26] Maersk Line, CMA CGM, Hapag-Lloyd, Mediterranean Shipping Company, Ocean Network Express, Evergreen Line, HMM, Marine Transport Corporation and Zim Integrated Shipping Services, covering 70 per cent of world trade. The consortium was first launched in November 2018.

[27] The main purpose of the International Safety Management Code is to provide an international standard for the safe management and operation of ships and for pollution prevention. It establishes safety management objectives and requires a safety management system to be established by "the Company", which is defined as the shipowner or any person, such as the manager or bareboat charterer, who has assumed responsibility for operating a ship. The company is then required to establish and implement a policy for achieving these objectives (www.imo.org/en/OurWork/HumanElement/SafetyManagement/Pages/ISMCode.aspx).

Part B, section 8.3 of the Code states that a ship security assessment should address, among others, the following elements on board or within the ship: "5. radio and telecommunications systems, including computer systems and networks, and 6. other areas that may, if damaged or used for illicit observation, pose a risk to persons, property or operations on board a ship or within a port facility".

With regard to cyberrisks, the IMO Assembly had as early as 2017 adopted a strategic plan that recognized the need to integrate new and advancing technologies into the regulatory framework for shipping (IMO, 2017b). In addition, to support effective cyberrisk management, two IMO committees, the Maritime Safety Committee and the Facilitation Committee, had adopted guidelines that provide high-level recommendations to safeguard shipping from current and emerging cyberthreats and vulnerabilities. These recommendations can be incorporated into existing risk management processes and are complementary to the safety and security management practices already established by IMO (that is to say, the International Safety Management Code and the International Ship and Port Facility Security Code) (IMO, 2017c). These guidelines present five functional elements: to identify, protect, detect, respond and recover.[28]

Other useful guidance, standards and regulations, adopted at the international, regional and national levels, are described below.

European Union Network and Information Security Directive (EU) 2016/1148 requires all Member States to protect their critical national infrastructure by implementing cybersecurity legislation by May 2018 (European Union, 2016). Inter alia, the Directive in chapter 2 lays down obligations for all Member States to adopt a national strategy on the security of network and information systems; creates a cooperation group to support and facilitate strategic cooperation and the exchange of information among Member States; establishes a computer security incident response teams network; sets security and notification requirements for operators of essential services and digital service providers; and spells out obligations for Member States to designate national competent authorities, single points of contact and computer security incident response teams. The Directive covers organizations in vital sectors that rely heavily on information networks and are referred to as "operators of essential services", including those in energy, transport, utilities, banking and finance, digital services and health care. As noted in preambular paragraph 10, in the water transport sector, security requirements for companies, ships, port facilities, ports and vessel traffic services under European Union legal acts cover all operations, including radio and telecommunication systems, computer systems and networks.

International standard 27001:2013 of the International Organization for Standardization and International Electrotechnical Commission, commonly known as ISO/IEC 27001:2013, specifies the requirements for setting up, implementing, maintaining and continually improving an information security management system within the context of an organization (International Organization for Standardization, 2013). It also includes requirements for the assessment and treatment of information security risks tailored to the needs of the organization. The requirements are generic and are intended to be applicable to all organizations, regardless of type, size or nature.

The Framework for Improving Critical Infrastructure Cybersecurity of the United States National Institute of Standards and Technology was prepared to assist companies with their risk assessments by helping them understand, manage and express potential cyberrisks internally and externally (National Institute of Standards and Technology, 2018).

The Code of Practice on Cybersecurity for Ships of the United Kingdom was drawn up to help companies develop cybersecurity assessments and plans, and mitigation measures, and to manage security breaches; it should be used along with ship security standards and other relevant IMO regulations (Institution of Engineering and Technology, 2017).

Guidelines on Cybersecurity on Board Ships offer guidance to shipowners and operators on procedures and actions to maintain the security of cybersystems in the company and on board ships (BIMCO et al., 2018).[29] Both the IMO guidelines and the United States National Institute of Standards and Technology framework have been taken into account. The guidance specifies, among others, that company plans and procedures for cyberrisk management should be incorporated into existing security and safety risk management requirements contained in the International Safety Management Code and the International Ship and Port Facility Security Code.

In the Asia-Pacific region, for instance, many countries have developed cybersecurity legislation and policy, elements of which are applicable across all industry areas; they have also set up relevant implementing bodies and entities both at the national and regional levels. However, sector-specific guidance and initiatives tailored to business needs, or the provision of methods to address unique risks or specific operations in certain sectors, including in the maritime sector, appear to be limited in the region (BSA/The Software Alliance, 2015; North Atlantic Treaty Organization Cooperative Cyberdefence Centre of Excellence, 2019).

At the national level, for instance, the China Classification Society in July 2017 issued guidelines on requirements and security assessments of ship cybersystems, offering

[28] For information on a platform aimed at helping shipowners and operators to better understand their vulnerabilities and improve their cybersecurity processes and systems ahead of the IMO deadline, see Safety4Sea, 2020a.

[29] For additional industry guidelines, see also Safety4Sea, 2018.

solutions for the increasingly serious threat to ship cybersecurity (China Classification Society, 2017). In February 2020, the Republic of Korea released guidelines based on international standards for type approval of maritime cybersecurity to help inspect the cybersecurity level and functioning of cybersystems, including remote access equipment, integrated control and monitoring systems on board ships (Safety4Sea 2020a).

Port cybersecurity

Ports are important to keep supply chains moving and economies across the world functioning. While they are becoming "smart", relying more on technologies and digitalization to become more competitive and optimize operations, ports are also facing increased cybersecurity challenges and threats. A recent report on port cybersecurity identifies the following good practices for terminal operators and officials responsible for cybersecurity implementation at port authorities (European Union Agency for Cybersecurity, 2019):

- Define clear governance concerning cybersecurity at port level, involving all stakeholders involved in port operations.
- Raise awareness of cybersecurity matters at port level and foster a cybersecurity culture.
- Enforce the technical cybersecurity basics such as network segregation, updates management, password hardening and segregation of rights.
- Consider security by design in applications, especially since ports use many systems, some of which are opened to third parties for data exchange. Any vulnerability in those systems can be a gateway to compromising port systems.
- Enforce detection and response capabilities at port level to react as quickly as possible to any cyberattack before it affects port operation, safety or security (see www.sauronproject.eu/).

Prompted by the Ryuk ransomware attack on enterprise environments in December 2019[30] (National Cybersecurity Centre, 2019; United States Coast Guard, 2019a) and by concerns that the maritime network is vulnerable to cybercrime (Riviera, 2019; United States Coast Guard, 2019b), the United States Coast Guard issued new guidelines for dealing with cyberrisks at Maritime Transportation Security Act regulated facilities (United States Coast Guard, 2020). According to the guidelines, regulated facilities must assess and document risks associated with their computer systems and networks in a facility security assessment and address them in a facility security plan or alternative security programme. Following this, owners and operators must demonstrate compliance. To allow time for owners or operators of such facilities to tackle cybersecurity vulnerabilities, the initial implementation period is 1.5 years with no further need to update a facility security assessment or an alternative security programme until 30 September 2021.

Similarly, the Department for Transport of the United Kingdom updated its 2016 cybersecurity guidance for ports and the wider maritime industry against cyberthreats. The guidance aims to help ports develop cybersecurity assessments and identify gaps in their security, while providing advice on handling security breaches and incidents and defining clear roles and responsibilities to deal with cyberattacks (Institution of Engineering and Technology, 2020).

COVID and maritime cybersecurity

Maritime digitalization has been an ongoing trend for some time both on board ships and ashore. The COVID-19 outbreak has heightened the need for digitalization and has brought maritime industry stakeholders closer through the collaborative use of digital technologies. These include video conferencing and other online platforms, as well as the sharing and remote monitoring of data to ensure that supply chains continue to function (Riviera, 2020a; Riviera, 2020b). At the same time, reports indicate an increase in shipping cyberattacks of 400 per cent between February and June 2020 (Splash, 2020a). According to cybersecurity systems provider Naval Dome, the ability of companies to sufficiently protect themselves has been reduced by travel restrictions, social distancing measures and economic recession. However, the primary reason behind this spike has been an increase in malware, ransomware and phishing emails exploiting the COVID-19 crisis (Marine Link, 2020).

With regard to ports, for instance, the COVID-19 crisis demonstrated that while some port communities had already digitalized their business processes and developed into smart ports, many others were lagging behind, relying heavily on personal interaction and paper-based transactions as the norm, for shipboard-, ship–port interface- and port–hinterland-based exchanges. As highlighted in a recent port industry policy statement, only 49 of the 174 IMO Member States have functioning port community systems (International Association of Ports and Harbours et al., 2020a). In these circumstances, the main shipping and port industry organizations have launched a call to action to accelerate the digitalization of maritime trade and logistics.[31] They have set the following priorities:

- Assess the state of implementation and find ways to enforce the already mandatory requirements defined

[30] Encryption was used to block access to systems, devices or files until a ransom was paid.

[31] BIMCO, Federation of National Associations of Ship Brokers and Agents, International Association of Ports and Harbours, International Cargo Handling Coordination Association, International Chamber of Shipping, International Harbour Masters Association, International Marine Purchasing Association, International Port Community Systems Association, International Ship Suppliers and Services Association, and the Protect Group.

in the IMO Convention on Facilitation of International Maritime Traffic, 1965 to support the transmission, receipt and response of information required for the arrival, stay and departure of ships, persons and cargo, including notifications and declarations for customs, immigration, and port and security authorities, through electronic data exchange.

- Ensure the harmonization of data standards beyond the aforementioned Convention to facilitate the sharing of port and berth-related master data for just-in-time operation of ships and optimum resource deployment by vessel services and suppliers, logistics providers, cargo handling and clearance, thereby saving energy, improving safety and cutting costs and emissions. This can be achieved by using the supply-chain standards of the International Organization for Standardization, the standards of the International Hydrographic Organization and the *IMO Compendium on Facilitation and Electronic Business*.
- Strive for the introduction of port community systems (www.ipcsa.international/) and secure data exchange platforms in the main ports of all Member States represented in IMO.
- Review existing IMO guidance on maritime cyberrisk management with regard to its ability to address cyberrisks in ports, developing additional guidance where needed.
- Raise awareness, avoid misconceptions and promote best practices and standardization on how port communities can apply emerging Internet technologies and automation; facilitate the implementation of such emerging technologies and other innovative tools to increase health security in port environments; and develop a framework and road map to facilitate the implementation and operationalization of digital port platforms that can connect with hinterland supply chains as well, and where data can be securely shared.
- Establish a coalition of stakeholders willing to improve transparency of the supply chain through collaboration and standardization, starting with the overdue introduction of the electronic bill of lading.
- Set up a capacity-building framework to support smaller, less developed and understaffed port communities, not only by providing technical facilities but also by training personnel (International Association of Ports and Harbours et al., 2020a).[32]

[32] For more information and a list of maritime technology initiatives that have been made available to help the industry deal with the disruption caused by the pandemic, see https://thetius.com/maritime-technology-initiatives-supporting-the-industry-covid-19-response. Also see International Association of Ports and Harbours et al., 2020b.

Given that digitalization and cyberrisks and vulnerabilities are growing during the ongoing COVID-19 crisis and its aftermath, related capacity-building will be required for many developing countries. On a more general note, in the developing world at large, the lack of reliable and affordable Internet services and a widespread digital divide continue to be a major concern, which needs to be effectively addressed (see Economic Commission for Asia and the Pacific, 2019).

2. Technological developments in shipping

Autonomous ships, navigation systems and drones

Work is advancing on the development of maritime autonomous surface ships, drones and navigation systems (see also UNCTAD, 2018; UNCTAD, 2019a). In 2019, it was announced that the Mayflower autonomous ship[33] would be attempting the world's first unmanned transatlantic crossing from Plymouth, United Kingdom, to Plymouth, Massachusetts, United States in the second half of 2020. This was described as a symbolic voyage, whereby a new Mayflower would set sail 400 years after the historic voyage, this time using artificial intelligence and other advanced technologies, providing for safer navigation and hazard avoidance (Safety4Sea, 2019b). The full-size, fully autonomous research ship was launched on 16 September 2020 and during its journey would spend six months gathering data about the state of the ocean (BBC News, 2020).

According to a report by technology and innovation consultancy Thetius, the market for maritime autonomous surface ships is worth $1.1 billion annually and will grow by 7 per cent each year to $1.5 billion by 2025. In addition, 96 per cent of almost 3,000 patents relating to autonomous shipping technology worldwide were registered in China. According to the report, this will lead other nations to develop and implement autonomous shipping within five years (Thetius, 2020). The report does not appear to include COVID-19-related considerations, however.

Global navigation satellite systems, used for the safe navigation of ships, and automatic identification system signals via satellites, tracking ships around the world, are considered critical to improve the safety of ship navigation and the reliability of data for vessel tracking and analytics, including for insurance purposes (also see chapter 3A). However, the safety of such systems can be compromised by jamming, spoofing or hacking, as evidenced by various incidents, which can be dangerous and may lead to grounding and collisions.

[33] Partners in this project are International Business Machines, Promare and the University of Plymouth, United Kingdom.

Automatic identification system tracking of ships may be occasionally disrupted, as some vessels switch off their devices when they enter zones in which they are legally prohibited from performing fishing or other illegal activities. Therefore, it is important to strengthen both global navigation satellite systems and automatic identification system communications, which both use satellites. For instance, the European Space Agency has started developing a solution to mitigate risks for its services in this area (Digital Ship, 2020).

Industry collaboration is continuing with respect to drones as well, including for instance, the launching in Singapore of a ship-to-shore pilot project by Wilhelmsen and Airbus, which worked to deploy drone technology in real-time port conditions, delivering a variety of small, time-critical items to vessels anchored in port (Splash, 2019), as well as the first commercial drone delivery to such vessels. Drone deliveries can help save costs, time and carbon-dioxide emissions compared with traditional shipping and have reduced unnecessary human contact during the pandemic. The drones that were used in the project could only deliver a maximum of 5 kg loads over 5 km, but the company was planning to complete the development of a drone that could carry 100 kg loads over 100 km, by the second half of 2021 (Splash, 2020b). In addition, in June 2020, the industry-first inspection by an autonomous drone, of an oil tank on a floating production, storage and offloading vessel, was completed. The drone uses light detection and ranging to navigate inside the tank, where reception of satellite signals for accurate positioning is unavailable in this enclosed space, and a three-dimensional map of the tank is created. As the technology matures, drones are expected to navigate more autonomously (Riviera, 2020c).

With regard to regulatory issues and intergovernmental meetings related to technology in shipping, the IMO Subcommittee on Navigation, Communications and Search and Rescue met in January 2020. It discussed advances in modernizing the Global Maritime Distress and Safety System – under the regulations in chapter IV of the International Convention for the Safety of Life at Sea, 1974, that is to say, performance standards for navigational and communication equipment. Interested parties were invited to give a progress report on updates to the document entitled “E-navigation strategy implementation plan: Update 1” (MSC.1/Circ.1595). The Subcommittee also reviewed issues related to the long-range identification and tracking system and testing and operating of maritime autonomous surface ships. The Subcommittee’s recommendations will be reviewed by the Maritime Safety Committee at its next meeting. The Committee was scheduled to meet in May 2020, but the meeting was postponed because of the COVID-19 crisis (IMO, 2020a).

Regulatory and other issues related to maritime autonomous surface ships were on the agendas of the IMO Legal Committee (scheduled for March 2020) and the IMO Facilitation Committee (scheduled for April 2020); both meetings also had to be postponed.[34]

Paperless bills of lading

Negotiable bills of lading are used for the carriage of goods by sea, particularly in containerized transport, which carries the world’s manufactured cargo. They are also used in the commodities trade in cost, insurance and freight terms (commonly known as CIF). Bills of lading must be physically presented to the carrier to obtain delivery, due to their documentary security function and their key role as a document of title in international trade (see Gaskell et al., 2000; UNCTAD, 2003). For various reasons, despite numerous attempts over the past decades, commercially viable electronic equivalents have only recently begun to emerge (UNCTAD 2003). The International Group of Protection and Indemnity Clubs provides indemnity insurance to about 90 per cent of the world’s ocean-going tonnage (International Group of Protection and Indemnity Clubs, 2020). The Group has recognized six electronic bill-of-lading systems or providers to date (United Kingdom Protection and Indemnity Club, 2017; United Kingdom Protection and Indemnity Club, 2020a; United Kingdom Protection and Indemnity Club, 2020b). Against this background, and in the light of the increased need for virtual interactions resulting from the ongoing COVID-19 crisis, recent developments and efforts to enable and promote paperless bill of lading solutions, including the following, are particularly worth noting.

The Digital Container Shipping Association announced plans to promote an initiative to enable the open collaboration necessary for achieving full electronic bill of lading adoption, based on the belief that an electronic bill of lading would be beneficial for all parties in container shipping (JOC, 2019). As part of this initiative, the Association aims to develop open-source standards for necessary legal terms and conditions, as well as definitions and terminology to facilitate communication among customers, container carriers, regulators, financial institutions and other industry stakeholders. In its view, carriers could reduce costs and inefficiencies associated with the manual creation of paper documents. If successful, ports and regulatory agencies would benefit from having access to the digital data within the electronic bills of lading, and irregular shipping patterns would be easier to identify.

According to research by the Association, paper bill processing costs three times as much as electronic

[34] IMO set a remote meeting plan for September–December 2020 (https://imo-newsroom.prgloo.com/news/imo-sets-remote-meeting-plan-for-september-december-2020).

bill of lading processing, which was determined to be an extra $4 billion annually in collective processing costs, at a 50 per cent adoption rate for the container shipping industry. With regard to the success of electronic air waybills for airfreight introduced by the International Air Transport Association in 2010, the Association suggests that a 50 per cent adoption rate may be feasible by 2030 if steps are taken now to begin standardizing electronic bills of lading (Digital Container Shipping Association, 2020c). This is an ambitious and worthwhile goal; however, air waybills, unlike negotiable bills of lading, do not serve as documents of title providing their holder with independent documentary security (UNCTAD, 2003). Therefore, there are fewer legal and regulatory problems associated with the use of electronic air waybills.

Progress is being made regarding acceptance of this technology by government authorities, banks and insurers, and this is likely to be accelerated as a result of the COVID-19 crisis. For instance, a number of Digital Container Shipping Association members had reported a sharp increase in electronic bill of lading adoption, in an effort to keep trade moving. As noted previously, the International Group of Protection and Indemnity Clubs has so far approved six electronic bill-of-lading systems or providers. As noted by the Association, in the case of negotiable bills of lading, the standard electronic bill of lading would likely have to be used in conjunction with new technologies, such as distributed ledger technology, peer-to-peer technology and blockchain technology, which offer potential solutions for eliminating the risk of a single catastrophic failure or attack that would compromise the integrity and uniqueness of an electronic bill of lading (Digital Container Shipping Association, 2020c; JOC, 2020).

Recently, Ocean Network Express, the world's sixth largest container line (see also chapter 2) became the latest shipping line to offer fully electronic bills of lading to their customers. The liner company recently announced that it had handled its first electronic negotiable bill of lading, using essDOCS's paperless document solution, CargoDocs, which is among the systems approved by the International Group of Protection and Indemnity Clubs (https://essdocs.com/). Ocean Network Express used this electronic bill of lading for a shipment of containerized synthetic rubber from the Russian Federation to China and is planning to allow customers to use electronic bills of lading on a regional and subsequently global basis commencing in the second quarter of 2020 as part of initiatives aimed at delivering an improved, digital customer experience (Ocean Network Express, 2020). Further, India is to integrate electronic bills of lading and digital documentation into the country's electronic port community system, incorporating the CargoX platform for blockchain document transfer into its infrastructure, to manage the secure exchange of data (Smart Maritime Network, 2020).

Given the number of earlier attempts to create commercially viable electronic alternatives to traditional paper-based bills of lading across the shipping industry, including, Bolero[35] and some other recent systems, such as essDOCS, the success of ongoing initiatives will remain to be seen. However, the COVID-19 crisis provides an added impetus for resolving long-standing legal and regulatory problems. The main challenge in efforts to develop electronic alternatives to the traditional paper bill of lading has been the effective replication of the document's functions in a secure electronic environment, while ensuring that the use of electronic records or data messages enjoys the same legal recognition as that of paper documents. For negotiable bills of lading, with the exclusive right to the delivery of goods traditionally linked to the physical possession of original document, this includes in particular, the replication, in an electronic environment, of the unique document of title function (UNCTAD, 2003). There are also concerns over legal enforceability, as not all Governments have legislative provisions to this effect in place.

Establishing the widespread use of a fully electronic equivalent to the traditional bill of lading will require much international cooperation and coordination to ensure that commercial parties across the world are readily accepting and using relevant electronic records, and that legal systems are adequately prepared. In addition, capacity-building may be required, particularly for small and medium-sized enterprises in developing countries that may lack access to the necessary technology or means of implementation. In this context, too, increasing cybersecurity and related capacity-building will be a matter of critical and strategic importance for the further development of international trade in an electronic environment.

The use of electronic trade documentation, including electronic bills of lading equivalents, has increased significantly in importance since the COVID-19 pandemic, and related physical distancing, teleworking and disrupted or suspended postal services have affected large parts of the world population. This matters, particularly since trade finance transactions typically require significant levels of in-person review and processing of hard-copy paper documentation. In these circumstances, international organizations and industry bodies have issued calls for Governments to remove restrictions on the use and processing of electronic trade documents and the need for any documentation to be presented in hard copy. For instance, the International Chamber of Commerce has called on all Governments to take two key actions without delay: as a temporary measure, void any legal requirements for trade documentation to be in hard copy and adopt the United Nations Commission on International Trade Law Model Law on Electronic Transferable Records (International Chamber of Commerce, 2020a; United

[35] See www.bolero.net and UNCTAD, 2003.

Nations Commission on International Trade Law, 2018; UNCTAD, 2017a).[36]

B. REGULATORY DEVELOPMENTS RELATING TO INTERNATIONAL SHIPPING, CLIMATE CHANGE AND OTHER ENVIRONMENTAL ISSUES

1. Developments under the auspices of the International Maritime Organization related to the reduction of greenhouse gas emissions from ships

Maritime decarbonization and the reduction of greenhouse gas emissions from ships have become a priority area for policymakers and industry to be achieved, among others, through the adoption of energy-efficient technologies, the optimization of ship operations and use of low- and zero-carbon fuels, as well as regulation. A number of measures are being adopted in these areas by Governments, often in collaboration with industry, both nationally and internationally.

The IMO Marine Environment Protection Committee has for some time been addressing greenhouse gas emissions from ships engaged in international voyages. The measures to improve the energy efficiency of international shipping were adopted under a new chapter of the International Convention for the Prevention of Pollution from Ships, 1973, as modified by the Protocol of 1978 relating thereto (MARPOL), annex VI. In force since 1 January 2013, these measures apply to ships of 400 gross tons and above that are engaged in international voyage. They make two key requirements mandatory: The energy efficiency design index for new ships and the ship energy efficiency management plan for new and existing ships.

The energy efficiency design index for new ships has become increasingly strict over time. In May 2019, the Marine Environment Protection Committee approved, for adoption at its next session (initially scheduled for April 2020, but postponed due to the COVID-19 pandemic), draft amendments to MARPOL annex VI. These aimed to significantly strengthen the phase 3 requirements of the index, bringing forward their entry into force date to 2022, from 2025, for several ship types, including container ships, gas carriers, general cargo ships and liquefied natural gas carriers.

The ship energy efficiency management plan for new and existing ships establishes a mechanism for improving the energy efficiency of ships, including by monitoring their energy efficiency performance, new practices and technologies. For instance, it is now mandatory for ships to collect and report ship fuel oil consumption data. Since 1 January 2019, flag States collect consumption data for each type of fuel oil used by ships of 5,000 gross tons and above, which are then transferred to the IMO ship fuel oil consumption database. Reports analysing and summarizing the data collected shall periodically inform the Marine Environment Protection Committee. Information from the reports also benefits analysis on emissions by flag or vessel type as presented in chapter 3.E of the Review.

Already in April 2018, the Marine Environment Protection Committee had adopted the Initial Strategy on reduction of greenhouse gas emissions from ships (IMO, 2018a, annex 1; UNCTAD, 2019a), which envisages a reduction of the total annual greenhouse gas emissions from international shipping by at least 50 per cent by 2050 as compared with 2008, while, at the same time, pursuing efforts towards phasing them out entirely. Candidate short-term measures, to be further developed and agreed upon by member States between 2018 and 2023, include technical and operational energy efficiency measures for both new and existing ships, such as speed optimization and reduction, the development of robust life cycle greenhouse gas and carbon intensity guidelines for all types of fuels to prepare for the use of alternative low-carbon and zero-carbon fuels, port activities and incentives for first movers.

Innovative emissions-reduction mechanisms, possibly including market-based measures, to incentivize greenhouse gas emission reduction – a controversial issue for a number of years – were included among candidate midterm measures. These are to be agreed and decided upon between 2023 and 2030, along with possible long-term measures to be undertaken beyond 2030 that would ultimately lead to zero-carbon or fossil-free fuels to enable the potential decarbonization of the shipping sector in the second half of the century (for more information, see UNCTAD, 2018). In October 2018, the Marine Environment Protection Committee approved a programme of follow-up actions of the Initial Strategy on reduction of greenhouse gas emissions from ships up to 2023. It is planned that a revised strategy on reduction of greenhouse gas emissions from ships will be adopted in 2023.

The Marine Environment Protection Committee Working Group on Reduction of Greenhouse Gas Emissions from Ships met for its sixth intersessional meeting in November 2019 and made progress on several issues, leading towards achieving the levels of ambition set out in the Initial Strategy (see IMO, 2019a). These include the following:

- Development of a draft resolution on national action plans to address greenhouse gas emissions

[36] For solutions that involve the use of electronic documents, scanned, faxed or emailed images and potential scenarios in the delivery of documents during the COVID-19 crisis, see International Chamber of Commerce, 2020b.

from international shipping. The development and update of relevant national action plans was envisaged as a candidate short-term measure in the Initial Strategy. The resolution suggests that national action plans could include, without being limited to, the following actions: improving domestic institutional and legislative arrangements for the effective implementation of existing IMO instruments; developing activities to further enhance the energy efficiency of ships; initiating research and advancing the uptake of alternative low-carbon and zero carbon fuels; accelerating port-emission reduction activities, consistent with resolution MEPC.323(74); fostering capacity-building, awareness-raising and regional cooperation; and facilitating the development of infrastructure for green shipping. Potential legal, policy and institutional arrangements to be put in place by Member States should be elaborated in accordance with national circumstances and priorities and relevant experiences shared with IMO.

- Consideration of various concrete proposals for mandatory short-term measures to further reduce greenhouse gas emissions from existing ships. Proposals of a technical nature included, for example, an energy efficiency existing ship index, which would require ships to make technical modifications, for example, mandatory engine power limitation, to improve their energy efficiency. Proposals for an operational approach included focusing on carbon-intensity-reduction targets using appropriate carbon-intensity indicators, including by means of strengthening the ship energy efficiency management plan based on regular energy audits of the ship. This approach could include measures to limit or optimize speeds for voyages. There was general agreement that a mandatory goal-based approach for both the technical and operational approaches would provide the needed flexibility and incentive for innovation.
- Assessment of impacts of the proposals on States, with particular attention to be paid to the needs of developing countries, especially the least developed countries and small island developing States.
- Consideration of the use of alternative fuels, in particular with regard to measures in the medium and long term. This is also important to encourage the uptake of low- and zero-carbon fuels in the shipping sector. The establishment of a dedicated workstream for the development of life cycle greenhouse gas or carbon-intensity guidelines (for example, from well to wake or tank to propeller) for all relevant types of alternative fuels was suggested. This could include, for example, biofuels, (renewable) electro- or synthetic fuels such as hydrogen or ammonia. The issue of methane slip, including enhanced understanding of the problem, how methane slip could be measured, monitored and controlled and which measures could be considered by IMO to address the matter, was discussed in relation to the uptake of methane-based fuels such as liquefied natural gas (IMO, 2019a).

Other recent IMO collaborative work to address greenhouse gas emissions from ships engaged in international voyage include the following:

- Fourth IMO greenhouse gas study. This study, published in August 2020, includes an inventory of current global emissions of greenhouse gases and relevant substances emitted between 2012 and 2018, from ships of 100 gross tons and above engaged in international voyages, as well as their carbon intensity, and projects scenarios for future international shipping emissions from 2018–2050. It builds on the third IMO greenhouse gas study, issued in 2014. The fourth study, mentioned above, indicates that the share of shipping emissions in global anthropogenic emissions increased from 2.76 per cent in 2012 to 2.89 per cent in 2018. Using a new voyage-based allocation of international shipping, the study indicates that carbon-dioxide emissions increased from 701 million tons in 2012 to 740 million tons in 2018 – a 5.6 per cent increase – but at a lower growth rate than that of total shipping emissions. Using the vessel-based allocation of international shipping taken from the third IMO greenhouse gas study, carbon-dioxide emissions grew from 848 million tons in 2012 to 919 million tonnes in 2018 – an 8.4 per cent increase. The study also notes that ship emissions are projected to rise from about 90 per cent of 2008 emissions in 2018 to 90–130 per cent of 2008 emissions by 2050. Thus, much work lies ahead to meet the IMO strategy goal of cutting greenhouse gas emissions from international shipping by at least 50 per cent from 2008 levels by 2050. Also, to phase out greenhouse gas emissions from the sector as soon as possible, regulations that encourage innovation and the widespread adoption of the cleanest, most advanced technologies are needed (International Council on Clean Transportation, 2020). Consideration and approval of the fourth IMO greenhouse gas study 2020 by the Marine Environment Protection Committee is still pending (IMO, 2020b).
- Multi-donor trust fund for reduction of greenhouse gas emissions from ships. This fund was established to provide a dedicated source of financial support to sustain IMO technical

cooperation and capacity-building activities to support the implementation of the Initial Strategy.

- Collaboration with UNCTAD on an expert review of the impact assessments submitted to the Intersessional Working Group on Reduction of Greenhouse Gas Emissions from Ships. The collaborative efforts aim to produce a review of the comprehensiveness of the impact assessments of the concrete proposals to improve the energy efficiency of existing ships submitted to the Working Group, taking into account the procedure for assessing impacts on States of candidate measures set out in MEPC.1/Circ.885 and the available data.

During the United Nations Climate Action Summit, held in New York in September 2019, many business leaders and local government representatives announced concrete actions to address climate change (United Nations, 2019). For example, the industry-led initiative "Getting to Zero Coalition", supported by UNCTAD, committed to the deployment of viable zero-emissions vessels by 2030 to further the achievement of the goals of the IMO Initial Strategy (United Nations, 2019).

With regard to the European Union and the European Economic Area, an important legal requirement is worth noting. Since 1 January 2018, large ships of over 5,000 gross tons that load or unload cargo or passengers at ports in the European Economic Area have been required to monitor and report their related carbon-dioxide emissions and other relevant information, in conformity with Regulation 2015/757, as amended by Delegated Regulation 2016/2071 (see https://ec.europa.eu/clima/policies/transport/shipping_en). As a result, since 2019, ships calling at ports in the European Economic Area must report under both the European Union regulation and the IMO data collection system. Every year, the European Commission publishes a report to keep the public abreast of trends in carbon-dioxide emissions and provides energy efficiency information concerning the monitored fleet (European Commission, 2020a; European Commission, 2020b).

2. Developments under the United Nations Framework Convention on Climate Change and related issues

The Conference of the Parties to the United Nations Framework Convention on Climate Change on its twenty-fifth session, held in Madrid, in December 2019, once again highlighted how much work lies ahead on both the domestic and international fronts with regard to climate action that is consistent with the goal of the Paris Agreement[37] of holding the increase in the global average temperature to well below 2°C above pre-industrial levels and to pursue efforts to limit the temperature increase to 1.5°C above pre-industrial levels (article 2). In respect of greenhouse gas emissions from international shipping, the Subsidiary Body for Scientific and Technological Advice is one of two permanent subsidiary bodies to the United Nations Framework Convention on Climate Change. The body, which supports the work of the Conference of the Parties by providing information and advice, including on emissions from fuel used for international aviation and maritime transport, did not reach agreement and postponed discussions until the next session, to be held at the twenty-sixth session of the Conference of the Parties in November 2021 (United Nations, 2020).

Documents and publications launched at the twenty-fifth session of the Conference of the Parties to assist countries in their efforts to implement the Paris Agreement include the following:

- A yearbook (United Nations Climate Change Secretariat, 2019).
- An online database in which a diverse range of stakeholders have registered their climate change mitigation and/or adaptation commitments, as well as a number of climate action pathways, developed by the Marrakech Partnership for Global Climate Action (United Nations Framework Convention on Climate Change, 2020).
- The Global Climate Action portal, formerly known as the Non-State Actor Zone for Climate Action, which outlines transformational actions and milestones in some key sectoral and cross-cutting areas, such as transport and resilience.

Also launched at the twenty-fifth session of the Conference of the Parties was a declaration on climate change by the World Association for Waterborne Transport Infrastructure, also known as PIANC (World Association for Waterborne Transport Infrastructure, 2019). The declaration highlights a number of priority actions to strengthen adaptation and resilience-building. These include inspection and maintenance; monitoring systems and effective data management; and risk assessments, contingency plans and warning systems. It also provides a focus on flexible and adaptive infrastructure, systems and operations, and engineered redundancy to improve resilience.

With regard to climate change adaptation and resilience-building for seaports, the transport pathway action table of the Marrakech Partnership for Global Climate Action includes two distinct action areas with a focus on adaptation for transport systems and transport infrastructure, respectively, as well as related milestones for 2020, 2030 and 2050 (Marrakech Partnership for Global Climate Action, 2019a). Inter alia, these milestones, which have also been integrated into the cross-sectoral resilience and adaptation pathway action table, envisage that, by 2030, "All critical transport infrastructure assets, systems/networks components are [made] climate resilient to (at least) 2050"; and,

[37] Ratified by 188 States. See https://unfccc.int/process/the-paris-agreement/status-of-ratification.

by 2050, "[A]ll critical transport infrastructure assets, systems/networks components are [made] climate resilient to (at least) 2100" (Marrakech Partnership for Global Climate Action, 2019b).[38] While this represents an important and timely ambition, a major acceleration of efforts will be required to put relevant measures in place.

Climate change adaptation and resilience-building is an increasingly important issue, in particular from the perspective of vulnerable developing countries that are at the forefront of climate change impacts, such as small island developing States.[39] Critical coastal transport infrastructure in these countries, notably ports and airports, are lifelines for external trade, food and energy security, and tourism, including in the context of disaster-risk reduction (UNCTAD, 2019b; UNCTAD and United Nations Environment Programme, 2019). These assets are projected to be at growing risk of coastal flooding, from as early as in the 2030s, unless effective adaptation action is taken (Intergovernmental Panel on Climate Change, 2018; Intergovernmental Panel on Climate Change, 2019; Monioudi et al., 2018). In the absence of timely planning and of the implementation of requisite adaptation measures, the projected impacts on critical transport infrastructure may have broad economic and trade-related repercussions and could severely compromise the sustainable development prospects of these vulnerable nations (Economic Commission for Europe, 2020; Pacific Community, 2019; UNCTAD, 2020a; UNCTAD 2020b;). However, there are still important knowledge gaps concerning vulnerabilities and the specific nature and extent of exposure that individual coastal transport facilities may be facing.[40]

A number of important issues have emerged as part of the related work of UNCTAD over the past decade. Thus, for the purposes of risk-assessment and with a view to developing effective adaptation measures, the generation and dissemination of more tailored data and information is important, as are targeted case studies and effective multi-disciplinary and multi-stakeholder collaboration. Successful adaptation strategies need to be underpinned by strong legal and regulatory frameworks that can help reduce exposure and/or vulnerability to climate-related risks of coastal transport infrastructure (UNCTAD, 2020a). Appropriate policies and standards also have an important role to play, particularly in the context of infrastructure planning and coastal zone management. Moreover, guidance, best practices, checklists, methodologies (for example, UNCTAD, 2017b) and other tools in support of adaptation are urgently required, and targeted capacity-building is going to be critical, especially for the most vulnerable countries.[41]

3. Protection of the marine environment and conservation and sustainable use of marine biodiversity

Relevant areas where regulatory action has recently been taken or is under way for the protection of the marine environment and conservation and sustainable use of marine biodiversity, are described below.

Implementing the 2020 sulphur limit of the International Maritime Organization

Sulphur oxides are known to be harmful to human health, causing respiratory symptoms and lung disease. They can lead to acid rain, which can harm crops, forests and aquatic species, and contribute to ocean acidification. Thus, limiting sulphur-oxide emissions from ships helps improve air quality and protect human health and the environment (IMO, 2020c). An IMO regulation limiting the sulphur content in ship fuel oil to 0.50 per cent, down from 3.50 per cent, entered into force on 1 January 2020 (UNCTAD, 2019a). In designated emission control areas, the limit remained even lower, at 0.10 per cent.[42]

To support consistent implementation and compliance and provide a means for effective enforcement by States, particularly port State control, IMO in October 2018 adopted an additional MARPOL amendment, which entered into force on 1 March 2020. The amendment prohibits not just the use, but also the carriage of non-compliant fuel oil for combustion purposes for propulsion or operation on board a ship, unless the ship is fitted with an approved equivalent method, such as a scrubber or exhaust gas cleaning system. Also, a comprehensive set of guidelines to support the consistent implementation of the lower 0.50 per cent limit on sulphur in ship fuel oil and related amendments to the Convention were approved in May 2019 (IMO, 2019b, annex 14).

[38] Key recommendations of technical experts, key industry stakeholders and international organizations participating in the ad hoc expert meeting entitled "Climate Change Adaptation for International Transport: Preparing for the Future", held by UNCTAD in 2019, are reflected in the Marrakech Partnership for Global Climate Action pathways on transport and on resilience (Marrakech Partnership for Global Climate Action, 2019a and 2019b). See https://unctad.org/en/pages/MeetingDetails.aspx?meetingid=2092.

[39] For further information and related work by UNCTAD, see https://SIDSport-ClimateAdapt.unctad.org; https://unctad.org/ttl/legal; https://unctad.org/en/pages/MeetingDetails.aspx?meetingid=2354.

[40] This is evidenced by recent port industry surveys and studies on climate change impacts and adaptation (Asariotis et al., 2018; Panahi et al., 2020).

[41] For further information on relevant practices and regulatory and policy approaches, see UNCTAD, 2020a. See also https://SIDSport-ClimateAdapt.unctad.org.

[42] The four emission control areas are as follows: the Baltic Sea area, the North Sea area, the North American area (covering designated coastal areas of Canada and the United States) and the United States Caribbean Sea area (around Puerto Rico and the United States Virgin Islands).

To support the enforcement of the carriage ban and the safe and consistent sampling of fuel oil being carried for use, in February 2020, the IMO Subcommittee on Pollution Prevention and Response made progress in preparatory work and various draft amendments and guidelines to be submitted to the next session of the Marine Environment Protection Committee with a view to their later consideration and adoption. The Subcommittee finalized draft guidelines that provide a recommended method for the sampling of liquid fuel oil intended to be used or carried for use on board a ship. It also finished its revision of the 2015 guidelines on exhaust gas cleaning systems (also known as scrubbers), with a view to enhancing the uniform application of the guidelines by specifying the criteria for the testing, survey, certification and verification of such systems under MARPOL annex VI, to ensure that they provide effective equivalence to the sulphur-oxide emission requirements of regulations. In addition, the Subcommittee agreed to recommend to the Marine Environment Protection Committee that its future work should look at the evaluation and harmonization of rules and guidance on the discharge of was water from exhaust gas cleaning systems into the aquatic environment, including conditions and areas. By way of background, some IMO members have expressed concern that several more factors must be taken into account when assessing the impact of wash water discharge from scrubbers operating in ports and coastal areas. It has also been suggested that open-loop systems currently in use and compliant with the 2015 guidelines may produce harmful impacts in certain coastal areas. A number of coastal States (China, Malaysia, Norway and Singapore) have announced a ban of open-loop exhaust gas cleaning systems in certain coastal areas (Safety4Sea, 2019c), and Egypt has banned the use of such systems when transiting the Suez Canal (IMO, 2020d; Seatrade Maritime News, 2020).

The implementation of the sulphur regulation as of 1 January 2020 was initially considered to be relatively smooth, and compliant fuel oil was reported to be widely available. However, some difficulties have arisen as a result of the disruptions caused by the pandemic. In March 2020, the ban on the carriage on non-compliant fuel oil entered force to support the implementation of the sulphur limit. However, it appears that its enforcement by port State control authorities was suspended, due to measures put in place to reduce inspections and contain the risk of spreading the virus (Heavy Lift, 2020).

Ballast water management

In February 2020, the IMO Subcommittee on Pollution Prevention and Response completed its work on the revision of a guidance document on the testing of ballast water management systems, intended to validate their installation by demonstrating that their mechanical, physical, chemical and biological processes are working properly. This guidance is expected to be adopted by the Marine Environment Protection Committee at its next session, as an amendment to regulation E-1 of the International Convention for the Control and Management of Ship's Ballast Water and Sediments, 2004, also known as the Ballast Water Management Convention, 2004.

Ballast Water Management Convention, 2004, has been in force since September 2017. By 31 July 2020, it had been ratified by 84 States, representing 91.10 per cent of the gross tonnage of the world's merchant fleet. The Convention aims to prevent the risk of the introduction and proliferation of non-native species following the discharge of untreated ballast water from ships. This is considered one of the four greatest threats to the world's oceans and a major threat to biodiversity, which, if not addressed, can have severe public health-related and environmental and economic impacts (UNCTAD, 2011; UNCTAD, 2015). From the date of the Convention's entry into force, ships have been required to manage their ballast water to meet standards D-1 and D-2; the former requires ships to exchange and release at least 95 per cent of ballast water by volume far away from a coast; the latter raises the restriction to a specified maximum amount of viable organisms allowed to be discharged, limiting the discharge of specified microbes harmful to human health. Currently, the regulatory focus continues to be on the effective and uniform implementation of the Convention.

Biofouling

While the Ballast Water Management Convention, 2004 aims to prevent the spread of potentially harmful aquatic species in ballast water, invasive species, such as marine animals, plants and algae, can attach themselves to the outside of ships (for example, ship hulls) and other marine structures. This is known as biofouling. When ships and structures move to new areas, these species can detach themselves, adapt to the new habitat, overcome local fauna and become invasive, with negative effects on the host ecosystem. Therefore, biofouling needs to be addressed as well. Biofouling has other negative effects – it increases the surface roughness of ship hulls and propellers, resulting in speed loss at constant power or power increase at constant speed and higher fuel consumption of up to 20 per cent (Riviera, 2020d; Riviera, 2020e).

Anti-fouling paints are normally used to coat the bottoms of ships to prevent sea life such as algae and molluscs attaching themselves to the hull, thereby slowing down the ship and increasing fuel consumption. The Convention for the Control of Harmful Anti-fouling Systems on Ships, 2001 defines anti-fouling systems as "a coating, paint, surface treatment, surface or device that is used on a ship to control or prevent attachment of unwanted organisms". It aims to prohibit the use of harmful organotin compounds in anti-fouling paints used on ships and establish a mechanism to prevent the potential future use of other harmful substances in anti-fouling systems. The Convention entered into force on 17 September 2008. As of 31 July 2020, 89 States parties, representing 96.09 per cent of the gross tonnage of the world's merchant fleet, had ratified the Convention. Annex 1 to the Convention states that as from 1 January 2003, all ships should not apply or re-apply organotin compounds, which act as biocides in anti-fouling systems, and as from 1 January 2008, ships either (a) shall not bear such compounds on their hulls or external parts or surfaces or (b) shall bear a coating that forms a barrier to such compounds leaching from the underlying non-compliant anti-fouling systems.

In July 2017, the Marine Environment Protection Committee started work on amending annex 1 to the Convention to include controls on the biocide chemical compound cybutryne, since scientific data had indicated that cybutryne causes significant adverse effects to the environment, especially to aquatic ecosystems. Work on this matter is ongoing in the Subcommittee on Pollution Prevention and Response, which in February 2020 finalized a proposed amendment to the Convention to include controls on cybutryne. The draft amendment will be presented to the Marine Environment Protection Committee at its next session for approval. The Subcommittee also began its review of the IMO Guidelines for the Control and Management Of Ships' Biofouling to Minimize the Transfer of Invasive Aquatic Species, also known as the Biofouling Guidelines (IMO, 2011), which provide a globally consistent approach to the management of biofouling (IMO, 2020d).

Marine pollution from plastics and microplastics

Marine debris in general, and plastics and microplastics in particular, give rise to some of the greatest environmental concerns today, along with climate change, ocean acidification and loss of biodiversity. These directly affect the sustainable development aspirations of developing States and small island developing States in particular, which, as custodians of vast areas of oceans and seas, face an existential threat from and are disproportionately affected by the effects of pollution from plastics. The issue of marine debris, plastics and microplastics in the oceans has been receiving increasing public attention and was the topic of the seventeenth meeting of the United Nations Open-ended Informal Consultative Process on Oceans and the Law of the Sea in 2016 (United Nations, 2016). Sustainable Development Goal 14.1, committing to prevent and significantly reduce marine pollution of all kinds, in particular from land-based activities, including marine debris and nutrient pollution by 2025, is particularly relevant in this context. Given the cross-cutting nature of the problem, plastics pollution is also relevant to other Sustainable Development Goals, including Goals 4 (education), 6 (clean water and sanitation), 12 (sustainable consumption and production patterns), and 15 (sustainable use of terrestrial ecosystems).

IMO is implementing an action plan to address marine plastic litter from ships, which contains measures to be completed by 2025, relating to all ships, including fishing vessels, and supports the IMO commitment to meeting the targets set in Goal 14 (IMO, 2018b). At its seventh meeting in February 2020, the Subcommittee on Pollution Prevention and Response prepared draft Marine Environment Protection Committee circulars on the provision of adequate facilities at ports and terminals for the reception of plastic waste from ships and on the sharing of results from research on marine litter and encouraging studies to better understand microplastics from ships. It also established a correspondence group to consider how to amend MARPOL annex V and the 2017 guidelines for the implementation of MARPOL annex V (resolution MEPC.295(71)) to facilitate and enhance reporting of the accidental loss or discharge of fishing gear and consider the information to be reported to Administrations and IMO, as well as reporting mechanisms and modalities (IMO, 2020d).

While the focus of this section of the Review is on developments related to plastic waste from ships, some considerations regarding plastics pollution arise in the context of the COVID-19 crisis. Various protective measures have been implemented as a priority over the past months with a view to controlling the spread of the virus. These include the wearing of surgical face masks and gloves and the frequent disinfection of hands, all of which involve the use of plastic. In addition, because of the threat of contamination, people may tend to use disposable or single-use plastic items such as food containers and utensils, rather than reusable ones. There is a risk for these items to end up as litter in the environment, including in the sea and along beaches, which in many countries are a mainstay of the local tourism industry. Short-term solutions to address an increase in plastics pollution arising from the ongoing pandemic may include imposing fines, placing labels on disposable items and making information on littering and recycling more available to the public. Public attention on plastics pollution is likely to increase, once the immediate COVID-19 health crisis is under control. In the meantime, researchers suggest recycling single-use plastic items, limiting food deliveries and ordering from grocery suppliers that offer more sustainable delivery packaging. In addition, wearing reusable face masks, disposing of single-use face masks correctly and buying hand sanitizer contained in ecologically sustainable packaging should also be considered (see https://earth.org/covid-19-surge-in-plastic-pollution/).

Safety considerations of new fuel blends and alternative marine fuels

To ensure compliance with the mandatory 0.50 per cent sulphur limit for fuel oil and meet the emission targets set out in the IMO Initial Strategy on reduction of greenhouse gas emissions, new fuels and fuel blends are being developed. At IMO, matters related to such fuels are considered by the Maritime Safety Committee in the context of discussions on the International Code of Safety for Ships using Gases or other Low-flashpoint Fuels. The Code, which entered into force in 2017, aims to minimize the risk to ships, their crews and the environment, given the nature of the fuels involved. It has initially focused on liquefied natural gas, but work is now under way to consider other fuel types.

In preparation for the next meeting of the Committee (scheduled for May 2020 but postponed due to the COVID-19 crisis), the Subcommittee on Carriage of Cargoes and Containers, at its sixth session in September 2019 took the following action:

- Finalized draft interim guidelines for the safety of ships using methyl or ethyl alcohol as fuel, for submission to the Maritime Safety Committee for approval.
- Made progress in developing draft interim guidelines for the safety of ships using fuel cell power installations.
- Agreed to develop amendments to the International Code of Safety for Ships using Gases or other Low-flashpoint Fuels to include safety provisions for ships using low-flashpoint oil fuels and established a correspondence group to continue this work.
- Approved in principle draft amendments to the Code, relating to specific requirements for ships using natural gas as fuel.
- Agreed to develop interim guidelines on safety provisions for ships using liquefied petroleum gas fuels.
- Completed draft guidelines for the acceptance of alternative metallic materials for cryogenic service in ships carrying liquefied gases in bulk and ships using gases or other low-flashpoint fuels, for submission to the Maritime Safety Committee for approval (IMO, 2019c).

Conservation and sustainable use of marine biodiversity of areas beyond national jurisdiction: Legally binding instrument under the United Nations Convention on the Law of the Sea, 1982

Areas beyond national jurisdiction hold unique oceanographic and biological features and play a role in climate regulation.[43] They provide seafood, raw materials and genetic and medicinal resources, which are of increasing commercial interest and hold promise for the development of new drugs to treat infectious diseases that are a major threat to human health – such as antibiotic-resistant infections and potentially, coronavirus disease. From the perspective of developing countries, access and benefit sharing, as well as the conservation of marine genetic resources, are of particular importance in this context (Premti, 2018).

The United Nations Convention on the Law of the Sea, 1982 sets forth the rights and obligations of States regarding the use of the oceans, their resources and the protection of the marine and coastal environment. However, it does not expressly refer to marine biodiversity or to the exploration and exploitation of resources within the water column in areas beyond national jurisdiction. Therefore, ongoing negotiations towards a new international legal instrument under the United Nations Convention on the Law of the Sea on the conservation and sustainable use of marine biological diversity of areas beyond national jurisdiction are particularly worth noting. Three sessions of the intergovernmental conference on the issue have taken place, the most recent, in August 2019 (see UNCTAD, 2019a for further information on discussions held). Discussions on a broad range of issues were expected to continue during the fourth session of the conference, scheduled to be held from 23 March to 3 April 2020, at United Nations Headquarters in New York, but were postponed due to the COVID-19 crisis.

One gap that the new international legally binding instrument aims to address is the establishment of marine protected areas. According to scientific evidence, these areas are effective tools for conserving and restoring oceans and their resources. However, under the current system of ocean management, there is no way to establish comprehensive marine protected areas for most parts of the high seas. A study was recently conducted to help determine which areas of the high seas should be protected first as ecologically or biologically significant (Visalli et al., 2020). It considered a variety of factors and conservation features and used a conservation prioritization tool to help select areas of the ocean that would include at least 30 percent of these conservation features, while minimizing overlap with areas that are already being heavily fished. This and other similar studies

[43] Maritime zones under the United Nations Convention on the Law of the Sea, 1982 include the following: the territorial sea, extending up to 12 nautical miles from the baseline (article 3); exclusive economic zones, extending from the edge of the territorial sea to 200 nautical miles from the baseline (article 57); the continental shelf, the natural prolongation of land territory to the outer edge of the continental margin, or 200 nautical miles from the baseline, whichever is greater (article 76); and areas beyond national jurisdiction, composed of "the Area" (article 1) and the high seas (article 86).

highlighting specific areas beyond national jurisdiction as high priorities for protection are expected to inform negotiations and decision-making on these issues at the United Nations.

C. OTHER LEGAL AND REGULATORY DEVELOPMENTS AFFECTING TRANSPORTATION

Extension of the European Union Consortia Block Exemption Regulation up to 2024

Article 101(1) of the Treaty on the Functioning of the European Union prohibits agreements between undertakings that restrict competition. However, article 101(3) of that treaty allows declaring such agreements compatible with the internal market, provided they contribute to improving the production or distribution of goods or to promoting technical or economic progress, while allowing consumers a fair share of the resulting benefits. Liner shipping is a highly concentrated industry, with 91 per cent of deep-sea maritime transport services controlled by 10 global operators (see chapter 2, table 10 of this report). In the European Union, liner conferences allowing their members to fix freight rates collectively and discuss market conditions were banned as of 2008 (Council (EC) Regulation 1419/2006). However, liner shipping consortia, as a form of operational cooperation, continue to enjoy a block exemption from European Union competition rules, set to expire on 25 April 2020. Given the international nature of liner shipping services and experience gained from the earlier initiatives of the European Union in this field (Premti, 2016), the impact of the European Union decisions goes beyond Europe and has a bearing on the container shipping markets in developing countries and other European Union trading partners.

In September 2018, the European Commission conducted an evaluation of the Consortia Block Exemption Regulation (European Commission, 2009), which included a consultation of stakeholders in the maritime liner shipping supply chain (for the results, see European Commission, 2019a). The aim was to assess the impact and relevance of that regulation in view of the general policy of harmonizing competition rules and recent important developments in the liner shipping industry and to determine whether it should be left to expire or to be prolonged, and if so, under which conditions. Allowing the Regulation to expire would not mean that consortia agreements become unlawful – but only that they would be examined under the general rules on competition just as cooperation agreements in other sectors (European Commission, 2019b). The first consortia block exemption regulation was adopted in 1995 and revised in 2009; since then, it has been prolonged every five years without modification.

The main stakeholders participating in the consultation were the carriers which apply the Regulation and their clients (shippers and freight forwarders), and port operators and their respective associations, including those in developing countries who may be affected by the freight rates and the quality and frequency of services resulting from a change in the European Union regulation.

Industry associations representing users of liner shipping services and service providers expressed their objection to the extension of the Regulation.[44] They argued that the evaluation criteria used by the European Commission were biased towards the interest of the carriers, that the 30 per cent market share threshold was difficult to monitor in practice due to missing data and that quality and choice, as well as service levels and schedule reliability, had decreased in recent years, while rate volatility had increased (Lloyd's Loading List, 2020) (see chapter 2). During the consultation, port operators expressed concerns, among others, about limited competition between individual lines that offered more or less equal service levels, and pointed out that any decrease in freight rates was a relatively small element of the total shipping costs (https://ec.europa.eu/competition/consultations/2018_consortia/index_en.html). In addition, representatives of transport workers were reluctant to prolong the Regulation, arguing that shipping companies were having a negative impact on the economic profitability of terminals and other service providers. Because of the increased size of ships, constant and significant investments from terminals were required, adversely affecting the working conditions and job security in ports.

On 24 March 2020, the European Commission announced an extension of the Consortia Block Exemption Regulation until 25 April 2024. According to the Commission, the Regulation results in efficiencies for carriers that can better use vessel capacity and offer more connections. Further, those efficiencies result in lower prices and better quality of service for consumers and a decrease in costs for carriers – in recent years, prices for customers have dropped by approximately 30 per cent (European Commission, 2020c).

D. STATUS OF CONVENTIONS

A number of international conventions in the field of maritime transport were prepared or adopted under the auspices of UNCTAD. The table below provides information on the status of ratification of each of those conventions as at 31 July 2020.

[44] European Association for Forwarding, Transport, Logistic and Customs Services; European Shippers Council; Global Shippers Forum and International Union for Road-Rail Combined Transport.

Table 5.1 Contracting States Parties to selected international conventions on maritime transport, as at 31 July 2020

Title of convention	Date of entry into force or conditions for entry into force	Contracting States
Convention on a Code of Conduct for Liner Conferences, 1974	6 October 1983	Algeria, Bangladesh, Barbados, Belgium, Benin, Burkina Faso, Burundi, Cabo Verde, Cameroon, Central African Republic, Chile, China, Congo, Costa Rica, Côte d'Ivoire, Cuba, Czechia, Democratic Republic of the Congo, Egypt, Ethiopia, Finland, France, Gabon, Gambia, Ghana, Guatemala, Guinea, Guyana, Honduras, India, Indonesia, Iraq, Italy, Jamaica, Jordan, Kenya, Kuwait, Lebanon, Liberia, Madagascar, Malaysia, Mali, Mauritania, Mauritius, Mexico, Montenegro, Morocco, Mozambique, Niger, Nigeria, Norway, Pakistan, Peru, Philippines, Portugal, Qatar, Republic of Korea, Romania, Russian Federation, Saudi Arabia, Senegal, Serbia, Sierra Leone, Slovakia, Somalia, Spain, Sri Lanka, Sudan, Sweden, Togo, Trinidad and Tobago, Tunisia, United Republic of Tanzania, Uruguay, Bolivarian Republic of Venezuela, Zambia **(76)**
United Nations Convention on the Carriage of Goods by Sea, 1978	1 November 1992	Albania, Austria, Barbados, Botswana, Burkina Faso, Burundi, Cameroon, Chile, Czechia, Dominican Republic, Egypt, Gambia, Georgia, Guinea, Hungary, Jordan, Kazakhstan, Kenya, Lebanon, Lesotho, Liberia, Malawi, Morocco, Nigeria, Paraguay, Romania, Saint Vincent and the Grenadines, Senegal, Sierra Leone, Syrian Arab Republic, Tunisia, Uganda, United Republic of Tanzania, Zambia **(34)**
International Convention on Maritime Liens and Mortgages, 1993	5 September 2004	Albania, Benin, Congo, Ecuador, Estonia, Honduras, Lithuania, Monaco, Nigeria, Peru, Russian Federation, Spain, Saint Kitts and Nevis, Saint Vincent and the Grenadines, Serbia, Syrian Arab Republic, Tunisia, Ukraine, Vanuatu **(19)**
United Nations Convention on International Multimodal Transport of Goods, 1980	Not yet in force – requires 30 contracting parties	Burundi, Chile, Georgia, Lebanon, Liberia, Malawi, Mexico, Morocco, Rwanda, Senegal, Zambia **(11)**
United Nations Convention on Conditions for Registration of Ships, 1986	Not yet in force – requires 40 contracting parties, representing at least 25 per cent of the world's tonnage as per annex III to the Convention	Albania, Bulgaria, Côte d'Ivoire, Egypt, Georgia, Ghana, Haiti, Hungary, Iraq, Liberia, Libya, Mexico, Morocco, Oman, Syrian Arab Republic **(15)**
International Convention on Arrest of Ships, 1999	14 September 2011	Albania, Algeria, Benin, Bulgaria, Congo, Ecuador, Estonia, Latvia, Liberia, Spain, Syrian Arab Republic, Turkey **(12)**

Note: For additional information, see UNCTAD Trade Logistics Branch, Policy and Legislation Section at unctad.org/ttl/legal. For official status information, see the United Nations Treaty Collection, available at https://treaties.un.org.

E. COVID-19 LEGAL AND REGULATORY CHALLENGES FOR INTERNATIONAL SHIPPING AND COLLABORATIVE ACTION IN RESPONSE TO THE CRISIS

1. Maritime health preparedness and response to the COVID-19 pandemic

Key shipping stakeholders, including international bodies and Governments, issued a number of recommendations and guidance which aimed to ensure, first of all, that seafarers were protected from the coronavirus disease, were medically fit and had access to medical care and that their ships met international sanitary requirements.[45]

Together with its industry partners and other international organizations, IMO developed and issued practical advice and guidance on a variety of technical and operational matters related to the pandemic. Given that IMO does not have an enforcement authority of its own, it cannot issue general exemptions from or delay implementation of the mandatory provisions of its relevant conventions or mandatory regulations for flag and port States. However, IMO issued a number of circular letters

[45] For a list of COVID-19-related communications on measures taken by IMO Member States and Associate Members (updated weekly), see www.imo.org/en/MediaCentre/HotTopics/Pages/COVID-19-Member-States-Communications.aspx. For a detailed list of recommendations by Governments and international bodies, see Safety4Sea, 2020b.

addressed to Member States, seafarers and shipping industry stakeholders and posted a compilation of its guidance and online resources from other international organizations and maritime industry on its website.[46] Circular letters included the following items:

- Information on the impacts of the pandemic on the shipping industry, including implementation and enforcement of mandatory IMO requirements, and a call for increased cooperation among flag and port States, taking a pragmatic approach to the uncertain COVID-19 situation (Circular Letter No. 4204/Add.1).
- Guidance relevant to all stakeholders, addressing global issues relevant to the health of seafarers, seagoing vessels and offshore infrastructure by establishing and implementing COVID-19 protocols for mitigating and preventing outbreaks at sea, following guidance from the European Commission, the International Chamber of Shipping, IMO and the World Health Organization on health and shipping in the context of COVID-19 (Circular Letters No. 4204/Add.1–Add.4).
- Recommendations for Governments and relevant national authorities on the facilitation of maritime trade during the pandemic (Circular Letter No. 4204/Add.6) and on ensuring the integrity of the global supply chain during the pandemic (Circular Letter No. 4204/Add.9).
- Guidance particularly relevant to shipbuilders, equipment suppliers, shipowners, surveyors and service engineers advising on newbuilding bulk carriers and oil tankers that were scheduled for delivery before 1 July 2020 (Circular Letter No. 4204/Add.7).
- European Commission guidelines on protection of health, repatriation and travel arrangements for seafarers, passengers and other persons on board ships (Circular Letter No. 4204/Add.11).
- World Health Organization information and guidance on the safe and effective use of personal protective equipment (Circular Letters No. 4204/Add.15 and Add.16).

On 20 February 2020, the European Union issued advice for ship operators on preparedness and response to the outbreak of COVID-19, which included a dedicated chapter on maritime transport and a focus on cargo ship travel (European Union, 2020a). Guidelines on the exercise of the free movement of the workers during the COVID-19 outbreak followed (European Commission, 2020d).

Representing the global shipping industry, the International Chamber of Shipping published new guidance for the industry to help combat the spread of the coronavirus disease. The guidance offered advice on managing port entry restrictions, practical protective measures against the disease for seafarers, including an outbreak management plan (International Chamber of Shipping, 2020a). The International Bunker Industry Association also adopted protective measures against the disease. Considering that in international shipping, the contact between ship and shore personnel during the bunkering process involved a possible risk of spreading the disease, it provided advice to mitigate the risk of infection during such process (Safety4Sea, 2020b). The International Association of Ports and Harbours adopted guidance on ports' responses to the pandemic, structured along a three-layered approach to present a methodology and a range of good practices on immediate measures addressing port operations, governance and communication; measures to protect the business and financial returns; and measures to support customers and supply chain stakeholders (International Association of Ports and Harbours, 2020b).

As part of its response to the COVID-19 outbreak, UNCTAD issued a call for action to keep ships moving, ports open and cross-border trade flowing (UNCTAD, 2020c). It also published a policy brief, highlighting a 10-point action plan to strengthen international trade and transport facilitation in times of pandemic (UNCTAD, 2020d). Related technical cooperation in collaboration with the United Nations regional commissions has already begun.[47] Moreover, the Secretaries-General of UNCTAD and IMO issued a joint statement in support of keeping ships moving, ports open and cross-border trade flowing during the pandemic (IMO and UNCTAD, 2020). To assist stakeholders in obtaining an overview of the multitude of COVID-19-related measures and responses, plan and potential implications thereof, UNCTAD drafted a technical note for ports and a non-exhaustive list of links to online resources from international organizations and industry groups that provide up-to date information about the ongoing developments in various countries.[48]

2. Maritime certification

Port State control regimes around the world, expressing solidarity with the shipping industry, also developed temporary guidance for their member authorities during the COVID-19 crisis.[49] In line with IMO efforts and circular letters related to the pandemic, port State control

[46] All COVID-19-related IMO circulars are available at www.imo.org/en/MediaCentre/HotTopics/Pages/Coronavirus.aspx.

[47] Transport and trade connectivity in the age of pandemics (project 2023X) (www.un.org/development/desa/da/da-response-to-covid-19/).

[48] See https://tft.unctad.org/ports-covid-19/ and https://etradeforall.org/unctad-repositories-of-measures-on-cross-border-movement-of-goods-and-persons/.

[49] Port State control is the inspection of foreign flag ships in national ports to verify their compliance with international rules on safety, security, marine environment protection and seafarers living and working conditions.

regimes developed temporary guidance on how they intended to deal with the impact of the pandemic. These included acceptance of extended periods of service on board for seafarers; extended periods for surveys, inspections and audits; and seafarers' certification, using a pragmatic and harmonized approach (see Indian Ocean Memorandum of Understanding on Port State Control Secretariat, 2020; Paris Memorandum of Understanding on Port State Control Secretariat, 2020; Secretariat of the Memorandum of Understanding on Port State Control in the Asia–Pacific Region, 2020).

Thus, as a general principle, the guidance adopted by the port State control regimes suggests that a pragmatic and risk-based approach regarding the above-mentioned issues be taken. In such cases, the active involvement of the flag State, and if appropriate, the recognized organization for the conduct of inspections and the issue of certification was expected. This would include examination of the available information on the ship and its history, as well as the performance of the ship's company. Whether an inspection took place remained the decision of the port State. Such temporary guidance might be reviewed, as appropriate, to keep aligned with the rapidly successive developments of the coronavirus disease and future initiatives by relevant stakeholders, including the International Labour Organization and IMO. In addition, recognized organizations,[50] including the American Bureau of Shipping,[51] Bureau Veritas,[52] DNV GL,[53] Indian Register of Shipping[54] and Lloyd's Register,[55] issued guidance for shipowners on how to apply for an extension of statutory certificates, including the Safety Management Certificate under the International Safety Management Code (International Convention for the Safety of Life at Sea, chapter IX); the International Ship Security Certificate under the International Code for the Security of Ships and of Port Facilities (International Convention for the Safety of Life at Sea, chapter XI-2); and the Maritime Labour Certificate (Maritime Labour Convention, 2006), or if possible, for remote surveys. A number of flag States also provided initial instructions on possible ways forward in cases where these certificates needed to be extended beyond the three months already suggested.[56]

Enabling the extension of the validity of licences and certificates leads to greater flexibility and legal certainty. These are necessary to maintain supply chains and ensure continued mobility at sea, while safeguarding safety and security. In this context, it is worth noting that in addition to the Safety Management Certificate and the International Ship Security Certificate, flag States are allowed to extend for up to three months the period of validity of the following certificates required under different mandatory IMO legal instruments (IMO, 2019d):

- Cargo Ship Safety Equipment Certificate.
- Cargo Ship Safety Construction Certificate.
- Cargo Ship Safety Radio Certificate.
- International Load Line Certificate.
- International Oil Pollution Prevention Certificate.
- International Pollution Prevention Certificate for Carriage of Noxious Liquid Substances in Bulk.
- International Sewage Pollution Prevention Certificate.
- International Air Pollution Prevention Certificate.
- International Certificate of Fitness for the Carriage of Dangerous Chemicals in Bulk or the Certificate of Fitness for the Carriage of Dangerous Chemicals in Bulk.
- International Certificate of Fitness for the Carriage of Liquefied Gases in Bulk.
- Passenger Ship Safety Certificate.
- Polar Ship Certificate.
- International Ballast Water Management Certificate.

As a general rule, according to IMO mandatory instruments, no certificate should be extended for a period longer than three months (IMO, 2019c), while according to the Maritime Labour Convention of the International Labour Organization (standard A5.1.3, paragraph 4), the flag State may extend the validity of the Maritime Labour Certificate for a period not exceeding five months. Therefore, it appears that due to the prevailing exceptional circumstances during the COVID-19 crisis, flag States should be able to extend the validity of all statutory certificates for a period of three months. If the normal operation of ports and travel of surveyors should continue to be restricted by the pandemic or eventual problems or delays created after the pandemic, alternative ways to address this would need to be found on a case-by-case basis, such as issuing short-term certificates based on remote surveys or use of alternative survey locations (Lloyds Register, 2020). On 8 April 2020, representatives of

[50] Those organizations responsible for carrying out surveys and inspections on behalf of Administrations.

[51] See https://ww2.eagle.org/en/news/abs-covid-19-update.html.

[52] See https://marine-offshore.bureauveritas.com/newsroom/covid-19-update-bureau-veritas-marine-offshore.

[53] See www.dnvgl.com/news/dnv-gl-maritime-response-to-the-coronavirus-covid-19-outbreak-166449.

[54] See www.irclass.org/covid-19/.

[55] See https://info.lr.org/l/12702/2020-02-27/8ntgzw.

[56] For example, Belgium, Denmark, the Marshall Islands, the Netherlands and Norway. For updated information, see Lloyds Register, 2020.

the 10 port State control regimes[57] that cover the world's oceans and the IMO Secretariat met in an online meeting. They reported that, while the number of physical on-board ship inspections had been reduced considerably to protect both port State control officers and seafarers, the regimes continued to work to target high-risk ships that might be substandard. They reported taking a pragmatic, practical and flexible approach, recognizing that exemptions, waivers and extensions to certificates had been granted by many flag States, and expressed a general desire for such practices to be standardized and harmonized (IMO, 2020e) (Circular Letter No. 4204/Add.8).

In addition, IMO addressed the certification of seafarers and fishing vessel personnel (Circular Letter No. 4204/Add.5/Rev.1), including medical certification, ship sanitation certification (Circular Letters No. 4202/Add.10 and Add.11), and certification of ships (Circular Letter No. 4204/Add.19/Rev.2), while the Special Tripartite Committee of the Maritime Labour Convention, 2006, as amended, in a statement on COVID-19 suggested extending the validity of seafarers' certificates for at least three months and adopting a flexible approach to ship certification (International Labour Organization, 2020). In addition, temporary measures were adopted in May 2020 at the European Union level, enabling the extension of the validity of certain certificates and licences in the road, rail and waterborne transport sectors (European Union, 2020b). An amendment was adopted to the Port Services Regulation (EU) 2017/352, which relaxed the rules on charging ships for the use of port infrastructures, providing flexibility on the reduction, deferral, waiver or suspension of port infrastructure charges as a response to the COVID-19 crisis, thus contributing to the financial sustainability of ship operators in the context of the pandemic (European Union, 2020c). Further, measures could be decided on a case-by-case basis by port-managing bodies. The temporary amendment could be applied for all measures taken as from 1 March 2020 until 31 October 2020 (European Union, 2020d).

Members of the International Association of Classification Societies, acting on behalf of flag States, also developed guiding principles for the provision of technical and implementation advice to such States when considering whether to permit statutory certificate extension beyond three months (Circular Letter No. 4204/Add.19). It was further clarified that the extension of the validity of certificates beyond the statutory maximum should only be considered in extraordinary circumstances and if no other alternative exists. The issuance of short-term certificates or other measures should be limited to specific situations caused by the pandemic, and relevant decisions should be made on a case-by-case basis. The guiding principles provide technical and implementation advice to flag States when considering whether to extend certificates beyond the three months allowed by the IMO treaty regime. They represent a six-step approach to an informed decision-making process that respects the existing regulatory regime and that can result in an evidence-based assessment for the justification of such an extension. Considering that port State control measures had been temporarily suspended to some degree by some port State control regimes, it is the responsibility of the flag State to issue clear statutory instructions and decisions to owners and recognized organizations regarding such extensions.

3. Crew changes and key worker status

Shipping and seafarers are vital to global supply chains and the world economy. Each month, a large number of seafarers need to be changed over to and from the ships they operate to ensure compliance with international maritime regulations for safety, crew health and welfare, and to prevent fatigue. Because of COVID-19-related restrictions, however, large numbers of seafarers had to have their service extended on board ships after many months at sea, unable to be replaced or repatriated after long tours of duty. The International Transport Workers' Federation estimated in July that approximately 300,000 seafarers were trapped working aboard ships due to the crew change crisis caused by government border and travel restrictions relating to the pandemic; the same number of unemployed seafarers, who were ashore, were waiting to join them. That makes 600,000 seafarers affected by this crisis (International Transport Workers' Federation, 2020a). This was considered unsustainable, both for the safety and well-being of seafarers and the safe operation of maritime trade (Marine Insights, 2020) (see also chapter 2 of this report).

During the implementation of border closures, lockdowns and preventative measures aiming to reduce the exposure to COVID-19 risk at ports and terminals, including the temporary suspension of crew changes and prohibition of crew disembarking at port

[57] Since the first regional port State control agreement (Paris Memorandum of Understanding on Port State Control) was signed in 1982, IMO has supported the establishment of eight other regional port State control regimes, achieving a global maritime network. The areas of responsibility cover the waters of the European coastal States and the North Atlantic basin from North America to parts of Europe and the north Atlantic (Paris Memorandum of Understanding on Port State Control); Asia and part of the Pacific Ocean (Memorandum of Understanding on Port State Control in the Asia–Pacific Region); Latin America (Latin American Agreement on Port State Control of Vessels); the Caribbean (Memorandum of Understanding on Port State Control in the Caribbean Region); West and Central Africa (West and Central Africa Memorandum of Understanding on Port State Control); the Black Sea (Memorandum of Understanding on Port State Control in the Black Sea Region); the southern part of the Mediterranean Sea (Mediterranean Memorandum of Understanding on Port State Control); the Indian Ocean (Indian Ocean Memorandum of Understanding on Port State Control); and the Persian Gulf (Riyadh Memorandum of Understanding on Port State Control). The United States Coast Guard maintains a tenth port State control regime.

terminals,[58] a major issue was the need for recognition by Governments and relevant national authorities of key-worker status for those operating essential services in maritime transport, including professional seafarers and marine personnel, regardless of nationality, when in their jurisdiction. This would give them the right to transit international borders and obtain medical attention ashore.[59] Another key issue was for Governments and national authorities to allow and facilitate crew changes and repatriation upon completion of their periods of service, permitting professional seafarers and marine personnel to disembark from ships in port and transit through their territory.

In cooperation with global industry associations representing various sectors of the maritime transport industry,[60] IMO adopted a number of general measures and protocols designed to address these issues and ensure that ship crew changes could take place safely during the pandemic (Circular Letter No. 4204/Add.14). Such protocols covered the travel and movement of seafarers to and from ships for the purpose of effecting ship crew changes, which included various locations (and potential locations) throughout the process of crew change and travel and the periods of time when there might be risks that needed to be managed and controlled in the process. The circular letter contained recommendations to maritime Administrations and other relevant national authorities, such as health, customs, immigration, border control, seaport and civil aviation authorities and outlined the roles of shipping companies, agents and representatives, including crew agencies and seafarers. The information was also extended to seaports, airports and airlines involved in travel operations for ship crew changes. Despite a gradual trend towards the easing of restrictions on crew changes by authorities, such easing was subject to conditions, mainly travel history and/or nationalities of crew on board. In many cases, full prohibition or closure of borders still remained.[61] Out of more than 102 countries surveyed in July 2020, 45 countries allowed crew changes, while 57 did not.

In a joint statement issued in May 2020, the International Civil Aviation Organization, IMO and the International Labour Organization recognized that for humanitarian reasons and the need to comply with international safety and employment regulations, crew changes could not be postponed indefinitely (Circular Letter No. 4204/Add.18). They advised that from mid-June 2020, around 150,000 seafarers a month would require international flights to ensure crew changeovers could take place. To facilitate crew change, they urged Governments and local authorities to designate the following personnel as key workers: seafarers, marine personnel, fishing vessel personnel, offshore energy sector personnel, aviation personnel, air cargo supply chain personnel and service provider personnel at airports and ports, regardless of nationality. They were urged to exempt them from travel restrictions to ensure the smooth changeover of crews, their access to emergency medical treatment and if necessary, emergency repatriation. The implementation included permitting seafarers, marine personnel, fishers and offshore energy sector personnel to disembark from and embark ships in port and transit through the territory of Governments and local authorities (that is to say, to an airport) for the purpose of crew changes and repatriation and the implementation of appropriate approval and screening protocols. Gradually, more and more reports of successful crew changes were being received (Splash, 2020c).

4. Commercial law implications of the COVID-19 crisis

As highlighted in an UNCTAD policy brief (UNCTAD, 2020d), the unprecedented disruptions associated with the pandemic and its massive socioeconomic consequences are giving rise to a plethora of legal issues affecting traders across the globe (for example, delays and performance failure, liability for breach of contract, frustration and force majeure). The effects of such issues may lead to large-scale economic losses and bankruptcies, in particular for small and medium-sized enterprises, including in developing countries, and in turn overwhelm courts and legal systems. Collaborative approaches by Governments and industry, policy coherence and synergy will be required to minimize adverse effects. Industry and traders need to be encouraged to waive some of their legal rights and agree on moratoriums for payments, performance and the like where appropriate, and Governments should consider where intervention or financial assistance may be necessary.

In all cases where performance is disrupted, delayed or becomes impossible, legal consequences arise, leading to the need for dispute resolution and potential litigation involving complex jurisdictional issues in a globalized context. Unless common approaches are found to reducing the incidence of disputes and facilitate their

[58] For COVID-19-related port restrictions on vessels and crew and an interactive map of ports around the world, see Wilhelmsen, 2020.

[59] For a draft template of letters of authorization from the International Chamber of Shipping and the International Transport Workers' Federation to help seafarers and authorities recognize key worker status, see International Chamber of Shipping and International Transport Workers' Federation, 2020.

[60] BIMCO, Cruise Lines International Association, Federation of National Associations of Ship Brokers and Agents, Intercargo, Interferry, InterManager, International Air Transport Association, International Association of Ports and Harbours, International Chamber of Shipping, International Federation of Shipmasters Associations, International Marine Contractors Association, International Parcel Tankers Association, International Transport Workers' Federation, Intertanko, Protection and Indemnity Clubs and World Shipping Council. See also International Transport Workers' Federation (2020b, 2020c and 2020d).

[61] For a list of countries that allow disembarkation for the purpose of crew change and related information on relevant restrictions, see BIMCO, 2020; S5 Agency World, 2020; and Waterfront Maritime Services, 2020.

resolution, for example by agreement on contractual extensions, restraint in terms of pursuing rights and legal claims, and efforts at mediation and informal dispute resolution, this could be on a scale overwhelming legal and administration of justice systems, with implications for global governance and the rule of law.[62] Coordinated government and collective industry action is required, as well as commercial risk-allocation through standard contractual clauses drafted to address contractual rights and obligations in the light of the circumstances associated with the pandemic. As part of its response to the COVID-19 crisis, UNCTAD has already begun lending technical assistance to provide related technical advice and guidance to small and medium-sized traders and policymakers, in particular in developing countries;[63] two related briefing notes are under preparation.

5. Need for systemic and coordinated policy responses at the global level

The urgent need for systemic and coordinated policy responses at the global level has prompted the United Nations Global Compact to issue a call to action that identifies recommendations for urgent political action to keep global ocean-related supply chains moving (United Nations Global Compact, 2020a). The recommendations were a consolidation of the work of the COVID-19 Task Force on Geopolitical Risks and Responses initiated by the Action Platform for Sustainable Ocean Business of the Global Compact (United Nations Global Compact, 2020b). The Task Force consists of representatives from leading international companies, industry associations, financial institutions, United Nations specialized agencies and academic institutions. The call to action recognizes that:

> The scale, complexity and urgency of the problem call for a comprehensive, systemic and coordinated approach at the global level. These issues cannot be effectively dealt with on a case-by-case basis, bilaterally or between a limited number of countries. An absence of decisive policy responses at the global level will likely trigger ripple effects which will reverberate through national economies and impede cross-border supplies of critical goods.

The call to action includes the following recommendations:

- Recognize the fundamental role robust international ocean-related supply chains play in the COVID-19 pandemic response.
- Pursue holistic and harmonized global cooperation and coordination to ensure the safety and integrity of ocean-related global supply chains.
- Ensure the continued cross-border flow of goods by sea to avoid disruptions to the integrity of ocean-related global supply chains.
- Adopt an internationally recognized key worker status system enabling unhindered movement, regardless of nationality, across international borders of personnel key to the safety and integrity of ocean-related supply chains.
- Implement measures to facilitate the safe and efficient cross-border movement of key personnel and flow of goods by sea.
- Adopt a uniform, evidenced-based and globally consistent approach to certification and classification procedures to ensure the safety and integrity of ocean-related global supply chains.
- Establish a system of metrics to gauge disruptions in the global ocean-related supply chains. (United Nations Global Compact, 2020a).

Detailed elaborations on the recommendations, along with suggestions for concrete actions to be taken, can be found in the annex to the aforementioned call to action.

F. SUMMARY AND POLICY CONSIDERATIONS

Technological advances, the COVID-19 pandemic and changes in the regulatory and legal environment provide a challenging environment for policymakers, who need to respond to these developments. Key issues presented and discussed above include the following.

1. Ensuring cybersecurity

The maritime industry is increasingly embracing automation, and ships and ports are becoming better connected and further integrated into information technology networks. Other trends affecting the industry are a growing shift towards digitalization and the development of smart navigation and advanced analytics. As a result, the implementation and strengthening of cybersecurity measures is becoming an essential priority for shipowners, managers and port operators. For ships, this becomes even more important regarding the need to implement IMO Resolution MSC.428(98), on Maritime Cyberrisk Management in Safety Management Systems, which encourages Administrations to ensure that cyberrisks are appropriately addressed in safety management systems, starting from 1 January 2021. Thus, in preparation for the implementation of the IMO resolution during 2020 – ahead of the first inspection by the International Safety Management auditors after 1 January 2021 – shipping companies need to assess their risk exposure and develop information technology policies for inclusion in their safety management systems. Owners who fail to do so are not only exposed

[62] Note in this context Sustainable Development Goal 16 on peace, justice and strong institutions and Goal 17 on the Global Partnership for Sustainable Development.

[63] Transport and trade connectivity in the age of pandemics (project 2023X), www.un.org/development/desa/da/da-response-to-covid-19/.

to cyberrisks but may have their ships detained by port State control authorities that would need to enforce this requirement.

The COVID-19 outbreak has brought maritime industry stakeholders closer in their efforts to ensure supply chains continue to function. Virtual platforms have played an important role in facilitating communication and operations during this time. However, an increase in shipping cyberattacks of 400 per cent was reported between February and June 2020, exacerbated by the reduced ability of companies to sufficiently protect themselves, in particular as a result of travel restrictions, social distancing measures and economic recession.

Cyberrisks are likely to continue to grow significantly, as a result of greater reliance on electronic trading and an increasing shift to virtual interactions at all levels; this heightens vulnerabilities across the globe, with a potential for crippling effects on critical supply-chains and services. Coordinated efforts towards developing appropriate protection mechanisms against cybercrime and attacks should therefore be pursued as a matter of urgency; this may require significant scaling up of investment and capacity-building for developing countries, including with respect to skilled human resources.

2. Using electronic trade documents

In the context of the pandemic, international organizations and industry have issued calls for Governments to remove restrictions on the use and processing of electronic trade documents and the need for documentation to be presented in hard copy. Governments have made significant efforts to keep their ports operational and speed up the use of new technologies, including digitalization. In addition, industry associations have been working to promote the use of electronic equivalents to the negotiable bill of lading and their acceptance by more government authorities, banks and insurers.

3. Reducing greenhouse gas emissions from international shipping and adapting transport infrastructure to the impacts of climate change

With regard to the reduction of greenhouse gas emissions from international shipping, progress was made at IMO towards achieving the levels of ambition set out in the Initial Strategy on reduction of greenhouse gas emissions from ships, including on ship energy efficiency, alternative fuels and the development of national action plans to address greenhouse gas emissions from international shipping. This includes the publication in 2020 of the fourth IMO greenhouse gas study. UNCTAD collaborates with IMO in a review of the impact assessments submitted to the Intersessional Working Group on Reduction of Greenhouse Gas Emissions from Ships. From the perspective of developing countries, many of which are particularly vulnerable to the growing risks of climate-change impacts, it is important that their legitimate interests be taken into account in the quest to reduce emissions from international shipping.

The twenty-fifth session of the United Nations Framework Convention on Climate Change, held in Madrid in December 2019, highlighted that much remains to be done on both the domestic and international fronts if climate action is to be achieved that is consistent with the long-term goal of the Paris Agreement under the Convention.[64]

In the context of climate-change adaptation and resilience-building for seaports – an issue of particular relevance to the developing world – the transport action table prepared by the Marrakech Partnership for Global Climate Action includes two distinct areas with a focus on adaptation, for transport systems and transport infrastructure, respectively, as well as related milestones for 2020, 2030 and 2050 (Marrakech Partnership for Global Climate Action, 2019a). These envisage, among others, that by 2030, all critical transport infrastructure will be climate-resilient to at least 2050. Relevant key actions and milestones for transport have also been integrated into the cross-sectoral resilience and adaptation action table, which highlights key actions and milestones for climate resilience-building (Marrakech Partnership for Global Climate Action, 2019b). UNCTAD actively contributed to the preparation of these documents. In the light of scientific projections, climate-change impacts and adaptation for critical transport infrastructure will remain key challenges, including during post-pandemic recovery.

4. Reducing pollution from shipping

There are several important areas where regulatory action has recently been taken or is under way for the protection of the marine environment and conservation and sustainable use of marine biodiversity. These are as follows: implementation of the IMO 2020 sulphur limit; ballast water management; action to address biofouling; reduction of pollution from plastics and microplastics; safety considerations of new fuel blends and alternative marine fuels; and the conservation and sustainable use of marine biodiversity of areas beyond national jurisdiction.

The implementation as of 1 January 2020 of the mandatory IMO limit of 0.5 per cent on sulphur content in ship fuel oil was considered to be relatively smooth at the outset; however, some difficulties have arisen as a result of the disruptions caused by the COVID-19

64 Paris Agreement, article 2.1(a): "Holding the increase in the global average temperature to well below 2°C above pre-industrial levels and pursuing efforts to limit the temperature increase to 1.5°C above pre-industrial levels…".

crisis. In March 2020, the ban on the carriage of non-compliant fuel oil entered into force to support the implementation of the sulphur cap. However, it appears that its enforcement by port State control authorities has been suspended, owing to measures put in place to reduce inspections and contain the risk of spreading the coronavirus disease. It will be important to ensure that any delay will not adversely affect the implementation of the sulphur cap regulation in the long term.

5. Responding to the COVID-19 pandemic

The spread of the coronavirus placed the entire world – and thus the international maritime industry – in an unprecedented situation. To slow the spread of the disease and mitigate its impacts, key shipping stakeholders, including international bodies and Governments, issued a number of recommendations and guidance that aimed to ensure, first of all, that port workers and seafarers were protected from the coronavirus disease, were medically fit and had access to medical care, and that ships met international sanitary requirements.

Seafarers in particular face major challenges stemming from the pandemic. Owing to COVID-19 restrictions, many seafarers had to have their service extended on board ships after many months at sea, unable to be replaced or repatriated after long tours of duty. This is a problematic state of affairs, both in terms of their safety and well-being and the safe operation of maritime trade. Therefore, calls have been issued to designate seafarers and other marine personnel as key workers, regardless of nationality, and to exempt them from travel restrictions, to enable crew changes. In addition, temporary guidance was developed for flag States, enabling the extension of the validity of seafarers and ship licences and certificates under mandatory instruments of the International Labour Organization and IMO. It has become more and more clear that due to the scale, complexity and urgency of the COVID-19 crisis, addressing these issues effectively calls for a comprehensive and coordinated approach at the global level.

In respect of the important and wide-ranging commercial law implications of the COVID-19 crisis and its aftermath, coordinated government and collective industry action will be required. Further, commercial risk-allocation through standard contractual clauses drafted to address legal rights and obligations will be necessary in light of the circumstances associated with the pandemic and to ensure that legal and administrative systems are not overwhelmed. In this regard, capacity-building and legal technical advice and guidance will be needed to support small and medium-sized enterprises, as well as policymakers in developing countries.

REFERENCES

Allianz (2019). Allianz: Shipping losses lowest this century, but incident numbers remain high. Press release. 4 June.

Asariotis R, Benamara H and Mohos-Naray V (2018). Port industry survey on climate change impacts and adaptation. Research Paper No. 18. UNCTAD.

BBC News (2020). Ship with no crew to sail across the Atlantic. 14 September.

BIMCO (2020). Coronavirus (COVID-19): Crew change challenges. 13 March.

BIMCO, Cruise Lines International Association, International Chamber of Shipping, Intercargo, InterManager, Intertanker, International Union of Marine Insurance, Oil Companies International Marine Forum and World Shipping Council (2018). The guidelines on cybersecurity onboard [sic] ships. Version 3.

BSA/The Software Alliance (2015). Asia–Pacific Cybersecurity Dashboard: A Path to a Secure Global Cyberspace.

China Classification Society (2017). CCS [China Classification Society] has issued the Guidelines for Requirement and Security Assessment of Ship Cyber System. 21 July.

Digital Container Shipping Association (2020a). DCSA *[Digital Container Shipping Association] Implementation Guide for Cybersecurity on Vessels v1.0*.

Digital Container Shipping Association (2020b). DCSA [Digital Container Shipping Association] publishes implementation guide for IMO cybersecurity mandate. 3 October.

Digital Container Shipping Association (2020c). DCSA [Digital Container Shipping Association] takes on eBill of lading standardization, calls for collaboration.

Digital Ship (2020). CYSEC helps ESA [European Space Agency] to protect ship tracking communications from cyberattack. 23 April.

Economic Commission for Asia and the Pacific (2019). Measuring the digital divide in the Asia–Pacific Region for the United Nations Economic and Social Commission for Asia and the Pacific. Asia–Pacific Information Superhighway (AP-IS) Working Paper Series.

Economic Commission for Europe (2020). *Climate Change Impacts and Adaptation for International Transport Networks* (United Nations publication. Sales No. E.20.II.E.23. Geneva).

European Commission (2009). Commission Regulation (EC) No. 906/2009 of 28 September 2009 on the application of article 81(3) of the Treaty to certain categories of agreements, decisions and concerted practices between liner shipping companies (consortia).

European Commission (2019a). Evaluation of the Commission Regulation (EC) No. 906/2009 of 28 September 2009 on the application of article 81(3) of the Treaty to certain categories of agreements, decisions and concerted practices between liner shipping companies (consortia). Commission Staff Working Document. 20 November.

European Commission (2019b). Consultation strategy for the evaluation of Consortia Block Exemption Regulation.

European Commission (2020a). Full-length report. Accompanying the document Report from the Commission: *2019 Annual Report on CO^2 [Carbon-dioxide] Emissions from Maritime Transport*. SWD(2020) 82 final. 19 May.

European Commission (2020b). *Report from the Commission: 2019 Annual Report on CO^2 [Carbon-dioxide] Emissions from Maritime Transport*. C(2020) 3184 final. 19 May.

European Commission (2020c). Antitrust: Commission prolongs the validity of block exemption for liner shipping consortia. Press release. 24 March.

European Commission (2020d). Communication from the Commission: Guidelines concerning the exercise of the free movement of workers during COVID-19 outbreak. 2020/C 102 I/03. 30 March.

European Union (2016). Directive (EU) 2016/1148 of the European Parliament and of the Council of 6 July 2016 concerning measures for a high common level of security of network and information systems across the Union.

European Union Agency for Cybersecurity (2019). *Port Cybersecurity: Good Practices for Cybersecurity in the Maritime Sector*. Athens.

European Union (2020a). Advice for ship operators for preparedness and response to the outbreak of COVID-19. Version 3. 20 February.

European Union (2020b). Regulation of the European Parliament and of the Council laying down specific and temporary measures in view of the COVID-19 outbreak concerning the renewal or extension of certain certificates, licences and authorizations and the postponement of certain periodic checks and periodic training in certain areas of transport legislation. PE-CONS 16/1/20 REV 1. 25 May.

European Union (2020c). Regulation of the European Parliament and of the Council amending Regulation (EU) 2017/352, so as to allow the managing body of a port or the competent authority to provide flexibility in respect of the levying of port infrastructure charges in the context of the COVID-19 outbreak. 25 May.

European Union (2020d). COVID-19 transport measures: Council adopts temporary flexibility for licences and port services. Press release. 20 May.

Gaskell N, Asariotis R and Baatz Y (2000). *Bills of Lading: Law and Contracts*. Informa Law from Routledge. LLP Professional Publishing. London.

Global Maritime Forum (2019). Getting to Zero Coalition. Available at www.globalmaritimeforum.org/getting-to-zero-coalition/members.

Heavy Lift (2020). Covid-19 shakes up Sulphur 2020. 3 April.

International Association of Ports and Harbours (2020a). IAPH [International Association of Ports and Harbours]–WPSP [World Ports Sustainability Programme] port economic impact barometer.

International Association of Ports and Harbours (2020b). Guidance on ports' response to the coronavirus pandemic. Version 2.0.

International Association of Ports and Harbours, BIMCO, International Cargo Handling Coordination Association, International Chamber of Shipping, International Harbour Masters Association, International Maritime Pilots Association, International Port Community Systems Association, International Ship Suppliers and Services Association, Federation of National Associations of Ship Brokers and Agents, and Protect Group (2020a). Accelerating digitalization of maritime trade and logistics: A call to action. 2 June.

International Association of Ports and Harbours, International Cargo Handling Coordination Association, TT Club Mutual Insurance and World Ports Sustainability Programme (2020b). Port community cybersecurity.

International Group of Protection and Indemnity Clubs (2020). Annual Review 2018/19. 14 January.

IMO (2011). The Guidelines for the control and management of ships' biofouling to minimize the transfer of invasive aquatic species. MEPC 62/24/Add.1. London.

IMO (2017a). IMO Resolution MSC.428(98) on maritime cyberrisk management in safety management systems. 16 June. MSC 98/23/Add.1. Annex 10..

IMO (2017b). IMO Resolution A.1110 (30), Strategic Plan for the Organization for the six-year period 2018 to 2023.

IMO (2017c). Guidelines on maritime cyberrisk management. MSC-FAL.1/Circ.3. 5 July.

IMO (2018a). Report of the Working Group on Reduction of greenhouse gas emissions from ships. MEPC 72/WP.7. 12 April.

IMO (2018b). Action plan to address marine plastic litter from ships. MEPC 73/19/Add.1. 26 October.

IMO (2019a). Note by the International Maritime Organization to the fifty-first session of the Subsidiary Body for Scientific and Technological Advice (SBSTA 51) Madrid, Spain, 2 to 9 December 2019: Agenda item 10(e)"Methodological issues under the Convention: emissions from fuel used for international aviation and maritime transport" – Update on IMO's Work to Address GHG [Greenhouse Gas] Emissions from Fuel Used for International Shipping.

IMO (2019b). Report of the Marine Environmental Protection Committee on its seventy-fourth session. MEPC 74/18. 9 June.

IMO (2019c). Subcommittee on Carriage of Cargoes and Containers. Sixth session. Report to the Maritime Safety Committee and the Marine Environment Protection Committee. CCC 6/14. 21 September.

IMO (2019d). Survey guidelines under the Harmonized System of Survey and Certification (HSSC). A31/Res.1140. 6 January.

IMO (2020a). Subcommittee on Navigation, Communications and Search and Rescue. Seventh session. Report to the Maritime Safety Committee. NCSR 7/23. 17 February.

IMO (2020b). Reduction of greenhouse gas emissions from ships: Fourth IMO greenhouse gas study 2020: Final report. MEPC 75/7/15. 29 July.

IMO (2020c). Sulphur 2020 – cutting sulphur oxide emissions. Available at www.imo.org/en/MediaCentre/HotTopics/Pages/Sulphur-2020.aspx.

IMO (2020d). Subcommittee on Pollution Prevention and Response. Seventh Session. Report to the Marine Environment Protection Committee. PPR 7/22. 24. London. April.

IMO (2020e). IMO and port State inspection authorities set pragmatic approach to support global supply chain. 10 April.

IMO and UNCTAD (2020). Joint statement in support of keeping ships moving, ports open and cross-border trade flowing during the COVID/19 pandemic. 8 June.

Indian Ocean Memorandum of Understanding on Port State Control Secretariat (2020). Indian Ocean MoU [memorandum of understanding] issuing guidance for dealing with impact of the outbreak of the COVID-19. Press release. 20 March.

Institution of Engineering and Technology (2017). *Code of Practice: Cybersecurity for Ships*. Queen's Printer and Controller of Her Majesty's Stationery Office. London.

Institution of Engineering and Technology (2020). *Good Practice Guide: Cybersecurity for Ports and Port Systems*. Elanders. Newcastle Upon Tyne.

Intergovernmental Panel on Climate Change (2018). Impacts of 1.5ºC global warming on natural and human systems (chapter 3). In: *Global Warming of 1.5°C: An IPCC [Intergovernmental Panel on Climate Change] Special Report on the Impacts of Global Warming of 1.5°C above Pre-industrial Levels and Related Global Greenhouse Gas Emission Path Ways, in the Context of Strengthening the Global Response to the Threat of Climate Change, Sustainable Development and Efforts to Eradicate Poverty*.

Intergovernmental Panel on Climate Change (2019). *IPCC* [Intergovernmental Panel on Climate Change] *Special Report on the Ocean and Cryosphere in a Changing Climate*.

International Chamber of Commerce (2020a). ICC [International Chamber of Commerce] memo to governments and central banks on essential steps to safeguard trade finance operations. 6 April.

International Chamber of Commerce (2020b). Guidance paper on the impact of COVID-19 on trade finance transactions issued subject to ICC [International Chamber of Commerce] rules.

International Chamber of Shipping (2020a). Coronavirus (COVID-19) *Guidance for Ship Operators for the Protection of the Health of Seafarers*. Version 1.0 – 3 March. Marisec Publications. London.

International Chamber of Shipping and International Transport Workers' Federation (2020). Facilitation certificates for international transport worker: Seafarer. 17 April. Available at www.ics-shipping.org/docs/default-source/resources/annex-1-finalcover-letter-seafarers.pdf?sfvrsn=4.

International Council on Clean Transportation (2020). New IMO study highlights sharp rise in short-lived climate pollution. Press release. 4 August.

International Labour Organization (2020). Statement of the Officers of the Special Tripartite Committee [of the Marine Labour Convention, 2006, as amended] on the coronavirus disease (COVID-19).

International Organization for Standardization (2013). IOS/IEC 27001:2013 [International Organization for Standardization/International Electrotechnical Commission standard 27001:2013]. Information technology – security techniques – information security management systems – requirements.

International Transport Workers' Federation (2020a). 300,000 Seafarers trapped at sea: Mounting crew change crisis demands faster action from governments. 16 July.

International Transport Workers' Federation (2020b). Message to G20 [Group of 20] leaders and ministers on facilitating essential movement of seafarers and marine personnel. International Chamber of Shipping and International Transport Workers' Federation joint statement. 7 April.

International Transport Workers' Federation (2020c). Crew changes and contractual rights and obligations. Joint statement from the International Transport Workers' Federation and the Joint Negotiating Group of Employers. 16 April.

International Transport Workers' Federation (2020d). COVID-19 and access to sanitation facilities for transport workers: Guidance for trade union negotiators. 17 April.

JOC.com (2019). Carriers driving container data standardization. 4 September.

JOC.com (2020). DCSA [Digital Container Shipping Association] targets elusive electronic bill of lading standard. 19 May.

Lloyd's Loading List (2020). Container shipping customer groups jointly urge EU [European Union] to re-evaluate BER [Block Exemption Regulation]. 21 January.

Lloyds Register (2020). Latest flag and port State instructions on COVID-19 (coronavirus).

Marine Insights (2020). IMO endorses new protocols designed to lift barriers to crew changes. 7 May.

Marine Link (2020). Surge in maritime cyberattacks reported. 15 June.

Marrakech Partnership for Global Climate Action (2019a). Climate action pathway: Transport. Action table.

Marrakech Partnership for Global Climate Action (2019b). Climate action pathway: Resilience and adaptation. Action table.

Monioudi I, Asariotis R, Becker A, Bhat C , Dowding-Gooden D, Esteban M, Feyen L, Mentaschi L, Nikolaou A, Nurse L, Phillips W, Smith D, Satoh M, O'Donnell Trotz U, Velegrakis AF, Voukouvalas E, Vousdoukas MI and Witkop R (2018). Climate change impacts on critical international transportation assets of Caribbean small island developing States (SIDS): The case of Jamaica and Saint Lucia. *Regional Environmental Change*. 18:2211–2225(2018). 31 May.

National Cybersecurity Centre (2019). *Advisory: Ryuk Ransomware Targeting Organizations Globally*. NCSC-Ops/17-19. 22 June.

National Institute of Standards and Technology (2018). Framework for Improving Critical Infrastructure Cybersecurity. Version 1.1. 16 April. Available at *www.nist.gov/cyberframework/framework*.

Nippon Yusen Kabushiki Kaisha Line (2019). NYK [Nippon Yusen Kabushiki Kaisha] Group accredited by ClassNK for cybersecurity management. 16 December.

North Atlantic Treaty Organization Cooperative Cyberdefence Centre of Excellence (2019). ASEAN cyber developments: Centre of excellence for Singapore, cybercrime convention for the Philippines and an open-ended working group for everyone.

Ocean Network Express (2020). Ocean network express issues its first electronic bill of lading and selects essDOCS to power its global bill of lading digitization initiative. 2 April.

Pacific Community (2019). Fourth pacific regional energy and transport ministers' meeting. 17 September. www.spc.int/updates/news/speeches/2019/09/fourth-pacific-regional-energy-and-transport-ministers-meeting.

Panahi R, Ng A and Pang J (2020). Climate change adaptation in the port industry: A complex of lingering research gaps and uncertainties. *Transport Policy*. 95:10–29.

Paris Memorandum of Understanding on Port State Control Secretariat (2020). Paris MOU [Memorandum of Understanding] guidance regarding the impact of COVID-19. Press release. 26 March.

Premti A (2016). Liner shipping: Is there a way for more competition? Discussion Paper No. 224. UNCTAD.

Premti A (2018). Conservation and sustainable use of marine biodiversity of areas beyond national jurisdiction: Recent legal developments. 29 October. UNCTAD Transport and Trade Facilitation Newsletter No. 80. Fourth quarter. Available at https://unctad.org/en/pages/newsdetails.aspx?OriginalVersionID=1905.

Riviera (2019). US [United States] 'not adequately addressing the problem' of maritime cyberthreats. 2 April.

Riviera (2020a). Covid-19 forces shipping to adopt digitalization. 28 April.

Riviera (2020b). Shipping adapts to coronavirus restrictions. 29 April.

Riviera (2020c). Altera drone test success paves the way for remote tank inspection. 10 June.

Riviera (2020d). Explained: Propeller fouling and 'the power penalty'. 11 February.

Riviera (2020e). Hull fouling can add 20% to fuel costs. 30 May.

S5 Agency World (2020). Service availability by port. Available at https://app.powerbi.com/view?r=eyJrIjoiN2I3MmJiYmYtYmYyNy00MGVkLWI0ZTktZDZmNTY4ZTBlMzM3IiwidCI6IjM5MGZkN2Q4LWUzMjktNDdiYy04MmY4LWM5NTY4NTg5MzYyYyIsImMiOjEwfQ%3D%3D.

Safety4Sea (2018). Cybersecurity library for shipping: Everything you need to know. 26 April.

Safety4Sea (2019a). Why underreporting is a major cyberthreat in the shipping industry. 2 July.

Safety4Sea (2019b). IBM joins Mayflower autonomous ship project. 16 October.

Safety4Sea (2019c). Scrubber discharges bans in ports: What you should know. 22 November.

Safety4Sea (2020a). KR [Korean Register] issues guidelines for type approval of maritime cybersecurity. 4 February.

Safety4Sea (2020b). How coronavirus affects shipping: Everything you need to know. 4 February.

Seatrade Maritime News (2020). Suez Canal clarifies open-loop scrubber confusion. 14 January.

Secretariat of the Memorandum of Understanding on Port State Control in the Asia–Pacific Region (2020). Tokyo MoU [memorandum of understanding] issuing guidance for dealing with impact of the outbreak of the COVID-19. Press release. 12 March.

Smart Maritime Network (2020). India moves forward with digital bills of lading. 11 June.

Splash (2019). Wilhelmsen and Airbus trial world's first commercial drone deliveries to vessels at anchorage. 15 March.

Splash (2020a). Number of shipping cyberattacks leaps 400 per cent since February. 5 June.

Splash (2020b). F-drones completes first commercial BVLOS [beyond visual line of sight] drone delivery in Singapore. 29 April.

Splash (2020c). Ship managers accelerate crew repatriation. 5 June.

Thetius (2020). Maritime autonomous surface ship market map.

United Kingdom Protection and Indemnity Club (2017). Legal briefing: Electronic bills of lading. May.

United Kingdom Protection and Indemnity Club (2020a). Electronic bills of lading: An update, part I. 26 March.

United Kingdom Protection and Indemnity Club (2020b). Electronic bills of lading: An update, part II. 1 April.

United Nations (2016). Report on the work of the United Nations Open-ended Informal Consultative Process on Oceans and the Law of the Sea at its seventeenth meeting. A/71/204. New York. 25 July.

United Nations (2019). *Report of the Secretary-General on the 2019 Climate Action Summit and the Way Forward in 2020*. 11 December. Available at http://sdghelpdesk.unescap.org/sites/default/files/2020-06/cas_report_11_dec.pdf.

United Nations (2020). Report of the Subsidiary Body for Scientific and Technological Advice on its fifty-first session, held in Madrid from 2 to 9 December 2019. FCCC/SBSTA/2019/5. 16 March.

United Nations Climate Change Secretariat (2019). *Yearbook of Global Climate Action 2019: Marrakech Partnership for Global Climate Action*. Bonn.

UNCTAD (2003). The use of transport documents in international trade. UNCTAD/SDTE/TLB/2003/3. 26 November.

UNCTAD (2011). The 2004 Ballast Water Management Convention – with international acceptance growing – the Convention may soon enter into force. Transport Newsletter No. 50.

UNCTAD (2015). The International Ballast Water Management Convention 2004 is set to enter into force in 2016. Transport and Trade Facilitation Newsletter No. 68.

UNCTAD (2017a). *Review of Maritime Transport 2017* (United Nations publication. Sales No. E.17.II.D.10. New York and Geneva).

UNCTAD (2017b). *Climate Risk and Vulnerability Assessment Framework for Caribbean Coastal Transport Infrastructure: Climate Change Impacts on Coastal Transportation Infrastructure in the Caribbean – Enhancing the Adaptive Capacity of Small Island Developing States*. UNCTAD/DTL/TLB/2018/1.

UNCTAD (2018). *Review of Maritime Transport 2018* (United Nations publication. Sales No. E.18.II.D.5. New York and Geneva).

UNCTAD (2019a). *Review of Maritime Transport 2019* (United Nations publication. Sales No. E.19.II.D.20. New York and Geneva).

UNCTAD (2019b). Outcome document: High-level panel discussion "Climate resilient transport infrastructure for sustainable trade, tourism and development in SIDS [small island developing States]". 10 December.

UNCTAD (2020a). *Climate Change Impacts and Adaptation for Coastal Transport Infrastructure: A Compilation of Policies and Practices* (United Nations publication. Sales No. E.20.II.D.10. Geneva).

UNCTAD (2020b). Climate change adaptation for seaports in support of the 2030 Agenda for Sustainable Development. TD/B/C.I/MEM.7/23. 10 February. Geneva.

UNCTAD (2020c). Coronavirus: Let's keep ships moving, ports open and cross-border trade flowing. 25 March.

UNCTAD (2020d). COVID-19: A 10-point action plan to strengthen international trade and transport facilitation in times of pandemic. Policy Brief No. 79. April.

UNCTAD and United Nations Environment Programme (2019). High-level panel discussion at COP 25 [the twenty-fifth Conference of the Parties to the United Nations Framework Convention on Climate Change], Madrid. Climate-resilient transport infrastructure for sustainable trade, tourism and development in SIDS [small island developing States]. 10 December. Available at https://unctad.org/en/Pages/MeetingDetails.aspx?meetingid=2354.

United Nations Commission on International Trade Law (2018). UNCITRAL [United Nations Commission on International Trade Law] *Model Law on Electronic Transferable Records* (United Nations publications. Sales No. E.17.V.5. New York).

United Nations Framework Convention on Climate Change (2020). Climate Action Pathways. Available at https://unfccc.int/climate-action/marrakech-partnership/reporting-and-tracking/climate_action_pathways.

United Nations Global Compact (2020a). COVID-19 Task Force on Geopolitical Risks and Responses: Call-to-action – Imminent threats to the integrity of global supply chains. Sustainable Ocean Business Action Platform.

United Nations Global Compact (2020b). Sustainable Ocean Business Action Platform. Available at www.unglobalcompact.org/take-action/action-platforms/ocean.

United States Coast Guard (2019a). Marine Safety Information Bulletin: Cyberattack impacts MTSA [Maritime Transportation Security Act] Facility Operations. 16 December.

United States Coast Guard (2019b). Marine Safety Information Bulletin: Cyber adversaries targeting commercial vessels. 24 May.

United States Coast Guard (2020). ACN 040/20 – March 2020 promulgation of navigation and vessel inspection circular No. 01-20: Guidelines for addressing cyberrisks at Maritime Transportation Security Act regulated facilities.

Visalli M, Best B, Cabral R, Cheung W, Clark N, Garilao C, Kaschner K, Kesner-Reyes K, Lam V, Maxwell S, Mayorga J, Moeller H, Morgan L, Ortuño Crespo G, Pinsky M, White T and McCauley D (2020). Data-driven approach for highlighting priority areas for protection in marine areas beyond national jurisdiction. *Marine Policy*. 28 March.

Waterfront Maritime Services (2020). COVID-19 overview: Crew change impact. 23 October.

Wilhelmsen (2020). COVID-19 global port restrictions map. Available at https://wilhelmsen.com/ships-agency/campaigns/coronavirus/coronavirus-map/.

World Association for Waterborne Transport Infrastructure (2019). PIANC [World Association for Waterborne Transport Infrastructure] Declaration on climate change.